Global Environmental Issues

Edited by

Frances Harris

John Wiley & Sons, Ltd

Copyright © 2004   John Wiley & Sons Ltd, he Atrium, Southern Gate, Chichester,
West Sussex PO19 8SQ, England

Telephone   (+44) 1243 779777

Email (for orders and customer service enquiries): cs-books@wiley.co.uk
Visit our Home Page on www.wileyeurope.com or www.wiley.com

Reprinted June 2005

This publication is designed to provide accurate and authoritative information in regard to the subject
matter covered. It is sold on the understanding that the Publisher is not engaged in rendering
professional services. If professional advice or other expert assistance is required, the services of a
competent professional should be sought.

Other Wiley Editorial Offices

John Wiley & Sons Inc., 111 River Street, Hoboken, NJ 07030, USA

Jossey-Bass, 989 Market Street, San Francisco, CA 94103-1741, USA

Wiley-VCH Verlag GmbH, Boschstr. 12, D-69469 Weinheim, Germany

John Wiley & Sons Australia Ltd, 33 Park Road, Milton, Queensland 4064, Australia

John Wiley & Sons (Asia) Pte Ltd, 2 Clementi Loop #02-01, Jin Xing Distripark, Singapore 129809

John Wiley & Sons Canada Ltd, 22 Worcester Road, Etobicoke, Ontario, Canada M9W 1L1

Library of Congress Cataloging-in-Publication Data

Global environmental issues/editor, Frances M.A. Harris.
    p.   cm.
  Includes bibliographical references and index.
  ISBN 0-470-84560-0 (alk. paper) — ISBN 0-470-84561-9 (pbk.:alk. paper)
  1. Environmental sciences.   I. Harris, Frances, M.A.

  GE105.G563 2004
  363.7—dc22   2003058347

British Library Cataloguing in Publication Data

A catalogue record for this book is available from the British Library

ISBN 10: 0-470-84561-9 (P/B)
ISBN 13: 978-0-470-84561-5 (P/B)
ISBN 10: 0-470-84560-0 (H/B)
ISBN 13: 978-0-470-84560-8 (H/B)

Typeset in 11/13pt Times by Integra Software Sevices Pvt. Ltd, Pondicherry, India
Printed and bound in Great Britain by Antony Rowe Ltd, Chippenham, Wiltshire.
This book is printed on acid-free paper responsibly manufactured from sustainable forestry
in which at least two trees are planted for each one used for paper production.

This book is dedicated to the memory of Sydney G. Harris, who encouraged me to take an interest in these topics, and to Thomas Lyon and those who will share his future.

Contents

Contents

Contributors

Doreen S. Boyd
School of Earth Sciences and Geography
Kingston University
Penrhyn Road
Kingston upon Thames KT1 2EE
UK
e-mail: D.Boyd@kingston.ac.uk

Giles M. Foody
Department of Geography
University of Southampton
Highfield
Southampton SO17 1BJ
UK
e-mail: G.M.Foody@soton.ac.uk

Frances Harris
School of Earth Sciences and Geography
Kingston University
Penrhyn Road
Kingston upon Thames KT1 2EE
UK
e-mail: F.Harris@kingston.ac.uk

Mike Hulme
Tyndall Centre for Climate Change Research
School of Environmental Sciences
University of East Anglia
Norwich NR4 7TJ
UK
e-mail: m.hulme@uea.ac.uk

Kenneth Lynch
School of Earth Sciences and Geography
Kingston University

Contributors

Penrhyn Road
Kingston upon Thames KT1 2EE
UK
e-mail: k.lynch@kingston.ac.uk

Kathy Morrissey
WSP Environmental Ltd
Buchanan House
24-30 Holborn
London ECIN 2HS
e-mail: kathy.morrissey@wspgroup.com

Patrick D. Nunn
Department of Geography
The University of the South Pacific
Suva
Fiji Islands
e-mail: nunn_p@usp.ac.fj

Nick Petford
School of Earth Sciences and Geography
Kingston University
Penrhyn Road
Kingston upon Thames KT1 2EE
UK
e-mail: N.Petford@kingston.ac.uk

Guy M. Robinson
School of Earth Sciences and Geography
Kingston University
Penrhyn Road
Kingston upon Thames KT1 2EE
UK
e-mail: G.Robinson@kingston.ac.uk

Ros Taylor
School of Earth Sciences and Geography
Kingston University
Penrhyn Road
Kingston upon Thames KT1 2EE
UK
e-mail: R.Taylor@kingston.ac.uk

Preface

Writing this book has coincided with the pregnancy and the birth of my first child. Writing and editing it has been a new experience and a welcome opportunity to broaden reading and research. All this has been 'supervised' by a growing child, a strong reminder of phrases such as 'intergenerational equity' and the importance of sustainable environmental management for the benefit of many future generations.

Acknowledgements

The structure of this book has been shaped by a lecture course at Kingston University. Doreen Boyd first suggested that the lecture course could be the basis of a book. Lyn Roberts, of Wiley, took up the idea and has guided me through the process of producing an edited volume. It would not have been possible without the effort of each of the contributing authors, for which I am very grateful. Claire Ivison has prepared most of the figures, and shown patience with all of us. The final stages of putting the whole together was made much easier by the assistance of Stella Bignold. Throughout, Fergus Lyon has been an excellent advisor and very supportive.

    The contributors would like to thank those who have given their permission to reproduce their figures and tables.

Frances Harris
School of Earth Sciences and Geography
Kingston University

Part One

Introduction

Chapter 1
Human–Environment Interactions

Frances Harris

1.1 Introduction

Environmental issues have been a concern for many years. Yet somehow they are problems that we have not been able to resolve, despite research, media attention, increased public awareness about environmental problems, campaigns by environmental pressure groups and international agreements. Our environment is dynamic, constantly changing and evolving in response to stimuli. Yet in the 20th century, it became apparent that mankind is having an increasing effect on the planet's ecosystems and biogeochemical cycles, so much so that our activities are now causing environmental change which is overriding the natural dynamism of the earth. Yet despite the evidence of environmental problems such as biodiversity loss, land cover change observable from satellite imagery, records of climate change and many examples of pollution, we still pursue activities which perpetuate the problems. As the world's population increases, and the *per capita* consumption of natural resources increases, we will have an even greater effect on these environmental problems, exacerbating them further.

Why are such problems so hard to resolve? There are three broad reasons: Firstly, the science of environmental problems is complex. We are dealing with many interrelated dynamic systems, within which and between which feedback mechanisms occur. Secondly, there are many stakeholders involved in both the causes and the solutions to environmental problems. Organising all of these stakeholders to act in a co-ordinated manner is difficult. Thirdly, resolving global environmental issues will require changes in our own consumption and pollution of natural resources, which will mean changes to lifestyles. This will require commitment at the personal level, which not everyone is willing to make.

Global Environmental Issues.   Edited by Frances Harris
© 2004 John Wiley & Sons, Ltd   ISBNs: 0-470-84560-0 (HB); 0-470-84561-9 (PB)

Human–environment interactions involve not only the question of resource use per person, but also our ability to understand the science of the environment, our ability to regulate our impact on the environment, our beliefs in the value of the environment, our attitudes to the future, particularly risk, and our ability to negotiate solutions at both the local and the global level. This book aims to discuss environmental issues from a scientific and socio-economic viewpoint, so that they are understood not only as contested science concerning natural resources, but also as political and social issues. In this way the reader gains a fuller understanding of the complexity of environmental issues and the challenges we are faced with in resolving them. 'The science of the environment is socially and politically situated, rather than unambiguous or separable from the subjective location of human perception' (Stott and Sullivan, 2000, p. 2).

1.2 Livelihoods and natural resources

Throughout the world, people earn their livelihoods through the use of whatever resources are available to them. Our livelihoods are ultimately natural resource dependent. Natural resources provide us with the land and water for agriculture (whether for subsistence needs or to serve a wider market), trees for firewood and timber, ocean and freshwater resources for fisheries, wildlife for meat, animal products, tourism, oil, gas and coal for energy, and also mineral resources (rocks, minerals, gems, coal...). Many economies are dependent on natural resources. At the household and community level, this can be in the form of agriculture or natural resource products gathered and sold (e.g. wild foods, honey). At the national level, most countries rely on their natural resource base to meet basic needs and provide the resources for economic development, for example through cash crops, forestry or mining. Globally, we rely on natural resources for ecosystem regulation.

Natural resources are irreplaceable. We have not devised a substitute for the global climate regulation mechanism. Neither can we in the short term undo all the effects of land cover change to recreate the natural environment which existed prior to land degradation and urban sprawl. Although we can save some seeds of plants, and keep some animals in zoos, recreating ecosystems is a much greater challenge. Therefore it is important to conserve the environment, both for its own sake and also because our livelihoods depend on it.

1.3 Population–environment theories

As national populations grow, and the demand for natural resources (particularly for food production and energy generation) increases, worries that we shall exceed the resources of the planet have been expressed by many over the years.

In 1798, Malthus predicted that human population growth would be checked by food supply. This argument has been further developed by several authors. Ehrlich (1968) argued that population growth rates at that time would exceed the world's resources. Furthermore, as most population growth, and also declining food production, was found to occur in developing countries, he advocated population control. However, as Bennett (2000) points out, 'there seems to be no evidence that our ability to produce food has been a lasting break on population growth'. In contrast, Dyson (1996) maintains that food production has increased and outstripped population growth in recent decades.

Such numbers-versus-resources calculations are far too simplistic because they fail to take into consideration variability in food production and food consumption across the globe. For example, Michaelson (1981, p. 3) stated that 'overpopulation is not a matter of too many people, but of unequal distribution of resources. The fundamental issue is not population control, but control of resources and the very circumstances of life itself'. Globally, sufficient food is produced to feed people. However, food shortages occur because of variations in land productivity, and also because of problems in food distribution, due to poverty, conflict and failing markets (Bennett, 2000). Problems of inequality and existing power struggles affect people's access to resources and people's entitlements to food and other natural resources (Sen, 1982; Leach, Mearns and Scoones, 1997) on which their livelihoods depend.

Numbers-versus-resources calculations also fail to consider changes in technology which can result in increased food production or more efficient use of existing resources. They assume a steady 'carrying capacity' of the earth. However, technology has changed so much since Malthus' time. As Boserüp (1965) argues, increasing populations can be the driving force for agricultural intensification, which increases food output per unit area of land. For example, the Green Revolution had an enormous impact on agricultural productivity, particularly that of rice and wheat. (Subsequently it was realised that the Green Revolution also created new social and environmental problems, as discussed in section 6.4, but its effect on the population–food debate remained.) Tiffen's work has followed on in Boserüp's line of thought. She argues 'increased population density can induce the necessary societal and technological changes to bring about better living standards' (Tiffen, 1995, p. 60). Simon (1981) also argues that more people bring positive change, as this results in more ideas, more experimentation and more technological innovation which can help resolve the problems of resource limitations.

Human–environment interactions are not just about meeting the global population's food needs, or even about meeting natural resource needs. The human population also affects the environment through what it leaves behind. The impact of the human population on the environment is seen as, among other things, land use change (forest clearance, reduced wildlife, changes in agricultural landscapes as farming systems intensify), urbanisation, pollution of water, seas and landscapes. In some cases, our impact is less visible, at least immediately, such as gaseous pollution and changing atmospheric composition. Harrison (1993) argues that it is the effect of pollution which will drive a 'third revolution' for change in the world. These arguments are summarised in the equation (Ehrlich and Ehrlich, 1990, section 9.3.3)

$$\text{Impact} = \text{Population} \times \text{Affluence} \times \text{Technology}$$

The impact of population on the environment is determined by the size of the population, its affluence (and hence consumption *per capita*) and the type of technologies used. Therefore an extremely large but poor population using low impact technology could have the same impact as a smaller but more affluent population using highly polluting technology. The impact depends not only on the size of the population, but also on whether the technology used is highly polluting or 'green' (i.e. reliant on renewable energy or non-polluting). It should also be remembered that in some cases, 'green' technology requires affluence, and hence is not necessarily associated with the developing world.

The arguments concerning population–environment theories range from debates based on numbers of people and food resources, more complex arguments concerning the effect of environment and technology on carrying capacity, to social and political factors affecting access and entitlement to natural resources. There are several important issues to bear in mind when considering the contrasting views on human–environment interactions:

- It is important to distinguish between naturally occurring long-term trends and human-induced changes in the environment. We live in a dynamic world which is constantly changing. While we need to be aware of anthropogenic and natural changes, we can only be held responsible for, and hope to reverse, anthropogenic environmental change.
- The fact that the world is dynamic means that models and predictions need to be constantly revised to match changing scenarios. We adapt to our changing environment, and this means predictions based on existing technologies or activities may become outdated or irrelevant. Population predictions are gradually being revised downwards in view of existing figures. The doom and gloom scenarios of Malthus and others have never been fulfilled (Bennett, 2000). As we increase our knowledge of the environment and develop less polluting technology, we overcome, or at least reduce, our effect on the environment (section 9.5). We should not underestimate mankind's capacity to adapt and respond to changing environmental circumstances, as argued by Tiffen, Mortimore and Gichuki (1994) and Boserüp (1965) in relation to agriculture, and by Adger (1999) (Box 2.1) in relation to climate change.
- Numbers-versus-resources calculations do not take into consideration factors which affect the choices and decisions humans make. What is a population's incentive to conserve resources and the environment? Does a

population or individual have guaranteed long-term ownership of a resource, and therefore feel certain that they can reap the benefits of environmental conservation? Are immediate needs for survival so acute that the longer term cannot be considered (e.g. land shortage precludes fallowing, even though it will result in declining yields over the longer term). In cases of shared or common property resources, environmental protection or conservation by one individual does nothing to ensure that other users rather than conserving the resources do not take advantage of the situation. This is a particular problem in the negotiation of inter-national agreements concerning global environmental change/carbon dioxide (CO_2) emissions and pollution (section 10.5).

- The impact of people on the environment depends not only on numbers, but also on lifestyle, consumption and technology.
- Our effect on the environment in terms of pollution may be more limiting than our consumption needs (Harrison, 1993).

1.4 Ecological footprints

The ecological footprint of a specified population or economy can be defined as the area of ecologically productive land (and water) in various classes – cropland, pasture, forests, etc. – that would be required on a continuous basis to (a) provide all the energy/material resources consumed, and (b) absorb all the wastes discharged by the population with prevailing technology, *wherever on Earth that land is located.*
Wackernagel and Rees, 1996, pp. 51–52

As such, ecological footprints are an 'accounting tool...to estimate the resource consumption and waste assimilation requirements of a defined human population or economy in terms of a corresponding productive land area' (Wackernagel and Rees, 1996, p. 9).

The concept of ecological footprints was developed recently, and has caught the attention of many due to the simplicity of the basic concept and the ability of the ecological footprint tool to be used in an educational manner to highlight and compare individual, community, regional or national effects on the environment. Ecological footprints link lifestyles with environmental impact. Ecological footprints are determined by calculating the amount of land and water area required to meet the consumption (food, energy, goods) of a population in a given area, and assimilate all the wastes generated by that population (Wackernagel and Rees, 1996). Obviously, such a calculation relies on the accuracy of the data provided, and of the 'conversion factors' used in determining agricultural productivity of the land providing food, and the forest area required to meet the energy needs. Indeed, there are those who have made serious criticisms of the method, like van den Bergh and Verbruggen (1999), some of which may be valid. However, as a comparative tool, it has its value in making individuals or societies think about the implications of their lifestyle on the environment. The following discussion focuses on national ecological footprints. Urban ecological footprints are discussed in section 8.3.2.

Obviously, many people are not 'living off the land', especially nearby land. Most people rely on some imported goods. International trade has gone on for centuries, and provides us with many of the staples we rely on. Sylva (1665, p. 48) stated that:

> the plains of North America and Russia are our [British] corn-fields; Chicago and Odessa our granaries; Canada and the Baltic are our timber-forests; Australasia contains sheep-farms, and in Argentina and on the western prairies of North America are our herds of oxen; Peru sends her silver, and the gold of South Africa and Australia flows to London; the Hindus and Chinese grow tea for us, and our coffee, sugar and spice plantations are all in the Indies. Spain and France are vineyards and the Mediterranean our fruit garden, and our cotton grounds, which for long have occupied the Southern United States, are now being extended everywhere in the warm regions of the Earth.

In the intervening centuries, world trade has increased, and in addition to food imports, we now rely on trade to provide many non-perishable goods and commodities. If we either import goods or change our environment to enable us to produce things locally (e.g. miles of hothouses to provide vegetables), we create an ecological footprint. However, our footprint is not only dependent on what we consume. We also generate waste, some of which must be dealt with locally or nationally (rubbish collected and deposited in landfill), and some of which is generated overseas in the creation of the imported goods (e.g. sugar refineries or leather tanneries). If we create demand for a waste generating product, then we are in some way responsible for the associated waste, even if it is not produced in our country. Furthermore, some waste, such as gaseous emissions, is dumped in the atmosphere: a global 'no man's land' whose degradation has implications for all of us. Dumping rubbish and waste in the world's seas and oceans is another problem. Pollution of the global commons is proving hard to regulate, and where funds are required to resolve problems of pollution of the global commons, there can be huge disagreement concerning who should bear responsibility and pay. Thus we may feel that we are living in an environmentally friendly way, because much of the pollution and waste generated by our consumption could be generated at a distance. We do not see the effect of imported goods on the ecological footprint (via pollution in production and transportation processes) but ecological footprint analysis does bring it into account.

In the western or developed world where we eat imported foods and use imported goods, our consumption is not just about the volume we consume, but also the source of the products we consume and transport. Figure 1.1 shows the ecological footprint of a sample of 23 countries. What is most apparent is that the countries with ecological footprints higher than the world's ecological footprint

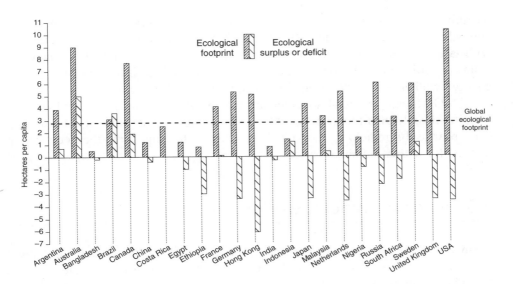

Figure 1.1 The ecological footprint and ecological deficit of 23 countries, compared with the global ecological footprint. (Compiled from data in Wackernagel *et al.*, 1997.)

are in the developed world, whereas those with lower ecological footprints are more likely to be in the developing world. The developing countries include nations with high population densities (Nigeria, China, India, Bangladesh), but the number of people does not seem to be the problem; rather it is the developed countries, where affluence is greater and technology in greater use, which have the big footprints.

Ecological footprints show that any change in the global ecological footprint will require a change in lifestyles, consumption and pollution from all nations, especially those in the developed world. There is considerable disparity between the north and south concerning environmental pressures and pollution. The ecological footprint analysis shows that the north's lifestyle (affluence, technology, consumption levels) is having a bigger impact than that of southern countries. If countries currently below the world average achieve living standards similar to those in the west, the global ecological footprint will be enormous. If the north is more responsible for some of the global environmental issues (e.g. climate change), should it bear more of the responsibility to overcome and resolve these problems, even if this means changing lifestyles to moderate its effect on the environment? This raises issues of ethics, justice and our relationship to distant others and future generations.

Figure 1.1 also relates a country's ecological footprint to the natural resources available to that country. An ecological deficit means that the needs of a country's population cannot be met from the resources within that country. Countries such as Australia and Brazil, with large, sparsely populated areas and large forest reserves, may have large ecological footprints, but can meet these from their own resources. This may partly be due to the fact that the mechanism whereby ecological footprints are calculated converts energy requirements into equivalent fuelwood (van den Bergh and Verbruggen, 1999), and so countries with large forested areas are able to compensate for high energy use, whereas countries without forests do less well in the calculation, even if they could provide energy by other renewable means such as hydroelectric power.

Of course, national statistics are the result of averages, and individual household ecological footprints could vary enormously. The ecological footprint concept is useful in helping individuals or societies to think about their contribution to global environmental issues.

1.5 The environment: worldviews

In any discussion of global environmental issues it is important to be aware that there are many different attitudes associated with valuing the environment. The value which individuals, communities and nations place on their environment is affected by cultural and religious values as well as economic and social systems. These are sometimes referred to as cultural filters (Pepper, 1986), and can affect the way we perceive the environment and 'scientific evidence' about it. Significantly, these attitudes underpin the development of strategies and priorities to conserve the environment.

1.5.1 Our relationship with nature

A fundamental issue is humanity's relationship with nature. Are we a part of nature, and one of many animals in a global ecosystem, or are we separate from nature, placed 'above' nature and entitled to control it and use it to further our own needs regardless of the effect on the remaining ecosystem? The answer to this question is fundamental as it affects how we treat the environment. The world's religions have addressed this issue. Stewardship is central to Judaic, Christian and Islamic beliefs. Religious texts can be cited stating that man can rule over and subdue the earth (Genesis 1:28–30), or that the world belongs to God, with humanity in the role of a servant or trustee, accountable to God concerning the stewardship of the earth (Attfield, 1999). The Great Chain of Being also places humans within a hierarchy, above nature, but below God. However, as each link of the chain is equally important, and mutually dependent, the Chain of Being also implies an equality between humans and nature (Pepper, 1986). Buddhism promotes respect for all forms of life, and encourages individuals to 'give back to the Earth what one has taken away' (National Environment Commission, Royal Government of Bhutan, 1998, p. 12). All the major religions believe that a judgement will be passed on acts in this life before progression to the next. Hence there is an incentive to follow religious teachings. For those who do not believe in an over-ruling God or religion, there are ethical arguments for stewardship of the environment based on our obligations to future generations (Attfield, 1999). Of course, people do not always live up to ideals. People of all religions may fall short of the teachings they profess to adhere to. Those who argue that we should conserve the environment on strictly moral and ethical (rather than religious) grounds may also fall short of achieving their ideal. Thus belief in the value of the environment does not necessarily translate into actions which conserve the environment. The assumption of human–environment duality underpins much of the writing on environmentalism in the West.

Religion and spiritual values are not the only factors which affect our attitudes to the environment; philosophical and political values can also have a strong influence. O'Riordan (1981) divides environmentalists into two broad groups: technocentrics and ecocentrics. Technocentrics have more faith in science and technology. They believe in man's dominance over nature, and furthermore are more optimistic that future scientific and technological developments will enable us to overcome environmental problems and constraints. Ecocentrics, on the other hand, believe in a greater degree of equality between humans and nature, and even the subordination of man to nature. As such, they believe we are just one part of a global ecosystem, which must be respected. Important issues shaping the extent to which someone is technocentric or ecocentric include their faith in the ability of science and technology to resolve environmental problems, and belief or scepticism regarding science and technology as driving forces in economic development.

How do these factors affect global environmental issues? The debates about global environmental issues and the sustainability of the planet are also debates about values and priorities of the populations relying on that environment. Any international debate about global environmental issues will include representatives of many cultures, political systems and values. Each may hope to impose their own views of human–environment interactions on others. Thus the predominantly Western technocentric view based on economic development as the pathway to development will be juxtaposed with more ecocentric views such as those of Bhutan, a nation whose national environmental strategy stresses the fact that 'socio-economic development and environmental and cultural integrity are not mutually exclusive, but are equally critical to the long-term viability of the Bhutanese nation' (National Environment Commission, Royal Government of Bhutan, 1998, p. 18). This more holistic approach is guided by Bhutanese culture and Buddhist values, and a belief that 'Gross National Happiness is more important than Gross National Product' (His Majesty King Jigme Singye Wangchuk, as quoted in National Environment Commission, Royal Government of Bhutan, 1998, p. 18).

1.5.2 Our attitudes to risk

Information is not always clear and unambiguous. Some research will be phrased in terms of 'likely outcomes', 'high probabilities', 'a possibility'… Such phrases mean that people need to consider the information and make decisions based on incomplete information or predictions, rather than absolute certainties. Decisions are made based on what we know, the likelihood of predictions or assumptions being true and the likely consequences if they are true. Thus an assessment of risk is made. Given the uncertainty surrounding much environmental information, some argue that 'risk and its management now occupy the centre-stage of environmental decision making' (Merrit and Jones, 2000, p. 77).

Perceptions of risk vary enormously, and are partly dependent on the character of the person. However, other factors also come into play as an individual or society considers the importance of an environmental problem. Location is an important consideration: is this an environmental problem which is nearby, and therefore a visible threat, or is it distant, on the far side of the world? (For example, sea level rise does not matter to continental or mountainous nations as much as to island nations or those with large, low-lying areas of coastland.) Is the problem likely to materialise in the near future, or in the distant future? Will it affect our lives, our children's lives, or those born many generations away? If the latter, how important is it to change activities now? Technocentrics may have faith that by that time, technology will have provided a solution to the problem. Finally, just how much will avoiding or overcoming an environmental problem impinge on individuals' and societies' livelihoods and lifestyles? Will it involve a personal sacrifice in terms of comfort or standard of living?

If the decision-makers are not absolutely sure that the science on which the decision is based is valid, they face a dilemma. Is it better to act on what we know and assume to be correct, or to delay action and wait for further information? Uncertainty is inevitable when dealing with science at the forefront of environmental knowledge. The precautionary principle, 'better safe than sorry' (Bennett, 2000), is cited in many of these debates to delay action. The aim of the precautionary principle is to prevent actions being taken when there is any uncertainty about the outcome. However, it can be misused, as when uncertainty is cited as an excuse for inaction in the face of predicted environmental catastrophe. (The precautionary principle is discussed in relation to climate change in section 2.5.2.)

1.6 The politics of science

Ideally, every decision concerning our environment is based on sound academic research, which is converted into sensible policy for protecting the environment. However, there are many factors which come into play and affect the conversion of up-to-date information about our environment into environmental, political, economic and social policies (Figure 1.2).

Global environmental issues are complex problems, and research involves a mixture of detailed investigations into specific problems as well as

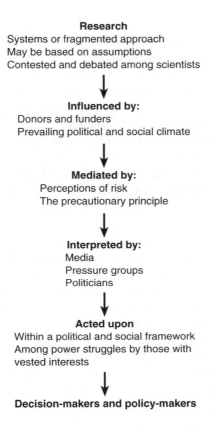

Research
Systems or fragmented approach
May be based on assumptions
Contested and debated among scientists

Influenced by:
Donors and funders
Prevailing political and social climate

Mediated by:
Perceptions of risk
The precautionary principle

Interpreted by:
Media
Pressure groups
Politicians

Acted upon
Within a political and social framework
Among power struggles by those with
vested interests

Decision-makers and policy-makers

Figure 1.2 Factors affecting the interpretation of scientific research to develop environmental, political, economic and social policies.

putting together the results from many investigations to try to understand larger systems (e.g. in the case of biodiversity, research on many species is used to develop models of ecosystems).

The scale at which research is carried out is important. For example, those who view land degradation at the local scale only may fail to see the relevance of global market conditions or terms of trade to local natural resource management. Likewise, those making decisions concerning global environmental problems need to consider the ramifications for national economies and local populations. Harris (section 5.4) discusses the range of stakeholders concerned with biodiversity, and shows that goals at different scales can vary widely. It is important to integrate natural and social sciences approaches in our investigations aimed at understanding the relationship between humans and the environment (Peterson, 2000). Consideration of global environmental issues through an inter-disciplinary framework is more likely to be of help in finding appropriate solutions to resolve environmental problems.

The need to find answers to pressing environmental concerns has increased the pace of environmental research. This has meant the development of new methodologies, new equipment and in some cases the acceptance of given assumptions as probably true, so that further research may be carried out. Little wonder then that sometimes research is contested and seen as right by some and wrong by others. There are strategies for assessing the validity of scientific research, mostly via scrutiny by other researchers. However, it should be remembered that research results are often debated amongst researchers, and usually only time will tell whether the uncertainty is well founded, or whether the research is subsequently proven right. In their hurry to address environmental problems, governments and policy-makers often want to have results before scientists are certain that they have carried out all the research necessary. For example, at the moment, much of the political debate about climate change is based on the findings of the Intergovernmental Panel on Climate Change (IPCC), but even IPCC documents contain lists of further research required (Parry, 2001). For some environmental issues, 'all that is certain is our uncertainty' (Harrison, 1993, p. 59).

Research is often influenced by the nature of funding available, which dictates what projects can and cannot be pursued. This may be directly controlled through corporate funding of research (e.g. medical research funded by pharmaceutical companies, research relating to fossil fuels and energy use), or by the fact that much government research funding is channelled through research councils, which themselves set priority or target areas of research. For example, several international organisations and the British Department for International Development are focusing development research on the 'poorest of the poor'. The targets or priority areas set by research councils often reflect the prevailing political and social climate at the time. In this subtle way, academic research is directed by the allocation of research funding: academics do not necessarily research the most pressing issue, but the most viable one.

Academic research is published in refereed academic journals. These are read by the academic community, but not by politicians and decision-makers. Research results take time to be published, and then must be disseminated to those it is hoped to influence. Key players in the dissemination process are often funding agencies, the individual researchers themselves and those for whom the research has a particular interest: pressure groups, particular sectors of the economy, etc. These groups promote the research results to a wider audience. The media can play a big role, either through television, radio, newspaper or magazines. Pressure groups often struggle to interest the media in presenting the research. The problem is that in the rush to simplify research into a short, easily understood sound bite or headline, it is often rephrased, and may be subject to reinterpretation by the disseminating groups. This can result in modification or alteration of the initial results. It most certainly results in simplification, so that caveats, nuances and conditions in which the results are held to be true are often disassociated from the results. A chain of reinterpretation ensues, along which the results may suffer from further distortion or misrepresentation. Furthermore, people can pick up part of the story, and add it to other information, to come to new, and not necessarily valid, conclusions. The Internet has been an incredibly important tool in this process. It is able to transfer information across the globe in very short periods of time. Anyone can set up an Internet site and post information on it, regardless of its authenticity. Individuals can search the global Internet for information on a particular topic, which will pick up valid and invalid information.

On the other hand, environmental pressure groups and the media are very effective at trumpeting news and bringing it to the attention of those who matter: politicians, decision-makers and the wider public who ultimately vote for politicians. The media can also play a role in educating the wider public on environmental concerns, and how their activities can influence sustainability.

When scientific information finally reaches those who are making political decisions which affect our environment, those decision-makers may not take actions that would seem sensible to the academic researchers. Decision-makers act upon new information within the political and social framework of the day. This includes the 'cultural filters' (Pepper, 1986) referred to earlier in this chapter. They juggle many priorities and urgent issues, and consider the implications of one decision on all the other concerns (such as the economy or social welfare). They are also influenced by so much incoming information, from other pressure groups, concerning international issues, concerning their own re-election, etc. Thus the likelihood of decisions being made which reflect the recommendations of the researchers is not high.

1.7 Conclusions

This introductory discussion aims to provide an analytical framework to the next nine chapters. As this book deals with environmental problems on a global scale, we need to step back from our particular cultural viewpoint and local or national needs to consider other points of view and priorities. Taken together, the previous sections on environment and culture, population–environment theories, ecological footprints, science and risk should help the reader to realise that environmental issues are not just scientific issues based on a global ecology; they are also political and social issues, framed by our cultural filters, political power struggles, aspirations for quality of life as well as the environment in which we live. We need to accept that there are many ways of valuing the environment and philosophies concerning how we, as humans, should interact with the environment.

The traditional population–natural resource debate has been adjusted over the years as we have acknowledged the adaptability of humans in the face of changing environmental conditions and the role that technology can play with regard to our impact on the environment (both good and bad). The interpretation of environmental science by non-scientists, and its translation into effective policy- and decision-making, is important in influencing how we react to global environmental issues. The science behind global environmental issues is rarely uncontested. How science is used also depends on the historical precedents which have resulted in patterns of access and entitlement to resources, and the distribution of power among the many different individuals, societies and nations which are affected by the global environmental issue (Bryant, 1998). Rees (1997) discusses the politics of negotiations concerning controlling global warming and CO_2 emissions. Nunn (Box 3.1) discusses the role of politics in influencing the interpretation of sea-level gauges in the South Pacific.

It is apparent that there are huge distortions in food availability and lifestyles/standards of living across the globe, which can be broadly seen as being a division between developed and developing countries or a north–south divide. This gross global inequity should have us all considering the difference between rights, needs, demands and desires. The adaptability of populations as a result of environmental change, and the willingness of populations to curb lifestyles and activities to avoid environmental damage, is important for our collective future. This is mediated by our perceptions of risk. As Newby (1991) states, the solutions to environmental problems rarely result from technical fixes alone, but rather from the interplay between technology and humans.

This book is divided into five parts. This part provides an introduction to human–environment interactions. The second considers four broad topics: climate change, fluctuations in sea level, changing land cover and conserving biodiversity. The third considers the challenges facing us as we seek to ensure food and energy supplies for the global population. The fourth considers our impact on the environment and how we cope with it. Urbanisation and pollution are the two topics focused on here. The final part of the book discusses the concept of sustainable development, and seeks to consider this within the framework set out in the introduction, and in the light of the information provided in other chapters. The book seeks to challenge readers to consider how choices made in our own environment affect livelihoods across the globe, and the ethics of current management of global environmental issues.

Given the scope of the book and the space available, the chapters do not seek to present all the scientific information concerning each global environmental issue. Instead, the aim is to provide a geographical perspective on environmental problems which are currently of global concern. Therefore, in addition to discussing the biophysical aspects of global environmental issues, each chapter will illustrate the interaction of environmental, technical, socio-economic and political factors in determining why and how people use and manage natural resources. This perspective considers how human–environment interactions affect global environmental issues.

Each global environmental issue has been presented in an individual chapter, but in reality they are interconnected. For example, changes in climate affect sea-level change, biodiversity and agriculture, and changes in energy affect climate change, sea-level rise and pollution. Any attempts to resolve global environmental issues and work towards sustainable development will need to take into consideration the interrelationships between global environmental issues and the ramifications of changes in one issue on all of the others. As we seek to achieve sustainable development, we shall need to be responsive to the dynamics of the global ecosystem as new developments in technology and resource exploitation, and changes in the distribution of resources, alter the global environmental system we are dealing with. Thus the concept of sustainable development is complex, linking environmental, ecological, social and political issues surrounding each global environmental issue, the interactions between global environmental issues, the role of society in adapting to, and causing, environmental change and the uncertainty of the future. Achieving sustainable development will require management of natural resources underpinned by good biophysical science alongside actions that confront social, political and economic issues as well as technological changes.

Further reading

Harrison, P. (1993) *The Third Revolution. Population, Environment and a Sustainable World*, London: Penguin.
A very readable book considering the implications of population growth, rising consumption and damaging technologies for the environment. It debates the effect of man's environmental impact on sustainable development.

O'Riordan, T. (1981) *Environmentalism*, 2nd edn, London: Pion.
Pepper, D. (1986) *The Roots of Modern Environmentalism*, London: Routledge.
These two books provide an excellent background to modern environmental ideology.

Sumner, D. and Huxham, M. (2000) *Science and Environmental Decision Making*, Harlow: Pearson Education Ltd.
Explores the use and limitations of science in environmental decision-making.

Wackernagel, M. and Rees, W.E. (1996) *Our Ecological Footprint: Reducing Human Impact on the Earth*. The new catalyst bioregional series, **9**. Philadelphia: New Society Publishers.
This book presents the concept of ecological footprints in detail, followed by examples for various regions.

Part Two

The Changing Surface
of the Earth

Chapter 2
A Change in the Weather? – Coming to Terms with Climate Change

Mike Hulme

2.1 Why is climate change a problem?

The climate of the earth has never been stable, least of all during the history and evolution of the human species. Glacial periods, globally, have been 4–5 °C cooler than now, and some interglacials have perhaps been 1–2 °C warmer. These changes in climate were clearly natural in origin (i.e. not related to human behaviour), occurring on a planet inhabited by primitive societies with far smaller populations than at present. Indeed, the regularity of the diurnal and seasonal rhythms of our planet has always been overlain by interannual, multidecadal and millennial variations in climate, over whatever timescale climate is defined. The works of great historical climatologists such as Ladurie (1972) and Lamb (1977) have made this very clear and shown that human history itself is partly driven by such (natural) variations in climate.

The causes of contemporary and future changes in climate, their rate and their potential significance for the human species, however, are all notably different from anything that has occurred previously in history or prehistory. The **causes** are now dominated by human perturbation of the atmosphere, the **rate** of warming already exceeds anything experienced in the last 10 000 years and is set to be more rapid, probably, than anything experienced in human history, and the **significance** for humanity is qualitatively different from the previously given ecological imprint made by our current and growing population of 6 billion and more. Yet at one level, these present and prospective changes in climate simply

Global Environmental Issues.   Edited by Frances Harris
© 2004 John Wiley & Sons, Ltd   ISBNs: 0-470-84560-0 (HB); 0-470-84561-9 (PB)

involve a geographical shift in weather phenomena from one part of the world to another and/or a change in the frequency or intensity with which the variety of weather events occur in a particular place. Climate change is **not** therefore causing the emergence of a new weather phenomena hitherto unknown to our species (Brönnimann, 2002). (In this regard, climate change is perhaps different from some other environmental concerns such as ozone depletion, or genetically modified organisms, where fundamentally novel chemical or organic species have emerged, or may emerge in the future, possibly to directly or indirectly cause damage to human health.) Neither is the experience of climate change novel for the reason suggested by Hulme and Barrow (1997). The exploration, and later colonisation, by European powers of the tropical world during the 17th to 19th centuries can in some sense be viewed as the discovery of new climates – the experience of climate change – and subsequent exploitation of these new climatic environments for the benefit of the trading companies and nations (if not for the indigenous inhabitants). Similarly, early travellers from places such as the Mediterranean, North Africa and India experienced new climates and environments during their travels.

So if climate change is not a new experience for humanity and if climate change is more a rearrangement of existing weather patterns than of unprecedented novelty of phenomena, why is the prospect of future climate change now overwhelmingly seen as a problem?

> We would be irresponsible to treat these [climate] predictions as scare-mongering … we cannot afford to ignore them.
>
> *Tony Blair, UK Prime Minister, 6 March 2001*

> Burning fossil fuels and using the atmosphere as an open sewer has turned out to be a recipe for disaster. The Earth is warming and the pace is quickening.
>
> *New Scientist, Global environment supplement, April 2001*

> We are constantly reminded about the need to respond to the challenges that climate change will bring.
>
> *Margaret Beckett, UK Secretary of State for Environment,*
> *Food and Rural Affairs, UK Third National Communication, October 2001*

It is a frequent refrain of politicians and environmentalists that climate change is the greatest environmental threat facing the earth. But is this the case? And, if so, why?

2.1.1 Climate and society

From micro-scales (gardens), to meso-scales (city pollution, acid rain), humans have long held the capacity to alter climate and weather advertently or inadvertently. The sheer numbers of people on the planet, their demand for (predominantly fossil-carbon) energy and the waste by-products of consumptive societies, have now, however, created the potential for humans to alter the global characteristics of the physical planetary system that delivers weather and climate to individual people and places. Whereas in the past, climates shaped and constrained societies, now global society itself can modify global and regional climate. The relationship between climate and society is now fully reflexive. A recognition of the nature of this relationship is what Thornes and McGregor (2002) demand in their redefinition of 'cultural climatology'.

The atmosphere delivers both resources (e.g. rain, sun, wind) and hazards (e.g. hurricanes, blizzards, droughts) to ecosystems and societies. Through autonomous and/or planned adaptation, species and individuals, ecosystems and societies are fashioned to a considerable extent by these climatic constraints. Our human cultures and economies are 'tuned' to the climate in which they evolve. Yet highly successful cultures develop in varying climates: cold/dry climates (e.g. Finland), cold/wet climates (e.g. Iceland), hot/dry climates (e.g. Saudi Arabia) and hot/wet climates (e.g. Hong Kong) (Figure 2.1). All societies have therefore evolved strategies to cope with some intrinsic level of climatic variability – for example, nomadic pastoralism, flood prevention, building design, weather forecasting, early warning systems and the weather hedging industry are all forms

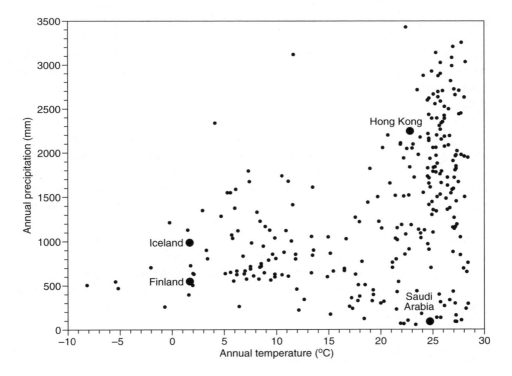

Figure 2.1 The climates of the world, by country – average national temperature plotted against average national precipitation. Each dot represents one nation. Averages are for the period 1961–1990. (Source: Tim Mitchell, Tyndall Centre.)

of human response to the variability of climate or the extremes of weather. Consequently, there exists some degree of variation in climate or some frequency or severity of weather extremes that can be 'accommodated' using existing strategies. Exactly **what** can be accommodated, however, varies greatly within and between societies, so that vulnerability to weather and climate change is strongly differentiated. For example, in developed, northern nations it is the elderly who die during heatwaves, and in developing countries it is often the poor and marginal who have their homes washed away in shanty towns built on flood plains.

2.1.2 The challenge of climate change

So the central concern is **not** that humans are altering climate – we have modified our environment to a marked extent throughout our history, although we have now increased the scale of this modification – but **whether** these changes in climate can be accommodated using our existing capacity to adapt, drawing upon our intellectual, regulatory, social or financial capital? An important supplementary question is whether we can consciously enhance this adaptive capacity, especially of the most vulnerable in our societies, to exploit the changing resources and minimise the changing hazards delivered to us by our (now) semi-artificial climate. Additional questions that flow from this perspective are: to what extent can we (need we?) predict future climates to assist this process of adaptation, and to what extent do we need (and desire) to reduce the size of the changes in the climate facing us to allow our adaptive potential to sustain an acceptable, dynamic equilibrium between climate and society?

This chapter addresses this sequence of questions with the view of using global climate change for a more holistic understanding of social and environmental change and interpreting its significance in the light of a dynamic view of the relationship that exists between climate and society. The chapter starts by summarising the physical nature of the problem, both the climate change that has been unambiguously observed and the climate change that is conditionally predicted. The consequences for society of this prospective change in climate, both direct and indirect, are discussed, paying particular attention to the nature of adaptation. Attention is then turned to the future prospects for global climate management, evaluating the constraints and frameworks within which any future strategy will have to develop. In conclusion, the importance is stressed of integrating climate change policies with the broader objectives of sustainable development.

2.2 The nature of the problem

Along with world population growth and global economic expansion over the past century, intensified human activities, particularly energy-intensive activities, have altered the properties of the earth's atmosphere by unlocking a vast quantity (~300 Gt) of underground fossilised carbon and emitting it into the atmosphere in the form of CO_2. Other greenhouse gases such as methane, nitrous oxide and a variety of halocarbons have also been injected into the atmosphere by humans, and land cover changes have impaired the capacity of parts of the biosphere to sequester atmospheric carbon into living biomass. This has altered the functioning of the global climate system. As concluded by the latest assessment report of the Intergovernmental Panel on Climate Change (IPCC, 2001a), the earth's climate system has demonstrably changed on both global and regional scales since the pre-industrial era, with at least some of these changes clearly attributable to human behaviour.

2.2.1 Observed changes in climate

The world has warmed since 1860 when the instrumental record started (Figure 2.2). During the 20th century, the earth's surface temperature increased by about 0.6 °C, with the 1990s being the warmest decade and 1998 the warmest year. Warming has continued into the 21st century, and the year 2002

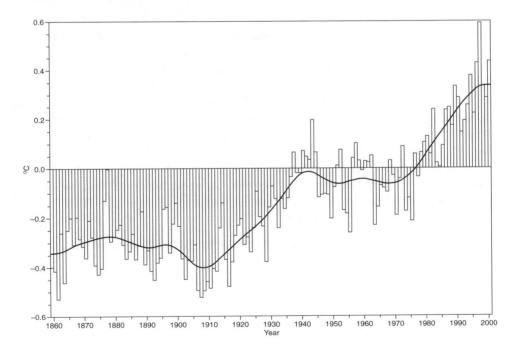

Figure 2.2 Globally averaged surface air temperature from 1860 to 2001. Anomalies (degrees Celsius) are from the 1961–1990 average. Smooth curve emphasises variations over 30 years and more. (Source: Phil Jones, Climatic Research Unit, University of East Anglia.)

is second only to 1998 for observed global warmth. Using proxy data across the northern hemisphere, annual temperature series have also been established for the past millennium (Jones, Osborn and Briffa, 2001). These reconstructions indicate that the 1990s is likely to have been the warmest decade in the past thousand years. As well as the rise in globally averaged temperature, changes in other features of global climate have also taken place: precipitation has very likely increased during the 20th century by between 5 and 10% over most mid- and high-latitudes of the northern hemisphere continents; globally averaged sea level rose by between 1 and 2 mm/year during the past 100 years; snow cover and sea ice extent have decreased in the northern hemisphere; and since the mid-1970s, warm episodes of the El Niño/Southern Oscillation (ENSO) phenomenon have become more frequent, persistent and intense compared with the previous 100 years.

2.2.2 Reasons for climate change

Global climate varies naturally, due to both 'internal variability' within the climate system and changes in external forcing unrelated to human behaviour – for example, changes in solar irradiance and volcanic activity. The reconstruction of temperature over the past thousand years suggests, however, that the warming over the 20th century is unusual and unlikely to be merely the response of the system to natural forcing. Indeed, detection and attribution studies consistently find evidence of an anthropogenic signal in the climate record of the last 35–50 years, despite uncertainties in forcing due to anthropogenic aerosols and natural factors (Mitchell and Karoly, 2001). Recent climate-model experiments show that natural causes of global temperature variability cannot, on their own, explain the observed surface warming of about 0.6 °C. On the other hand, when these experiments are repeated using rising historic concentrations of greenhouse gases and shifting distributions of sulphate aerosols, much better agreement between observed and modelled global patterns of temperature change is achieved. Hence, the IPCC Third Assessment Report concluded that:

> there is new and stronger evidence that most of the warming observed over the last 50 years is attributable to human activities.
>
> *IPCC, 2001a, p. 5*

2.2.3 Future changes in climate

This human-induced climate change will almost certainly continue in the decades and centuries to come. Actions to mitigate climate change may yet slow the rate of climate change, but will almost certainly not stop it. To prepare our societies better for the changes in climate ahead, and to identify possible critical thresholds in the climate system, considerable efforts have been made to project the likely regional and global climatic consequences of a range of plausible socio-economic development pathways.

Taking into account a wide range of possible future greenhouse gas emissions (IPCC, 2000, Figure 2.3), and embracing the uncertainty implicit in climate modelling, globally averaged surface air temperature is estimated to rise by between 1.4 and 5.8 °C over the period 1990–2100 (IPCC, 2001a). About 0.3 °C of this warming has already been observed to date (2002). This rate of change is about two to ten times larger than the warming observed over the 20th century and based on palaeoclimate data is very likely to be without precedent during at least the last 10 000 years. Corresponding to the above range of global warming,

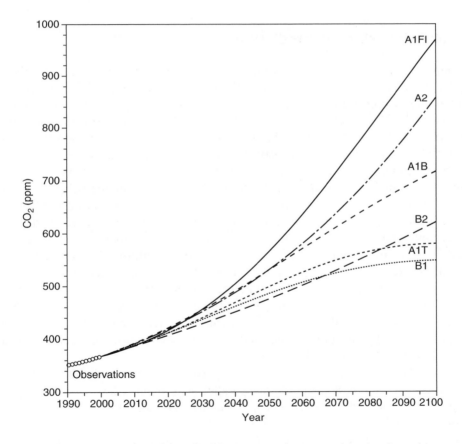

Figure 2.3 Estimated future atmospheric CO_2 concentrations assuming six alternative energy scenarios (A1B, A1T, A1FI, A2, B1, B2) specified in the Special Report on Emissions Scenarios of the IPCC (2000). Observed concentrations at Mauna Loa (Hawaii) are shown up to year 2000. (Redrawn from Bolin and Kheshgi, 2001.)

globally averaged sea level is estimated to rise by between 9 and 88 cm between the years 1990 and 2100. Globally averaged precipitation is also estimated to increase during the coming century as a direct consequence of this planetary warming (Allen and Ingram, 2002), although such increases will be far from geographically uniform. Changes in some climatic variables will be substantially different from the global average at the regional scale, and for some variables such as precipitation, cloud cover and relative humidity, the sign of change might well be different from region to region. Apart from changes in average climatic features, the accumulation of atmospheric greenhouse gases will also change the duration, location, frequency and intensity of extremes of weather (e.g. heatwaves, intense rainfall events, tropical cyclones, etc.; Easterling *et al.*, 2000), which in turn will alter and/or disrupt the functioning of many natural biophysical and human socio-economic systems.

2.2.4 Reflexivity and prediction

When discussing estimates of future climate change, it is important to distinguish between that part of future climate which is independent of human behaviour (which we represent by our (imperfect) ability to model the biogeophysics of the planetary system), and that part which we **can** influence through our perturbation of the atmosphere and of the land surface (which we represent through imagined scenarios of the future). The act of making a prediction or predictions of future climate makes no difference to the way the planet functions, but it clearly **does** alter the way in which humans behave, in turn altering the assumed scenarios of greenhouse gas emissions or land cover. This 'reflexivity' of social behaviour in relation to the real or perceived threat of climate change is itself a clear demonstration of the coevolution of climate and society (Berkhout, Hertin and Jordan, 2002). Future climate, at least in the longer term (30 years and more), is not therefore predetermined, nor is it 'predictable' using natural science methods alone. Humans can indeed shape the climate of future generations, if not the climate of their own, by the decisions they make over the coming years; and these decisions will at least in part be influenced by the predictions that we make about how our actions affect global climate.

The inadvertent biogeophysical experiment we are conducting with planetary climate is therefore resulting in a parallel, semi-managed, socio-technological experiment. The latter experiment is about whether we can consciously shape, globally, our long-term energy-technology-lifestyle development path in such a way that it minimises the dangers posed by climate change and maximises the opportunities to restructure our economies into more sustainable forms. Such a conscious global social experiment has not been attempted before and our institutions are poorly designed to manage it.

2.3 The direct impacts of climate change

A warming of global climate, of whatever magnitude, will inevitably alter the characteristics of weather experienced in all regions and localities. The climate system is fully integrated and warming it by 1 °C or more will alter, for example, the distribution and magnitude of precipitation over the earth's surface, the frequency, severity and distribution of storms around the world, and the nature of thermal regimes, especially extreme heat and extreme cold. Human modification of the global climate will also induce changes in the behaviour of natural, large-scale oscillatory phenomena in the climate system, such as the ENSO and the North Atlantic Oscillation (NAO). Changes in the attributes of these large-scale drivers of interannual climate variability will have repercussions for the functioning of ecological systems, the behaviour of species and hence, ultimately, for biodiversity (Stenseth *et al.*, 2002).

In relation to extreme weather events, however, the primary concern should not be to try and identify which weather events have been caused or influenced by global warming – from now on **all** of our weather is being influenced to a greater or lesser extent by human alterations to the global atmosphere. A more important question to answer is: Can we predict, and with what confidence, the future changes in regional and local weather characteristics that follow from this change in global climate? Estimating these changes in future weather characteristics, and how responsive they are to human actions, may prove crucial for redesigning our resource management systems, our social institutions and our policy regulations (Figure 2.4). Such pre-emptive actions offer the prospect of allowing society to be better prepared for the changes in climate that are ahead of it.

Many attempts have been made to estimate and quantify the future direct physical impacts of climate change on a range of environmental and social systems. These studies often focus on water resources (e.g. Arnell, 1999), food productivity (e.g. Parry *et al.*, 1999), forest distributions, coastal flooding (e.g. Nicholls, Hoozemans and Marchand, 1999), human health (e.g. Martens *et al.*, 1999), energy demand, tourism and the like. These resource systems and human responses are all to a greater or lesser degree sensitive to climate, and hence to climate change, and it is entirely appropriate that we undertake work aimed at quantifying just what the scope of climate change impacts might be. There are some potentially serious disruptions that will be induced by a warming climate; for example, greater risks of riverine and coastal flooding, or changes in the distribution of certain disease vectors and pathogens.

It must also be recognised, however, that future changes in climate will occur in a world that will be changing in nearly every other dimension as well (Berkhout and Hertin, 2000) and, as with climate, probably changing more

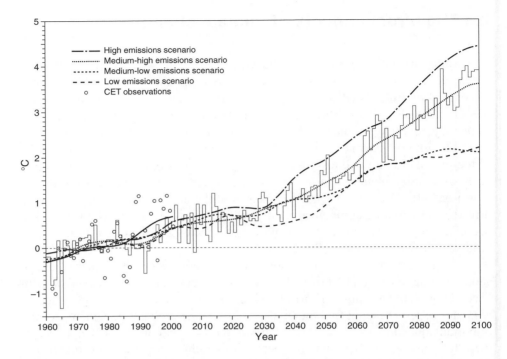

Figure 2.4 Scenarios of future annual warming over the UK, derived from four alternative emissions scenarios interpreted for regional climate using the HadCM3 climate model. These four scenarios (called the UKCIP02 scenarios) are used by the UK Government as the basis for planning adaptation strategies in the UK. Observed annual temperature anomalies are shown for 1961–2001 from the Central England Temperature record. Anomalies are with respect to the 1961–1990 average. The curves emphasise variations over 30 years and more; the annual histogram shows individual future years for the medium-high emissions scenario.

rapidly than in the past. Some of these changes are likely to exacerbate stresses or dislocations induced by climate change, but some of them may well ameliorate the envisaged climate impacts or provide opportunities for new forms of wealth creation to emerge or new abilities to manage changing environments. Customers, communities, corporations and countries will not be passive observers of these changes in climate, or dumb victims of their impacts; they will be active agents, in some cases pre-empting changes in climate through precautionary adaptation and in most cases by, at least, reacting to changes in climate and altering behaviour patterns, management strategies, investment plans and regulatory policies.

Some of the more dire predictions of the impacts of climate change (e.g. Parry *et al.*, 2001) therefore need to be taken with some caution, especially where no consideration has been given to the nature and rapidity of social change or the capacity of systems to adapt. In many senses, such studies lead to rather pedagogical statements that draw necessary attention to the potential dangers associated with climate change. But as predictions of future reality they are poorly founded if they do not recognise the broader context of social change and development or the reflexivity of the systems being analysed.

2.3.1 Adaptation

Over the history of evolution, natural and human systems have adapted to spatial and temporal variations in climate. Many social and economic systems – including agriculture, forestry, settlements, industry, transportation, human health and water resource management – have evolved to accommodate some deviations from 'normal' conditions, although there exist fewer examples of successfully accommodating the extremes of weather. Nevertheless, this capacity of systems to accommodate variations in climatic conditions from year to year is referred to as the 'coping range' or 'resilience'.

This ability of human society to adapt to, and cope with, climate change is a function of wealth, technology, information, skills, infrastructure, institutions, equity, empowerment and ability to spread risk (Adger, 2001). Groups and regions with adaptive capacity that is limited along any one of these dimensions are more vulnerable to climate change, just as they are more vulnerable to other stresses. Adaptive capacity in natural systems tends to be more limited than adaptive capacity in human systems. Many species have limited ability to migrate or to change behaviour in response to climate change.

Adaptive measures are necessary to enhance the resilience of ecological and socio-economic systems to future climate change. Adaptation to climate change can take many forms. These include actions taken by people with the intent of lessening impacts or utilising new opportunities, and structural and functional changes in natural systems made in response to changes in pressures. In terms of adaptive actions taken by human societies, the range of options includes reactive adaptations (actions taken concurrent with changed conditions and without prior preparation) and planned adaptations (actions taken either concurrent with or in anticipation of changed conditions, but with prior preparation). Adaptations can be taken by private entities (e.g. individuals, households or business firms) or by public entities (e.g. local, state or national government agencies). Box 2.1 illustrates a case study which explored the factors contributing to the vulnerability of individuals and social groups to climate extremes, and the roles of socio-economic and institutional changes in the process of adapting to climate extremes in part of Vietnam.

Numerous possible adaptation options for responding to climate change have the potential to reduce adverse, and enhance beneficial, impacts of climate change. Yet these options will usually incur cost. This reveals a major challenge for policy-makers: how do they plan these costly adaptive strategies against a threat that may be likely, but is uncertain in the precise way that it will manifest?

Box 2.1: Assessing social vulnerability and institutional adaptation: a case study in Vietnam

Conventionally, climate change vulnerability assessments have followed a physical approach, that is, focusing on the physical impacts of climate change on natural systems. The social aspects of climate change impacts have often been under-emphasised. In this case study, a framework for analysing social vulnerability was developed to elaborate the complex nature of social vulnerability and the importance of the context of the political economy. Poverty, resource dependency and social resilience, inequality and institutional effectiveness are identified as the key determinants of social vulnerability. Employing individual-level and collective-level vulnerability indicators (e.g. poverty indices, dependency and stability, gross domestic product (GDP) *per capita*, relative inequality, etc.), the framework was applied to determine the vulnerability of the population of Xuan Thuy District in coastal northern Vietnam.

In general, the population exhibits resilience through its use of available natural resources, but the liberalisation process has had, at best, an ambivalent impact on vulnerability as a whole by undermining some institutional practices which acted as security and coping mechanisms in terms of stress. In Xuan Thuy, the privatisation of mangrove forests and their conversion into aquaculture is the major cause for increasing inequality in income. This concentration of resources in fewer hands constrains entitlements to use and disposal of assets under coping strategies in times of stress. Institutional adaptation appears to have offsetting influences on the vulnerability of Xuan Thuy District, given present patterns of land use, land ownership and control and the role of the state in resource and risk management. The reduction in power and autonomy of state institutions associated with collective measures for protection from the impacts of coastal storms is one major accentuation of vulnerability observed in the case study. The agricultural co-operatives no longer play a primary role in allocating labour and resources to collective action for water and irrigation management or for coastal defence. The atomisation of agricultural decision-making has, however, contributed to increased marketed production of agricultural commodities, thereby contributing to the rising incomes in the District, but this has been at some cost to collective security. Offsetting these impacts associated with the rolling back of the state has been the re-emergence of informal, social coping mechanisms associated with both entrepreneurial and community activities.

Source: Adger, 1999

The problem is particularly acute for developing countries. Many developing nations are vulnerable to current climate hazards (Smit and Pilifosova, 2001). Most have highly variable climates and all are limited in their capacity to adapt to extremes of weather. Despite limited contributions of historical greenhouse gas emissions, developing countries are highly vulnerable to the impacts of projected climate change (Apuuli *et al.*, 2000). Hence, adaptation to climate change is a development issue that competes for resources with other development issues such as food security, social equity, education and health (Rayner, 2000). The final section of the chapter returns to this theme.

2.4 The indirect effects of climate change

Quite distinct from the possible direct physical impacts of climate change are what might be called the 'indirect effects' of climate change. These indirect effects result from changes to economic and social policy that follow from the adoption of some form of climate change management strategy and the consequences of these policy changes for individuals, communities, corporations and nations. These indirect effects of climate change are already being experienced in the UK and some other countries.

At a global level, climate change management has taken on the form of an international convention, the UN Framework Convention on Climate Change (UNFCCC). Most of the world's nations have signed this Convention and it came into force in 1994. The ultimate objective of the Convention expressed in Article 2 states:

> to stabilise greenhouse gases concentrations in the atmosphere at a level that would prevent dangerous anthropogenic interference with the climate system...
>
> *UNFCCC, 1992a*

What level is implied by this objective is of course subject to political negotiation and may well change over time. There are a number of different approaches that can be taken to defining 'dangerous' (O'Neill and Oppenheimer, 2002). The text of the Convention provides some further direction for such negotiation by stating that such stabilisation should be achieved:

> within a time frame sufficient to allow ecosystems to adapt naturally to climate change, to ensure that food production is not threatened, and to enable economic development to proceed in a sustainable manner.
>
> *UNFCCC, 1992a*

Dangerous climate change might also be defined more radically, however, according to the future viability of certain sovereign states, such as Pacific atoll nations, whose very existence might be threatened by sea-level rise (Barnett, 2001).

There is also an important distinction to make between defining dangerous climate change using external indicators, such as millions of people at risk (Parry *et al.*, 2001) or destruction of coral reefs, and the internal perception of danger or insecurity as experienced by individuals or communities. These latter perceptions may in the end prove more powerful in influencing agreed international targets than externally imposed biophysical or economic definitions of what constitutes dangerous climate change.

If the ultimate objective of the current climate change management strategy is difficult to define, how any such objective will be **achieved** is even less clear. The Kyoto Protocol (Box 2.2), drafted in 1997 but only likely to be ratified sometime after 2003, contains the first attempt to lay out a set of procedures and mechanisms that start the process of delivering on the objective, that is, on starting to consciously reduce global greenhouse gas emissions below levels that would otherwise have occurred. Despite the tortuous path towards ratification and the obstacles in the path of achieving full global ownership, the terms of the Protocol have already triggered a stream of actions and reactions at both national and corporate levels that should already be seen as indirect effects of climate change and part of the reflexive relationship between climate and society. In the UK, for example, these effects include the introduction of a Climate Change Levy in 2001 (basically an energy tax applied in the non-domestic sector) and the commencement of a carbon emissions trading regime in 2002.

Box 2.2: The Kyoto Protocol

Five years after the adoption of the UNFCCC, on 11 December 1997, governments took a further step forward and adopted the landmark Kyoto Protocol (see section 7.2.1). Building on the framework of the UNFCCC, the Kyoto Protocol broke new ground with its legally binding constraints on greenhouse gas emissions and its innovative 'mechanisms' aimed at cutting the cost of curbing emissions (UNFCCC, 1997). At the heart of the Kyoto Protocol lies a set of legally binding emissions reduction targets for industrialised countries, the so-called Annex I countries. These reductions amount to a total cut among all Annex I Parties of 5.2% from 1990 levels by the first commitment period, 2008–2012. The total cut is shared unequally so that each Annex I Party has its own individual emissions reduction target. These individual targets were decided upon in Kyoto through intense negotiation rather than through any formal allocation principle, and allowed some nations – for example Norway, Australia and Iceland – to actually **increase** their emissions relative to 1990 because of the particular historical or prevailing economic or energy technological circumstances of these nations.

Continued on page 35

Continued from page 34

Table 2.1 Annex I countries and their emissions reduction targets as stated in the Kyoto Protocol (Source: UNFCCC secretariat)

Country	Target (1990*–2008/2012)
EU-15, Bulgaria, Czech Republic, Estonia, Latvia, Liechtenstein, Lithuania, Monaco, Romania, Slovakia, Slovenia, Switzerland	−8%
US**	−7%
Canada, Hungary, Japan, Poland	−6%
Croatia	−5%
New Zealand, Russian Federation, Ukraine	0
Norway	+1%
Australia	+8%
Iceland	+10%

* Some economies in transition (EIT) have a baseline other than 1990.
** The US has indicated its intention not to ratify the Kyoto Protocol.

The 15 member States of the European Union will take advantage of a scheme under the Protocol, known as a 'bubble', to redistribute their collective 8% reduction target amongst themselves.

The Protocol broke new ground with its three innovative additional mechanisms which can be used to achieve emissions reduction: joint implementation (JI), the clean development mechanism (CDM) and emissions trading. JI refers to emissions reduction schemes in other industrialised nations, CDM refers to emissions reduction schemes in non-industrialised nations, while emissions trading involves buying and selling carbon-emissions certificates between industrialised nations. These additional mechanisms aim to maximise the cost-effectiveness of climate change mitigation by allowing Parties to pursue opportunities to cut emissions, or enhance carbon sinks, more cheaply abroad than at home. The cost of curbing emissions varies considerably from region to region as a result of differences in, for example, energy sources, energy efficiency and waste management. It therefore makes economic sense to cut emissions, or increase removals, where it is cheapest to do so, given that the impact on the atmosphere is the same wherever the carbon emissions are avoided or sequestered. However, there have been concerns that the mechanisms could allow Parties to avoid taking climate change mitigation action at home, confer a 'right to emit' on certain Parties, or lead to exchanges of fictitious credits which would undermine the Protocol's environmental goals. Projects implemented under these mechanisms therefore need to be evaluated to ensure the Protocol and Convention's environmental goal is attained and its integrity preserved.

Source: Grubb, Vrolijk and Brack, 1999; Victor, 2001

This will be followed by the introduction of a European carbon trading regime in 2005. These policy measures are altering the way in which some businesses, especially those in the energy sector, assess their corporate policy and long-term strategic objectives. They have also set in motion a reassessment by investment-fund managers about which companies are most likely to offer a good return on investment in a future carbon-constrained world. In terms of business risk and management these are real and measurable (indirect) effects of climate change.

2.5 The future of global climate management

Given the reality of climate change, society's reflexive and adaptive relationship with climate and the existing global climate change management regime expressed through the UNFCCC, what does the future hold for our efforts to consciously manage climate and its impacts on society? Our newly found, but largely unwanted, ability to alter the climate of our planet presents human society with a profound challenge. As a species we substantially adapt to our ambient climate(s); we depend intimately upon the resources this climate delivers to us; and we attempt to protect our assets against the climatic hazards experienced. Changing climate in far-reaching and unpredictable ways disrupts this sometimes delicate relationship between climate and society, and introduces a new obstacle to achieving sustainable development and economic growth. On the other hand, changes in climate may also present new opportunities for different forms of economic activity and for improving climate hazard management, if these can be grasped.

2.5.1 Shaping climate policy

The scale of this challenge is hard to overestimate. With prospective global warming of between 1.4 and 5.8 °C over the coming century – a two to tenfold increase in the historical rate of warming observed – we need actions that both reduce the size of the problem and allow society to cope sustainably with whatever residual change in climate will occur. To design and implement these actions we need an overall global climate management strategy that recognises a number of fundamental constraints. Three of these are mentioned here.

First, we must recognise that mitigation and adaptation policies need designing in parallel. Efforts to reduce the magnitude of climate change and efforts to enhance society's coping capacity are not alternatives; they need to be tackled simultaneously, identifying appropriate synergies between respective measures. Mitigation efforts can buy time for adaptation, whilst adaptation can raise the thresholds of environmental or social tolerance that need to be avoided through mitigation efforts. For example, our future building-stock needs designing **both** for a warmer climate with a different regime of weather extremes **and** at the same time to be less carbon intensive in its materials and energy demands.

Second, we must recognise the truly global nature of the challenge. For example, we **need** a (near) comprehensive international framework for designing effective climate policies, such as the Kyoto Protocol, in order to make meaningful collective progress. The present (and past) imbalance in *per capita* carbon emissions between different world regions highlights both the inequity in our use of a global 'good' – the atmosphere – and the extent of the challenge, since 80% of the world's population presently live in nations not participating in the Protocol (Figure 2.5). This need for a global framework for effective climate management reveals the parallel need for appropriate global forms of environmental governance (O'Riordan, 2001). These are presently lacking, in stark contrast to the growing power and influence of the World Trade Organisation as a body that regulates global trade.

Third, we must recognise that there are widely differing values, interests and perceptions existing both within and between societies that will hinder attempts at reaching a global compromise in relation to climate policy. On the one

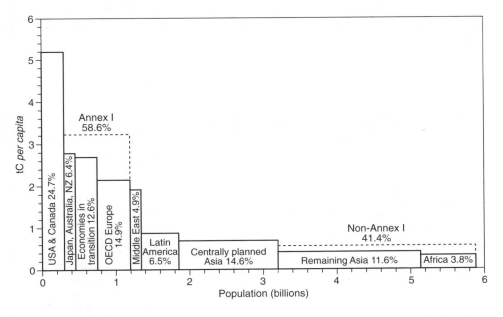

Figure 2.5 *Per capita* fossil carbon emissions in 1999 averaged for nine geographical regions. Height of bars gives the average *per capita* annual emissions of each region. Width of bars gives the population. The percentages given indicate the fraction of 1999 global emissions attributed to each region. Less than 20% of the world's population resides in countries that have formal emissions reductions targets under the Kyoto Protocol (the so-called Annex I countries). (Redrawn from Bolin and Kheshgi, 2001.)

hand, this is expressed through the wording of Articles 4.8 and 4.9 of the UNFCCC, where it is stated that measures should be taken to reduce not only the direct impacts of climate change but also the impacts of emissions-reduction policies on the economy of energy exporting countries, such as Organisation of Petroleum Exporting Countries (OPEC) (Barnett and Dessai, 2002). On the other hand, it is obvious that individual citizens within a given nation sometimes express deeply differing levels of awareness of and concern about climate change when asked to express their views. For example, individual citizens polarise into a typology of 'deniers', 'disinterested', 'engagers' and 'disengagers' (Figure 2.6), these differing psychologies inevitably influencing the way in which climate policy measures are perceived and accepted. Reconciling such differences may well be beyond any single climate policy measure, but understanding such 'discourse coalitions' and how they operate in society may well be essential for policy effectiveness (Gough and Shackley, 2001).

Given these, and many other, constraints, what intellectual frameworks exist within which a global climate management strategy might be developed and operationalised, and what demands do they each make on our knowledge base? Three such overarching frameworks are suggested, one of which has been predominant to date (the precautionary principle), one of which is likely to remain unattainable (optimisation) and one of which needs further development and considerable additional input from both science and society (risk assessment and management).

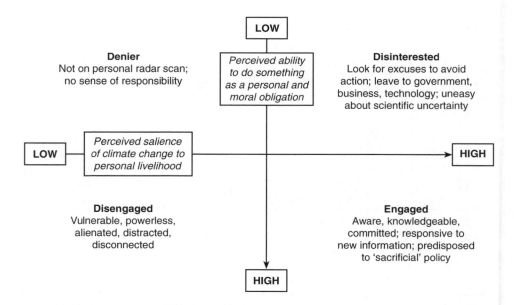

Figure 2.6 A typology of people's attitudes to climate change derived from questionnaires and focus groups in the Norwich area, UK. (Source: Irene Lorenzoni, University of East Anglia, unpublished.)

2.5.2 The precautionary principle

The precautionary principle and its application in recent years to a variety of environmental problems have been well analysed (e.g. Harremoës *et al.*, 2002). There is no consensual definition of the precautionary principle, but one suggestion is:

> a general rule of public policy action to be used in situations of potentially serious or irreversible threats to health or the environment, where there is a need to act to reduce potential hazards before there is strong proof of harm.
>
> *Harremoës et al., 2002*

In effect, precaution underlies the construction of the UNFCCC and its objective of stabilising atmospheric concentrations of greenhouse gases at a level that avoids dangerous climate change. This objective was drafted in 1992, well before anyone had any firm basis for establishing what was dangerous and what was not. As we have already seen, 10 years on and the notion of dangerous climate change is still notoriously difficult to make operational through consensus international politics, and may remain so for the foreseeable future. In effect, therefore, the precautionary principle delivers us an arbitrary target for emissions reduction that is either the outcome of the **real politik** of international negotiations (cf. Kyoto's 5.2% reduction target for Annex I nations) or else that reverts to the widely quoted, but poorly founded, stabilisation level for CO_2 in the atmosphere of 550 ppm (roughly equivalent to a 60% reduction in global emissions). This level equates roughly to a 2 °C global warming and yet its origins can be traced back, almost 20 years, to one of the first international assessments on climate change and is largely a function of estimated rates of ecosystem migration. Futhermore, not only does the precautionary principle give little helpful steer towards the establishment of a sound and defendable target for mitigation policy, but it allows little formal role for the processes of social adaptation.

2.5.3 An optimised benefit/cost analysis

A fundamentally different framework within which climate management strategies might possibly be devised is that of economic benefit/cost analysis. Not only is this framework espoused by certain economists, it was recently adopted by the statistician Bjørn Lomborg and it provided him with the basis for arguing that the Kyoto Protocol was based on poor science and on bad economics and should be abandoned (Lomborg, 2001). In essence, applying this framework requires costing the full economic costs associated with different levels of climate change, together with the costs associated with different mitigation targets, and using benefit/cost analysis to identify the optimal mix of mitigation and adaptation interventions. On this basis, Lomborg argues that the most beneficial strategy is to abandon all international mitigation efforts driven by the

Kyoto Protocol and to invest the saved resource in improving basic human needs in developing countries.

At first sight, this argument, and indeed the economic benefit/cost framework, has an appeal. However, the range of demands placed on our knowledge base in order to operationalise such a framework suggests that it remains a flimsy basis for devising climate management strategies. For example, three well-debated deficiencies in our knowledge base in this area are: (a) the full economic, including non-monetary, costs of different levels of climate change, (b) the full economic costs of achieving different CO_2 stabilisation levels and (c) representing the processes of adaptation and learning in both (a) and (b) over a 50 or 100 years time span. Many more could be cited. Economic benefit/cost analysis has some serious flaws in relation to its use for long-term global climate management, and it places massive, and probably unattainable, demands on the scientific, social and economic knowledge base. Yet it remains an often-implicit intellectual pillar for many who are opposed to the current Kyoto policy regime.

2.5.4 A risk framework

This framework perhaps offers the prospect of progressing beyond a rather flabby and arbitrary application of the precautionary principle, which can too easily be used to support almost any climate policy depending on the predilection of the proposer, and yet does not require us to go so far as to pretend we know enough to apply classical benefit/cost analyses to the problem of global climate change. In adopting a risk framework, we need, *inter alia*, to identify critical climate thresholds in natural and social systems, to estimate their probabilities of being exceeded, to understand the psychology of risk held by different communities and

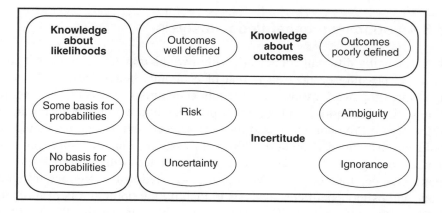

Figure 2.7 A simple typology of the difference between risk and uncertainty. (Source: SPRU, 1999.)

cultures, and to allow political leaders to work with this raw material to lead societies in a direction in which climate change policies are acceptably balanced against the objectives of national development and the aspirations of individual citizens. This framework also places exacting demands on our knowledge base (Figure 2.7), but ones that are perhaps more attainable. This framework also recognises the complementary domains of risk assessment (science), risk perception (society) and risk management (government).

2.6 Climate change and sustainable development

Whatever framework, or combination of frameworks, come to dominate the development of long-term global climate management, it will need to increasingly recognise the pre-eminent objective of sustainable development. (Chapter 10 provides a fuller discussion of the meaning of sustainable development and the complexity of achieving this objective.) Sustainable development is a clearly stated goal of both the UNFCCC and the Kyoto Protocol (e.g. the preamble and Articles 2 and 3 of UNFCCC and Articles 2 and 10 of the Kyoto Protocol). The pursuit of sustainable development is integral to lasting climate change mitigation; equally, combating climate change is vital to the pursuit of sustainable development.

The Third Assessment Report of the IPCC concluded that climate change is expected to impact negatively on development, sustainability and equity (IPCC, 2001b). People in developing countries are generally expected to be exposed to relatively higher risks of adverse impacts from climate change, affecting human health, water supplies, agricultural productivity, property and other resources, than are people in developed countries. Poverty, lack of training and education, lack of infrastructure, lack of access to technologies, lack of diversity in income opportunities, a degraded natural resource base, misplaced incentives, inadequate legal framework, struggling public and private institutions, debt and unfair trade systems create conditions of low adaptive capacity in most developing nations. Exposure to climatic hazards, combined with low capacity to adapt, makes populations in developing countries generally more vulnerable than populations in the developed world (Adger *et al.*, 2002) (see Box 3.2 on climate change and sea-level rise). Inequities in health status and in access to essential resources are likely to be exacerbated by climate change, affecting disproportionately both the developing countries and the poor persons within all countries who are already vulnerable to current climate variability and change.

Policies aimed at reducing greenhouse gas emissions can have positive side effects on society, irrespective of the benefits of avoided climate change. Climate change mitigation strategies offer a clear example of how co-ordinated and harmonised policies can take advantage of the synergies between the implementation of climate mitigation options and the broader objectives of development.

Energy-efficiency improvements, including energy conservation, switching to low carbon-content fuels, using renewable energy sources and the introduction of more advanced non-conventional energy technologies, will all have significant impacts on curbing greenhouse gas emissions. Similarly, adoption of new technologies and practices in agricultural and forestry activities, as well as adoption of clean production processes, has the potential to contribute to climate change mitigation. Depending on the specific context in which they are applied, these options may entail positive side effects, or 'double dividends', which in some cases are worth undertaking whether or not there are climate-related reasons for doing so.

In spite of recent progress, it remains very challenging to develop quantitative estimates of the ancillary effects, benefits and costs of climate mitigation or adaptation policies. Despite these difficulties, the short-term ancillary benefits of climate mitigation policies under some circumstances can form a significant fraction of private (direct) costs of mitigation, and in some cases they may be comparable. Ancillary benefits may be of particular importance in developing countries, especially in relation to human health benefits of avoided air pollution (Seip *et al.*, 2001). The exact magnitude, scale and scope of these ancillary (co-) benefits will vary with local geographical and baseline conditions.

2.6.1 Integrating climate change policies with development plans

The impacts of climate change, climate policy responses and associated socio-economic development will all affect the ability of countries to achieve their sustainable development goals. The pursuit of those goals will, in turn, affect the opportunities for, and success of, implementing climate change policies. In particular, the socio-economic and technological characteristics of different development paths will strongly affect greenhouse gas emissions, the subsequent rate and magnitude of climate change and associated climate change impacts, and the ability of societies to adapt to climate change and their capacity to subsequently implement further mitigation options. The effectiveness of climate policies can therefore be enhanced when they are integrated with broader strategies designed to make national and regional development paths more sustainable.

The process of integrating and internalising climate change and sustainable development policies into national development agendas requires new problem-solving strategies and decision-making approaches. This task implies a twofold effort. On the one hand, sustainable development discourse needs greater analytical and intellectual rigour (methods, indicators, etc.) to make the concept advance from theory to practice. On the other hand, climate change discourse needs to be aware of the restrictive set of assumptions underlying the tools and methods applied in climate change analyses, and also the social and political implications of scientific constructions of climate change (Demeritt, 2001).

2.7 Conclusion

We face certainly continuing, probably accelerating and possibly unprecedented changes in the earth's climate over the coming years and decades. These changes in such a fundamental resource for society and such a powerful influence on environmental, economic and cultural development will introduce new challenges to the way we live with and influence climate. Some of these challenges may be broadly foreseeable, many of them may not. Some of the risks associated with a rapidly changing climate may be quantifiable, many of them may not. What should be our response?

As evidence is emerging that some physical and biological systems are already reacting to this human-induced change in climate, and as we know that at least for some regions and for some communities climate variability has already imposed huge costs, doing nothing is unlikely to be the best option. Society needs to develop and implement appropriate strategies to reduce the risks associated with a changing climate – to ensure that these changing climatic resources are appropriately exploited and that the adverse impacts of changing climatic hazards are minimised.

Mitigation measures are required to reduce global greenhouse gas emissions with the intention of eventually stabilising atmospheric concentrations at some level at which an acceptable dynamic equilibrium could be sustained between climate, ecosystems and human society. What this level ought to be, however, remains poorly known and increasingly contested. On the other hand, due to the inertia of both the climate system and our energy structures, greenhouse gases accumulated and accumulating in the atmosphere since the pre-industrial era will continue to affect global climate long into the future. Together with the existing exposure of many communities and assets to extremes of weather, adaptive measures become essential in order to enhance the coping abilities of valued ecosystems, vulnerable communities and exposed infrastructures.

These crucial perspectives on climate change need to be integrated fundamentally into the full range of policy measures that are demanded by our drive towards sustainable development, an argument equally valid for the nations of the south as for the nations of the north. We all need to come to terms with climate change.

Further reading

Claussen, E. (ed.) (2001) *Climate Change: Science, Strategies and Solutions*, Leiden/Boston/Köln: Brill Publishers, USA.
A perspective, written largely by American authors, on climate change addressing both science and policy issues.

Grubb, M., Vrolijk, C. and Brack, D. (1999) *The Kyoto Protocol: A Guide and Assessment*, London, UK: RIIA.
A definitive guide to the negotiations leading to the 1997 Kyoto Protocol and an assessment of its basic principles; it doesn't cover the more recent developments.

Harvey, L.D.D. (2000) *Global Warming: The Hard Science*, Essex, UK: Prentice Hall.
A good primer on the science of climate change. A better tutorial than that the IPCC could give you.

IPCC (2001) *Climate Change 2001: Synthesis Report. A Contribution of Working Groups I, II, and III to the Third Assessment Report of the Intergovernmental Panel on Climate Change*, R.T. Watson, and the Core Writing Team (eds), Cambridge, UK: Cambridge University Press.
The 'Bible' on our considered view of anthropogenic climate change, this synthesis report is a required reading to get a panoramic view of our current knowledge of climate change and its implications for society.

Jepma, C.J. and Munasinghe, M. (1998) *Climate Change Policy: Facts, Issues and Analyses*, Cambridge, UK: Cambridge University Press.
Written largely from a developing country perspective, this book will introduce to you the dilemmas facing climate policy developers and the relationship between climate and development policy.

Chapter 3
Understanding and Adapting to Sea-Level Change

Patrick D. Nunn

3.1 Introduction: sea-level changes

The surface of the ocean is never still. In most parts of the world, every few seconds the wind whips up waves. Then there are the daily changes experienced as tides. There are seasonal oscillations and interannual changes that are sometimes manifested as decadal-to-century scale periods of overall sea-level rise or fall. Then there are long-term sea-level changes, changes that are often obscured over timescales like that of a human lifespan by shorter-term changes.

When we peer deep into the earth's past history, we cannot see such shorter-term changes clearly. Instead we see evidence only for the longer-term changes, evidence which includes often spectacularly elevated or sunken shorelines. During the European Renaissance, philosophers were divided into those who thought that the presence of seashells high in the Alps was evidence for the Deluge described in the Christian Bible, and those who considered this idea fatuous and regarded such shells as having probably fallen from the hats of pilgrims. Both in their way were wrong.

For centuries the subject of sea-level change was one that evaded the public interest, considered exclusively a topic for academics or fisherfolk, of interest to those engaged in petroleum exploration or mariculture, for example, but hardly front-page news. All this changed in the 1970s and 1980s, as the spectre of sea-level rise loomed up suddenly. The world, it seemed, faced an unprecedented catastrophe, as water levels around coasts were apparently set to rise because of the rapid industrialisation during the past 150 years or so. When newspapers in the Cook Islands splashed headlines in the late 1980s talking of 10 m of sea-level rise in the

Global Environmental Issues. Edited by Frances Harris
© 2004 John Wiley & Sons, Ltd   ISBNs: 0-470-84560-0 (HB); 0-470-84561-9 (PB)

next 10 years, people there were understandably alarmed. Such misinformation was fuelled by the many uncertainties involved in 'predicting' the rate and magnitude of future sea-level rise; early authoritative opinions talked of 3.5 m of sea-level rise by AD 2100, and many of the first impact studies were based on this figure. Since the establishment of the Intergovernmental Panel on Climate Change (IPCC) in 1988 'estimates' of 21st-century sea-level rise have been revised downwards quite considerably, yet, within the same period, we have also begun to explore the multifarious effects of sea-level rise on human societies.

3.2 Past sea-level changes

In the distant past, most long-term changes in sea level (or eustatic changes) appear to have been slow, monotonic and unidirectional. Often the growth of an underwater mountain range or the gradual deformation of an ocean basin caused these changes – both examples of **tectonic eustasy**. At those times during the history of the earth when supercontinents existed, the area of continental lithosphere was comparatively small and the oceans spread over comparatively wide areas. Such times were thus marked by low sea levels. The ensuing periods of supercontinent break-up were therefore associated with slowly rising sea level, a good example being the sea-level rise during the Mesozoic Era which ended in the later part of the Cretaceous Period with a comparatively rapid sea-level fall of perhaps as much as 200 m (Hallam, 1984).

Tectonic eustasy was not the only cause of ancient sea-level changes. Sometimes lateral shifts in material within the upper layers of the earth caused changes in the geoid surface which in turn brought about massive shifts in sea level – this is **geoidal eustasy**. The precise contribution of this cause of sea-level change is still uncertain although Mörner (1981) argued, from his recognition of spatial changes in contemporary sea level, that it was an important cause of Cretaceous sea-level change.

Some sea-level changes were also caused by changes in the volume of land ice – the so-called **glacial eustasy**. Such changes have often been linked to changes in earth surface temperatures and although glacial-eustatic changes occurred in pre-Cenozoic times, it is widely accepted that they are the major cause of later Cenozoic sea-level change, specifically that which occurred within the last 30 million years or so.

During the Quaternary there may have been 20–25 glacial-eustatic oscillations driven primarily by changes in the earth's orbit around the sun. These oscillations were of shorter duration in the early Quaternary but began to last longer around 0.85 Ma (million years ago), a result of the dominance of the orbital-eccentricity cycle. During the Last Interglacial (128–111 thousand years ago (ka)) the sea level was close to its present (interglacial) level, perhaps even rising rapidly some 6 m higher towards the end of this stage as a result of West Antarctic ice sheet collapse (Hearty *et al.*, 1999).

Then began the long sea-level fall which characterised most of the Last Glacial (111–12 ka). During this stage the sea level reached a maximum 120–130 m below its present mean level at a time (22–17 ka) when land ice was at its maximum extent and temperatures were at a minimum. As temperatures rose in the late Last Glacial, so too did sea level. As for all Quaternary glacial cycles, the sea-level rise at the end of the Last Glacial was much faster than the sea-level fall which marked its first part.

Owing to its rapidity, this 'postglacial sea-level rise' disrupted the activities of the growing coastal populations worldwide and transformed coastal environments in ways which are still abundantly manifest. The next section details the nature and effects of Holocene sea-level changes.

3.3 Holocene sea-level changes

Between approximately 15 000 and 6000 years ago, the level of the ocean surface rose a net 120–130 m largely as a result of land-grounded ice melting as a consequence of rising temperatures. The result was that every coastline in the world drowned and ecosystems responded both to sea-level rise and to the attendant climate changes (Tooley, 1994).

Humans were displaced in most parts of the world. Along the western rim of the Pacific, the vast deltaic and alluvial lowlands of the Huanghe–Yangtze were inundated and the early agriculturalists in these areas were compelled to move out. Many apparently took to the sea and began a process of island colonisation within the Pacific Islands that was unrivalled on the planet at the time. The ancestors of modern Pacific Islanders crossed the entire ocean from west to east – a distance of some 12 000 km – well before Europeans even knew that this ocean existed.

Elsewhere in the world, Holocene sea-level rise isolated previously connected human communities. Examples include the flooding of the 'land bridges' between mainland Australia and Tasmania and New Guinea, and the flooding of the Bering Strait which the first humans to colonise North America are believed to have crossed sometime during the Last Glacial (Nunn, 1999). Postglacial sea-level rise had major effects on low-lying continental margins; in Germany, for example, the shoreline moved landward 250 km between 8600 and 7100 years ago (Streif, 1995).

3.3.1 Early Holocene sea-level rise

The early Holocene sea-level rise was neither unbroken nor monotonic (Figure 3.1). There were periods of comparatively abrupt sea-level fall, associated with episodes of cooling, such as the Younger Dryas (11–10 ka) and the '8200-year event'

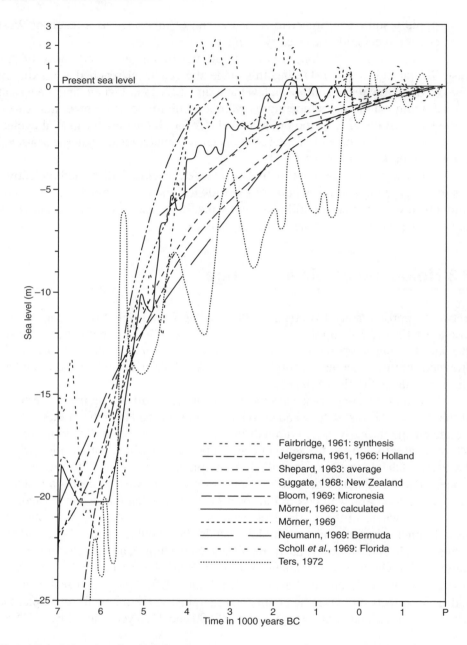

Figure 3.1 Composite record of Holocene sea-level changes. (Reproduced with permission from Jelgersma and Tooley, 1995; published by Coastal Education and Research Foundation.)

(Edwards *et al.*, 1993; Alley *et al.*, 1997). There were also periods of rapid sea-level rise associated with discharging of subglacial and proglacial lakes; rates of 24 mm/year about 12 ka, and 28 mm/year about 9.5 ka were calculated by Fairbanks (1989). A global survey suggested that Holocene sea-level rise was at the rate of 10–30 mm/year (Cronin, 1983).

Early Holocene sea-level rise both destroyed ecosystems and created opportunities for new ones to develop. The relationship between coral reefs and rising Holocene sea level is instructive in this context. Around the start of the Holocene, sea-surface temperatures became warm enough in most parts of the tropics for hermatypic corals and other reef-building organisms to become re-established on many shallow-water surfaces. As sea level rose, these nascent coral reefs began growing upwards. In those areas of the oceans where oceanographic conditions were optimal for coral growth, many such reefs managed to 'keep up' with rising sea level. Elsewhere, coral reefs were able only to 'catch up' with sea level once this stabilised during the middle Holocene (section 3.3.2). In some cases coral reefs had to 'give up' because the rate of sea-level rise was too fast (Neumann and MacIntyre, 1985), and this may account for the 'drowned atolls' which are widespread in the Indian Ocean (e.g. Stoddart, 1973).

Many of the world's coastlines which were drowned by postglacial sea-level rise had been, as many are today, marked by river valleys and alluvial-coastal flats. When these were drowned, many straight coasts became embayed as a result of the lower parts of river valleys being inundated. The development of sheltered embayments created new environments for particular organisms. A good example is the mangrove, various species of which colonised tropical Australian coasts around 6000 years ago (Woodroffe *et al.*, 1985). Another example comes from southwest Europe where estuarine environments became important centres for biological diversification during the early Holocene and attracted human groups for that reason (Straus, 1996).

3.3.2 Middle Holocene sea-level stability

Around 6000 years ago, sea level became effectively stable. In many parts of the world this happened at a level 1–2 m above the present sea level. This middle Holocene period of higher-than-present sea level coincided with a period of warmer and generally wetter-than-present climate known as the Holocene Climate Optimum.

For most of the world's coastlines, the middle Holocene was thus a time of lateral shoreline development. This ranged from the development of broad erosional shore platforms along many rocky coasts to the progradation through peripheral sediment deposition of many others, particularly around the mouths of large rivers. Good examples come from southern Spain (Lario *et al.*, 1995).

Fixing the elevation and the timing of the maximum middle Holocene sea level has implications for understanding many aspects of Holocene/interglacial climate evolution. This is true particularly for discerning, through the medium of lithospheric response to changing ice and water loads, some of the rheological properties of the outer layers of the solid earth. For example, the ICE-4G model of lithospheric response to changing loads, which assumes that the contribution of melting ice to Holocene sea-level rise ended 6000 years ago, has been tested with empirical data from many of the world's coastlines. It can now be used to predict (actually, retrodict) the level to which the sea surface would have reached at its Holocene maximum in different parts of the world (Peltier, 1998).

3.3.3 Late Holocene sea-level fall

Sea level has been falling in many parts of the world for the last 3000 years or so. Of course, it is sometimes difficult to separate real sea-level changes from sea-level changes overprinted with tectonic changes, but the generalisation seems correct both theoretically and empirically. The late Holocene sea-level fall has been associated with cooling and, in some regions of the world, with increasing aridity. The late Holocene was also the time when human societies burgeoned, both in their complexity and in their influence on the environment. This was due at least in part to the new opportunities presented to humans by emerging coastlines (Nunn, 1999).

Late Holocene sea-level fall has to some extent countered the earlier effects of drowning even though the magnitude of the latter was far less than that of the transgression. For example, along many coastlines, the shore platforms and the alluvial-marine flats created during the middle Holocene emerged during the late Holocene. This has created more flat land along the world's coastlines, a process which has played an important role in the development of human societies in such areas (Nunn, 1994).

Sea level did not fall everywhere during the late Holocene. Sometimes the fall was obscured by uplift, sometimes the sea level has apparently risen continuously throughout the Holocene. The reason for the variation has principally to do with the rheological response of the earth's lithosphere to changing postglacial ice and water loads (Peltier, 1998).

Examples of sea-level data for various parts of the world are given in Table 3.1. Note how the late Holocene trend is generally falling and the tide-gauge trend, usually no more than a 100-year record, is rising.

In most parts of the world it seems that sea level had reached close to its present level around 1200–1100 years ago. Although many scientists have assumed that sea-level changes subsequently were negligible, recent work suggests otherwise (Nunn, 2000). The last 1200 years were marked by a period of rising sea level called the Little Climatic Optimum followed by a period of lower-than-present sea level termed the Little Ice Age. About AD 1800 began the period of sea-level rise within which we are still living and which, some argue, is now being driven largely by global warming associated with the human-enhanced greenhouse effect.

Table 3.1 Selected data for late Holocene sea-level change and tide-gauge records (from data in Gornitz, 1995)

Station	Late Holocene trend (mm/year)	Tide-gauge trend (mm/year)
Aberdeen, Scotland	−0.47	0.76
Bergen, eastern	−1.95	−0.96
North America	0.97 ± 1.94	2.48 ± 1.81
Point-au-Pere, Canada	−2.5	0.51
Sydney, Australia	−0.3	1.31

3.3.4 Recent sea-level rise

Within the last 200 years or so, sea level has been rising. As with earlier periods, this sea-level rise has been neither continuous nor monotonic. It has been marked by considerable 'noise' about the 'signal', noise that may involve alternately periods of sea-level rise more rapid than the signal and periods of sea-level fall (Figure 3.2). Separating the signal from the noise is difficult and several decades of data are needed before one can begin to feel confident of having succeeded in doing so (Douglas, 2001).

The instrumental records of sea-level change have been exhaustively analysed (Table 3.2). Several studies of proxy sea-level data, such as studies of coral-growth rates, have also been used to extend such trends back before the start of the instrumental record (Quinn *et al.*, 1993; Nunn, 1999).

In some parts of the world where a long-term monitored sea-level record is not available, proxy data have been gathered. An example comes from the

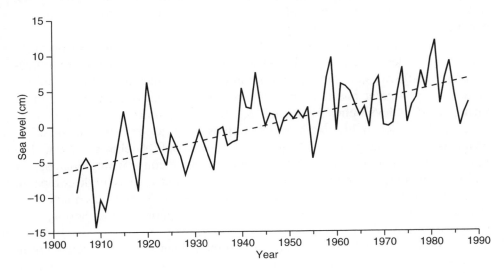

Figure 3.2 Recorded sea-level change from Honolulu, Hawaii (from Nunn, 1994, Reproduced by permission of Blackwell Publishing Ltd.).

Table 3.2 Results of selected analyses of recent sea-level data (Adapted from Warrick and Oerlemans, 1990; Gornitz, 1995)

Rate (mm/year)	Comments	Source
1.1 ± 0.8	Many stations, 1807–1939	Gutenberg (1941)
1.2	Selected stations, 1900–1950	Fairbridge and Krebs (1962)
1.2 ± 0.1	195 stations, 1880–1980	Gornitz et al. (1982)
1.75 ± 0.13	84 stations, 1900–1986	Trupin and Wahr (1990)
1.15 ± 0.38	655 stations, ~1807–1990	Nakiboglu and Lambeck (1991)

Pacific Islands where elderly long-term residents of long-established coastal settlements have been questioned about how the shoreline has changed since the time when they were young.

There seems little doubt that recent sea-level rise has been driven largely by the contemporaneous temperature rise. Rising ground-surface temperatures cause the melting of land-based ice, which ends up in the oceans and causes sea level to rise. Rising sea-surface temperatures cause a slight yet significant expansion of ocean-surface waters, also leading to sea-level rise. Most of the sea-level rise within the past 40 years has been a result of this thermal expansion (Cabanes et al., 2001).

The magnitude of recent sea-level rise varies from coast to coast, due in part to hydroisostatic effects, coastal lithology, tectonics (including anthropogenic ground subsidence) and instrumental imprecision. The overall rate for the past 200 years or so has been around +1.5 mm/year. This does not sound very great, particularly when placed alongside some future projections (Table 3.3), but we can measure its 'greatness' only by examining its effects.

The effects of recent sea-level rise are not easy to quantify along many coasts in the 'developed' world because of the long history of shoreline protection and modification associated with maintaining or increasing the economic value of a particular coast. Conversely, recent sea-level rise effects are more visible in the 'developing' world where often the resources to counter them have not been so readily available. Examples are found throughout much of southeast Asia and parts of Africa (Leatherman and Nicholls, 1995). The effects of recent sea-level rise are even more conspicuous on single islands or within archipelagos where the ratio of coast to land area is high compared with continents. In many such places, recent sea-level rise has manifestly affected entire island populations in one way or another (Leatherman and Beller-Simms, 1997).

In the past 30 years or so, there is some evidence for an acceleration in sea-level rise. This acceleration has been heralded by some scientists as empirical proof that the effects of the human-enhanced greenhouse effect are finally 'kicking in' to the sea-level record (Gornitz and Lebedeff, 1987).

3.3.5 Future sea-level rise

Due to the warming associated with the human-enhanced greenhouse effect, it seems likely that, irrespective of what actions are taken now, sea level will rise throughout this century at a significantly faster rate than that at which it rose within the last century (Table 3.3).

Widespread anxiety at the looming spectre of accelerated 21st-century sea-level change (and climate change) coupled with the appreciation that this was an issue which would not respect national boundaries led the United Nations to establish in 1988 the IPCC. One of the things with which the IPCC was charged was to come up with the best possible projections of future sea-level change. Although some of the early projections are now regarded as excessively large, a significant acceleration of the rate of recent sea level rise appears inevitable. The IPCC Reports (2001) state that sea level is projected to increase 8–88 cm between 1990 and 2100. As with anything involved in predicting the future with an optimal degree of certainty, there is considerable controversy surrounding the question of future sea-level rise (and climate change). Some of this controversy is explored in section 3.4.

3.4 Predicting 21st-century sea-level rise

Sea-level rise during the 21st century is likely to have major negative effects on the world's coasts and the 21% of the world's people who occupy them (Gommes *et al.*, 1997). Direct inundation will be perhaps the most severe (Table 3.3) but also the longest to accomplish, therefore leaving opportunities for adaptation. In contrast, increased storm-surge hazards associated with sea-level rise may leave little time for adaptation. It has been estimated that currently around 46 million people per year are affected by flooding associated with storm surges. A 1 m sea-level rise would increase this to about 118 million, or more if population growth is taken into account (Watson *et al.*, 2001).

3.4.1 Imperatives, challenges and hindrances

Sometimes in the field of sea-level studies, science appears to lose touch with its purpose. Unless science can ultimately be of some practical use to humankind, then perhaps it is of no use. Sea-level scientists sometimes expend so much energy arguing about the precise magnitude of future sea-level changes that they forget that the purpose of this knowledge is to ascertain the impacts of these changes on the earth's environments. It matters little to persons living in Bangladesh, or on a Pacific atoll, whether sea level rises by 35 or 55 cm by the year 2050 because the effects of this on the physical environment and on the livelihoods of its occupants will be virtually indistinguishable.

It is perhaps more helpful for scientists to concentrate on the probable effects of sea-level rise (Table 3.3) rather than continuing to argue about its precise magnitude. In other words, formulating and implementing appropriate adaptation options are of more value than debating the precise nature of that to which we need (or will need) to adapt. People living in Bangladesh or in the central Pacific atoll nation of Kiribati know that they are vulnerable. They also know that this vulnerability is associated with sea level and that this has been increasing. With good reason they fear that this trend will continue into the future, and they want to do something about it. It is imperative that they are helped in this regard.

In 'developed' countries, most coastal-management policies are formulated at national level, and enforced effectively through a 'top-down' process. Thus governments of such countries can develop approaches to future sea-level rise which combine mitigation and adaptation, and mix intergovernmental and national initiatives (mitigation deals with the causes of climate change and sea-level rise, adaptation deals with their effects).

In 'developing' countries, the situation is different. While mitigation is usually something that can be pursued only at government level, adaptation is something in which everyone can be involved. Many coastal communities, particularly those outside urban centres, need to help themselves adapt to future sea-level rise. The challenge is persuading them to do so, giving them the right information and tools, and then ensuring that their adaptation response is part of a co-ordinated regional effort.

A final challenge is for the IPCC and those who regard as plausible its projections of 21st-century sea-level (and climate) change to convince the doubters. There are numerous scientists who are outspoken in their opposition to certain assumptions made by IPCC modellers. If their views cannot be critically evaluated by government decision-makers, then it is possible that some governments will take 'fringe' views seriously, particularly to justify committing scarce revenue to adaptation strategies.

It is obvious that 21st-century sea-level rise will be no respecter of national boundaries. Yet some countries are in better positions to adapt than others. Often countries that are better able to adapt are those in the 'developed'

Table 3.3 Primary and secondary impacts of future sea-level rise (Reproduced with permission from Nicholls and de la Vega-Leinert, 2000; published by Asia pacific Network for Global Change Research.)

Primary impacts	Secondary impacts
Increased erosion	On livelihoods and human health
Inundation of coastal wetlands and lowlands	On infrastructure and economic activity
Increased risk of flooding and storm damage	Displacement of vulnerable populations
Salinisation of surface and ground waters	Diversion of resources to adaptation responses
	Political and institutional instability, social unrest
	Threats to particular cultures and ways of life

world that are widely perceived as having caused the problem of accelerated sea-level rise. Conversely those which are going to be most affected by this are those which did not contribute to it, at least not significantly. These are the ingredients of one of the greatest hindrances to a co-ordinated and coherent international response to the issue of sea-level rise.

The 'victims' want the 'perpetrators' to pay in terms of both mitigation and adaptation, while at the same time often compounding the problem by their own actions. The victims of sea-level rise, like the people of the Marshall Islands or Bangladesh, do not have the resources to adapt successfully to the projected 21st-century sea-level rise. They would like the perpetrators, such as those countries that became industrialised more than a hundred years ago and which are now reaping the benefits, to help them. But then even the self-confessed perpetrators find it difficult to justify such help when many of the victims, particularly China and India, are industrialising rapidly and now emitting far more greenhouse gases than the perpetrators, thereby ensuring that sea level will continue to rise long after the year 2100. Agreements to limit emissions are thus imperative, but to reach them in the light of such arguments, presented here very superficially, is a huge challenge.

Hard as it may be to convince someone living off the land in a highly vulnerable coastal region, the issue of future sea-level rise is a political as well as an environmental one. Most of our lives are controlled to some extent by politicians, many of whom are at heart unconcerned about long-term issues (such as sea-level rise) on which action will not pay short-term dividends to their electorate. It is therefore no surprise to find that most international initiatives for coping with future sea-level rise (and climate change) are driven by intergovernmental organisations, and that national preparedness in many 'developing' countries depends in large part on non-governmental organisations and aid donors. A more unified approach would help address the problem more effectively. Sometimes, as discussed in Box 3.1, politics can extend a malign influence into the science and reporting of sea-level rise.

Box 3.1: Case study: the politics of sea-level rise

In the late 1980s, in response to increasing concern from its Pacific Island neighbours about threats from rising sea levels, the Australian Government announced an ambitious aid scheme. It would fund state-of-the-art tide gauges in 11 Pacific Island nations that would help those nations monitor rising sea levels. Now there were already tide gauges in the region, and the overall trend of sea-level rise was well established (Wyrtki, 1990). Further, such a huge amount of money might have been better spent on more pragmatic assistance, such as enabling appropriate artificial (seawall) and natural (mangrove) shoreline protection schemes. But at the time, it seemed a good idea to know more about the nature of sea-level rise and the Australian offer was roundly applauded.

Continued on page 56

Continued from page 55

At around the turn of the present century, when the debate on how the global community should respond to 21st-century sea-level rise was becoming increasingly polarised between those advocating drastic reductions in greenhouse gas emissions and those questioning the scientific basis of the IPCCs projections, world attention became focused on the South Pacific nation of Tuvalu. With a total population of around 9000 living on nine low-lying atolls, Tuvalu was one of the countries thought likely to feel the worst effects of future sea-level rise (Roy and Connell, 1989).

Then in 2000, at a climate change conference in the Cook Islands, the National Tidal Facility (NTF) – the Australian organisation charged with overseeing the operations of the Pacific tide-gauge network – announced that sea level had actually fallen by 8.69 cm around Tuvalu over the preceding 7 years during which the tide gauge had been functioning. This report was promulgated worldwide, expanded and amplified. Governments like those of Australia and the United States, which had been loud in their resistance to emissions reductions targets, appeared to take heart from the Tuvalu tide-gauge record, pointing out that their suspicion that 'predictions' of future sea-level rise were scientifically flawed must indeed be correct.

The reality is different. Firstly, the NTF conclusion in 2000 that the sea level around Tuvalu was falling has now been revised as more data have been collected (Figure 3.3). The latest statement from the NTF concluded that in the 104 months to January 2002, the sea level around Tuvalu had been rising at 0.9 mm/year (Mitchell, 2002). Yet even this fails to acknowledge that the

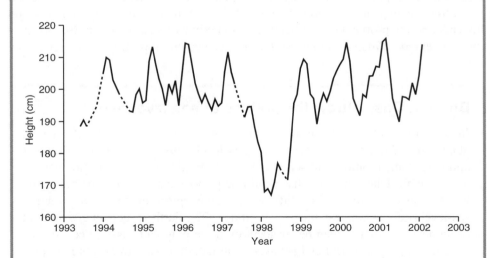

Figure 3.3 The tide-gauge record which caused a global furore: the 1995–2001 Funafuti, Tuvalu record collected by the NTF of Australia.

Continued on page 57

Continued from page 56

uncertainty in their sea-level trend is ±13.7 mm/year and that a more realistic estimate based on the NTF gauge and earlier data is +1.2 ± 0.8 mm/year (Hunter, 2002). The Government of Tuvalu is also critical of NTF science and its public pronouncements on sea-level change around Tuvalu (Laupepa, 2002).

The Australian Government no doubt seeks a return on its multimillion-dollar investment in state-of-the-art tide gauges in the South Pacific, a return by which Australian voters will understand the importance of this 'aid'. As the agency charged with overseeing the tide-gauge network, the NTF is keen to demonstrate that useful results are coming from these tide gauges, even though it is well understood that the signal of sea-level rise will not necessarily be visible until the gauges have been operating for 30–50 years or more. The Australian Government would presumably rather not wait so long and would like to see island states say something useful about the data generated from the tide gauges thus far – hence the report by the NTF talking about a 7-year sea-level fall.

However, the empirical justification for failing to cut greenhouse gas emissions by the Australian government would appear to be flawed. Every spring tide, water floods into the Tuvalu Meteorological Office premises in Funafuti. Every king tide (storm surge), it reaches farther. The Government of Tuvalu publicly calls on the Government of Australia to discuss plans for accommodating 'environmental refugees' from Tuvalu. The Government of Australia appears to doubt the seriousness of the problem because of the supposed 'evidence' for sea-level fall around Tuvalu.

In a meeting in Suva, Fiji, in March 2002, attended by delegates (including the author, from whom this account comes) from many Pacific Island countries, NTF Deputy Director, Bill Mitchell, criticised the Government of Tuvalu for having raised the fate of environmental refugees from their country with the Australian Government. 'There will be no such problem', stated Mitchell, 'because sea level has been falling not rising around their country.' Protests from the Tuvaluan representatives that their observations over the past few years contradicted this conclusion were drowned in the sea of data presented by the NTF team.

The NTF team went on to say that the data being generated by the Australian-funded tide-gauge network was showing that the 'predictions' of the IPCC for future sea-level change were wrong. Just how a decade or less of empirical data might contradict the broad projections of future sea-level rise by the IPCC was left unexplained, but the message was clear: 'sea level is not rising as predicted, and so our regional response, particularly in emissions reductions, can be toned down'. So it might be argued that Australians – among the highest *per capita* producers of greenhouse gases in the world – will be spared cutbacks while Tuvalu might disappear.

Continued on page 58

Continued from page 57
Cynics might argue that any global problem which involves identifying perpetrators and victims will end up being tackled less than adequately
because of political considerations. That issue is irrelevant to people inhabit-
ing Tuvalu. Their problems are real. The sea is yearly, monthly, invading
their living space and reducing their livelihood options. No amount of data
analysis will change that.

3.5 Adapting to sea-level rise

Accelerated sea-level rise will pose major problems for most of the world's
coastal environments. To understand the nature of these problems, it is helpful to
distinguish the 'vulnerability' and 'resilience' of particular coasts. Vulnerability is
defined as the potential for the attributes of a system to respond adversely to the
occurrence of hazardous events. Resilience is defined as the potential for the
attributes of a system to absorb the impacts of hazardous events without significant or adverse response (Yamada *et al.*, 1995).

In September 1991, the Common Methodology for Assessing Vulnerability
to Sea-Level Rise was launched by the IPCC so that Vulnerability Assessments
carried out for different countries might be readily compared (IPCC, 1991). The
basis of the 'Common Methodology' was that a country's vulnerability to sea-
level rise involves the susceptibility of coastal systems (natural and social) to
change, and the ability of their inhabitants to adapt to reduce the negative impacts
of this change. The Common Methodology thus sought to mesh the physical
vulnerability of a country with its ability – technological and financial – to
respond. This approach has worked well in many parts of the world but the
Common Methodology required detailed data that many developing countries
found difficult to acquire. The shortcomings of the Common Methodology for
subsistence and customary communities, in which applying quantitative ('dollar')
values to every piece of coastline was unworkable and inappropriate, were
discussed by Kaluwin (2001).

Vulnerability and resilience can be quantified either loosely or rigorously.
Loose quantification can be a very useful tool for rapid appraisal of hazards,
as shown in Figure 3.4. The composite index based on vulnerability and resilience
of settlements along the coasts of Ovalau and Moturiki islands in central Fiji
was developed on the basis of interviews with long-term residents of these
settlements.

Rigorous quantification involves the assignment of scores (typically -3
to $+3$, most vulnerable to most resilient) to a range of natural, human, infra-
structural, economic, institutional and cultural systems. The IPCC Common

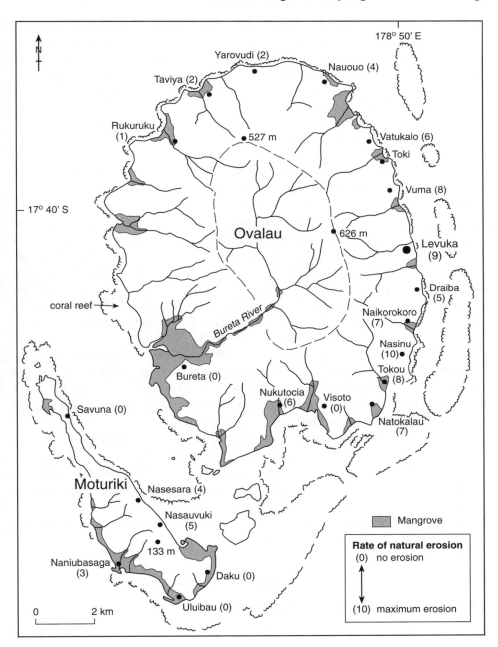

Figure 3.4 Loose quantification of the coastal erosion hazard around Ovalau and Moturiki islands in central Fiji. Erosion is greatest along exposed windward (southeast) coasts, least along leeward shores. Note the low scores for Bureta and Visoto villages on Ovalau, where traditional taboos have prevented the cutting-down of the mangrove fringe that probably existed along the entire lengths of these islands' coasts before and during the early period of human occupation.

Methodology is one example. A contrasting example for South Pacific countries was developed by Yamada *et al*. (1995). The principal advantage of rigorous quantification is that it allows the vulnerability and resilience of large stretches of coastline to be evaluated consistently. This in turn allows planners to identify the most vulnerable areas and to take appropriate action.

3.5.1 Adaptation responses

The IPCC Coastal-Zone Management Subgroup (IPCC, 1991) identified three main options by which coasts could adapt to sea-level rise: retreat, accommodate and protect.

Retreat involves no attempt to protect the land from sea-level rise, but instead requires that everything of value along the coast is moved inland to a place where it is no longer likely to be affected by sea-level rise. This option is viable in low-lying coastal areas with a higher-level hinterland which could house what has been displaced. Good examples are the implementation of building setback codes and the encouragement of upland over lowland agriculture, but retreat is obviously not a viable option for communities along atoll-island coasts or in densely populated deltas, for example.

Accommodation also involves no attempt to protect the land from sea-level rise, but requires that the people living there and the activities carried out there continue in different forms which acknowledge the higher sea level. An example would be where lowland agriculture is replaced by aquaculture, or where buildings formerly at ground level are elevated on piles.

Protection involves the construction of hard artificial structures (such as seawalls and dykes) and/or soft solutions (such as landfilling or revegetation) to protect the land from the sea, and allow existing uses of that land to continue. Caution must be exercised with hard structures; sometimes they exacerbate or transfer the problems they are designed to solve. Soft solutions are generally preferable in 'developing' countries where the costs of constructing and maintaining hard structures are often prohibitive.

Box 3.2: Case study: Bangladesh

Bangladesh, a country crowded around the subsiding delta of the Ganges, Brahmaputra and Meghna rivers, is one of the countries on earth most vulnerable to sea-level rise. Most of its >130 million people live in areas only 1–2 m above mean sea level.

The vulnerability of Bangladesh to sea-level rise has been demonstrated on many occasions in the past. Following the 1970 storm surge in

Continued on page 61

Continued from page 60

the Bay of Bengal, 200 000 people drowned (Flierl and Robinson, 1972), while 15 000 died as a result of the 1985 surge (Maddox, 1985); the May 1991 surge was even greater in magnitude than that of 1970 (Tooley, 1994). On average, 1.5 storm surges affect Bangladesh yearly and reach as far as 160 km inland of the shoreline.

Were the sea level to rise by, say, 1 m by the end of the 21st century, a huge area would be inundated and an even larger one would become vulnerable to direct storm surge effects. The potential loss of life and land would likewise be enormous. A 1 m rise in sea level 'could displace more than 13 million people (or 11% of the national population) from their homes' (Huq *et al.*, 1995, p. 48). The vulnerability of the whole country to storm surges and sea-level rise is being exacerbated by the subsidence of the delta, an effect which is amplified by groundwater extraction (Milliman *et al.*, 1989).

Over 85% of the population of Bangladesh depend on agriculture; a 1 m sea-level rise would have severe impacts (Table 3.4). Much of the lowest area being farmed is already affected by saltwater intrusion and this will increase in the future. The vulnerability of the agricultural sector will increase if damming of the rivers feeding the delta takes place: 'if the Bengal rivers are effectively dammed, it is conceivable that nearly the entire nation could be affected by the intrusion of salt water' (Broadus *et al.*, 1986, p. 175). Salinisation of groundwater also affects the supply of potable water, most of which is drawn from underground aquifers.

In terms of physical resilience, the nature of the shoreline is important. The 5770 km² Sundarbans on the southwestern coast is a barely inhabited complex of mangrove and nepa palm, and is an effective buffer against storm surges in the area. Its removal would increase vulnerability. Unfortunately it is already affected by saltwater intrusion, made worse by the reduction in freshwater inputs from the Ganges associated with a dam at Farakka in India (Huq *et al.*, 1995).

Table 3.4 Direct effect of a 1 m sea-level rise on the main crops in Bangladesh (Reproduced with permission from Huq *et al.*, 1995; published by Coastal Education and Research Foundation)

Crop	Area liable to inundation (km²)	Total for the whole of Bangladesh (%)
Aman (monsoon) rice	1 280 000	21
Aus (summer) rice	40 000	12
Boro (winter) rice	102 000	8
Jute	13 800	2

Continued on page 62

Continued from page 61

Bangladesh cannot accommodate the effects of a 1 m sea-level rise by the end of this century without massive disruption to its economy and many of its inhabitants. Minimal physical protection against this sea-level rise will cost around 1 billion dollars (Huq *et al.*, 1995) but that may not halt an increased frequency and extent of flooding and groundwater salinisation.

3.5.2 Prospects for adapting successfully to future sea-level rise

Schneider (1988) stated that 'not to decide' about addressing the problem of future climate change and sea-level rise 'is to decide'. For the first time in human history, we can predict the future with a high degree of certainty and can therefore adapt to minimise the disruptions to our lifestyles. Such anticipatory adaptation is risky because the future can never be predicted with absolute certainty. Despite this, many governments around the world have recognised the long-term financial and social advantages of anticipatory adaptation and have enacted legislation accordingly. Thus new bridges in Canada are now being built a metre higher than they apparently need to be to take into account 21st-century sea-level rise. Building projects along many parts of the United States coastline are now subject to strict 'setback' laws in anticipation that many existing coastal areas will be inundated in the future as a direct result of sea-level rise.

Yet many other countries, particularly in the 'developing' world, see the costs of anticipatory adaptation as prohibitive, given their future development plans. Yet in deciding not to anticipate the effects of future sea-level rise (and climate change) those countries are inviting an abrupt escalation of their environmental problems at some point during the 21st century. The example of Alexandria, Egypt, is given in Table 3.5. Another example is provided by Nigeria (French *et al.*, 1995).

Table 3.5 Impacts of projected sea-level rise on the Alexandria Governorate, Egypt, assuming 'No Protection' (Reproduced with permission from El-Raey *et al.*, 1995; published by Coastal Education and Research Foundation)

Sea-level rise (m)	Population displaced		Area loss (km²)	Economic losses (billions of 1990 US dollars)
	1990 Population	**2030 Population (projected)**		
0.5	1.81	4.02	1017.2	40.2
1.0	2.02	4.50	1237.8	44.9
2.0	2.30	5.11	1518.7	52.9

Other good examples are of those where coastal vegetation, which had hitherto protected the shoreline from erosion, has been removed; examples include the conversion of mangrove forests for aquaculture in East Asia (Furukawa and Baba, 2000). In a similar vein, the removal of sandy beaches, often a result of mining, lowers the natural resilience of the shoreline to erosion. The degradation of coral reefs has the same effect; more than 30% of the 27 000 km^2 coral reefs in the Philippines are in poor condition, largely as a result of preventable human impacts (Perez, 2000). These are all problems that could have been reduced by effective enforcement of environmental legislation.

Partly because of the uneven response to the threat of sea-level rise (and climate change), much of the global response has been driven by international treaties. Two of the more important treaties which relate to sea-level rise are the United Nations Framework Convention on Climate Change (UNFCCC), adopted in New York, 1992, and the Kyoto Protocol to the UNFCCC, adopted in Kyoto, 1997 (Box 2.2). The former laid the ground for data to be gathered and the adaptation options to be explored by signatories. The latter, not yet in force, sought to reduce emissions of greenhouse gases to levels where the rate of sea-level rise (and warming) at some unspecified time in the future would significantly decelerate.

In some countries, often those in the 'developing' world, initiatives to counter the effects of sea-level rise are both 'top-down', as discussed above, and 'bottom-up'. The latter, often community-driven, are frequently the most effective ways of addressing the problem in countries with large, rural coastal populations, such as archipelagic nations.

Often in such places it is implied that if the 'wrong' adaptation option is somehow implemented, then adaptation will fail. A good example is the 'seawall mindset' in the Pacific Islands, whereby community leaders, following advice from 'experts' and observing urban solutions to shoreline erosion, believe that hard artificial structures are the only possible adaptation option in rural areas. The result has been that millions of scarce funds have been poured into constructing seawalls that invariably create new, unanticipated problems and collapse after a year or two. Far cheaper and more effective in the long run in the tropical Pacific Islands is the restoration of the mangrove fringe which is thought to have existed along most island coasts up until a few hundred years ago. A 30 m broad buffer zone of mangroves can effectively reduce shoreline erosion and storm impacts to a minimum. Yet in the Pacific Islands such initiatives are being driven largely by non-governmental organisations or 'green' aid donors and are popularly perceived as less effective than 'hard' solutions like seawalls.

In a similar vein, adaptation runs the risk of failing if it is too specific, a fact lost on many scientists and planners who often suggest to the general public that precise data are needed to identify the optimal adaptation option and that any other will fail. The uncertainties of the projected 21st-century sea-level rise are often sidelined in such calculations. In this author's view it is more sensible to adapt thoroughly than to what is perceived as only sufficiently, not least to try and negate any 'unpleasant surprises in the greenhouse' (Broecker, 1987).

3.6 Conclusion

Our understanding of sea-level rise is good. We understand, at least in general terms, how and why sea level changed in the past, and how and why it may rise in the future. While there is considerable scientific merit in refining our understanding, and improving our knowledge of the magnitudes of past and future sea-level change, those goals should not perhaps be the priorities in this field of study. Far less secure is our understanding of what effects sea-level rise had and will have on coastal environments and peoples. Without such knowledge they cannot effectively adapt.

Finally, it is often forgotten that sea-level rise is not a problem in isolation. It may be convenient to study it as a separate entity but, as a 21st-century impact on human societies, it is inextricably bound up with other factors such as changes in various climate and human parameters. Sometimes global initiatives to reduce the impacts of sea-level rise forget this, implying that if this problem is somehow solved, then all other environmental concerns will be simultaneously nullified. It is likely that the future will see more emphasis placed on integrated studies that are able to balance the contribution of sea-level rise with the contributions of other factors to a particular problem. Only then will we be able to address effectively the 21st-century environmental imperatives.

This chapter has sought to explain the main causes of sea-level rise (and fall) in the past, focusing on the Holocene – the interglacial period in which we are currently living. Although sea level has fallen for much of the late Holocene, the last 200 years or so have seen a net increase at a rate (within the last 100) of about 1.5 mm/year. The next hundred years are likely to see a more rapid rate of sea-level rise which will disrupt coastal communities worldwide.

Further reading

Devoy, R.J.N. (ed.) (1987) *Sea Surface Studies: A Global View*, London: Croom Helm.
Tooley, M.J. and Shennan, I. (eds) (1987) *Sea-Level Changes*, Oxford: Blackwell.
Douglas, B.C., Kearney, M.S. and Leatherman, S.P. (eds) (2001) *Sea-Level Rise: History and Consequences*, San Diego: Academic Press.
These are all good studies on sea-level changes.

Pugh, D.T. (1987) *Tides, Surges and Mean Sea-Level*, Chichester: Wiley.
This book explains the science of short-term sea-level change.

van de Plassche, O. (ed.) (1986) *Sea-Level Research: A Manual for the Collection and Evaluation of Data*, Norwich: Geo Books.
This book looks at the various ways in which past sea-level changes can be assessed.

Chapter 4
Changing Land Cover

Doreen S. Boyd and Giles M. Foody

4.1 Introduction

Land cover is a critical variable linking the human and physical environments. Moreover, these three variables namely land cover, the human environment and physical environment are intimately and interactively linked, with change in one impacting on the others. Most of the world's habitable land is now used and managed in some way by humans. Hannah *et al.* (1994) estimate that of the total land surface, 83.25% is habitable land. Of this habitable land, 27% remains undisturbed (i.e. primary vegetation cover, very low population density), 36.7% is partially disturbed (i.e. shifting or extensive agriculture, secondary but naturally regenerating vegetation, livestock density over carrying capacity or other evidence of human disturbance) and 36.3% is human dominated (i.e. permanent agriculture or urban settlement, primary vegetation removed, current vegetation differs from potential vegetation, desertification or other permanent degradation). Given the inextricably close interaction between humans and the environment that now exists, the issues surrounding land cover and its change are of particular importance, both to the humans and the environment. This chapter aims to provide a brief overview of land cover and its dynamics with particular regard to its influences on basic environmental processes. The principal causes of land cover change will be presented, as will the practices of acquiring land cover data and monitoring of land cover change, with particular emphasis on regional to global scales. Finally, the ways in which humans have responded to the consequences of land cover change will be discussed; given the profound environmental consequence of land cover change, it is these responses that determine the impact of environmental changes to come.

Global Environmental Issues.   Edited by Frances Harris
© 2004 John Wiley & Sons, Ltd   ISBNs: 0-470-84560-0 (HB); 0-470-84561-9 (PB)

4.1.1 Land cover and land cover change: terms and context

At the outset it is important to understand what the term land cover means. Here, land cover relates to nothing more than the surficial covering of the earth's surface. This is typically classified thematically into classes such as forest, grassland and urban. The classes required are often application-specific (DeFries and Los, 1999). The classes are often broad, akin to those depicted in Figure 4.1, since they are the concern within the context of global environmental change. Whatever classes are used, land cover relates simply to the nature of the earth's surface. It is, therefore, quite different to land use, which relates to the activity associated with a tract of land. The difference between land cover and land use is critical as the same tract of land can have the same cover (e.g. grassland) but be put to different uses (e.g. recreation, agriculture, transport) and *vice versa* and these can vary spatio-temporally across the globe. Unfortunately, much of the literature uses the terms land use and land cover synonymously, and so do some land classification systems (e.g. the United States Geological Survey (Anderson *et al.*, 1976)), hence the meaning used in any article should be carefully deciphered.

Land cover exerts a large influence on many basic environmental processes and consequently a change in land cover can have a marked impact on the environment. Two broad types of land cover change occur. The first, and most widely studied, is land cover conversion. A conversion of land cover involves

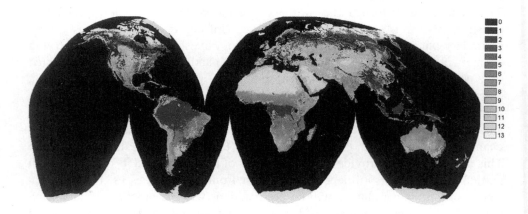

Figure 4.1 An example of a global land cover map. The map shown is that derived by DeFries *et al.* (1998) from NOAA AVHRR data. Although full appreciation requires colour, the map indicates the nature of land cover maps typically derived from satellite remote sensor data. The 13 land cover types shown are Evergreen needleleaf forests (1); Evergreen broadleaf forests (2); Deciduous needleleaf forests (3); Deciduous broadleaf forests (4); Mixed forests (5); Woodlands (6); Wooded grasslands/shrublands (7); Closed bushlands or shrublands (8); Open shrublands (9); Grasses (10); Croplands (11); Bare (12); and Mosses and lichens (13). For more information on the land cover types, refer to DeFries *et al.* (1998). (Reproduced with permission from DeFries *et al.*, 1998; published by Taylor & Francis Ltd and website http://www.tandf.co.uk/journals.)

a complete alteration in the cover of a tract of land. One commonly encountered land cover conversion of tremendous importance within the context of environmental change is deforestation. In many regions, forested land is cleared of tree cover to facilitate agricultural use of the land and the removal of the forest has enormous environmental impacts, some of which will be discussed throughout this chapter. The second type of land cover change is land cover modification. A land cover modification involves a change in the character or condition but not the type of land cover. Thus, while deforestation may involve the complete clearance of tree cover and replacement with, for example, grassland for pasture, a forest could be modified by reducing, but not removing, tree cover. Forest managers commonly thin stands of forest to promote growth of commercially valuable timber and although the thinning process can involve the removal of a considerable amount of trees, the tract of land remains a forest after thinning and so has been modified rather than converted. In most cases, once the land has been modified or converted from its prior state further change is likely to occur. Moreover, a combination of poor practices in using the land cover may have cumulative and subtle consequences. This may be evident during desertification, which is often attributed to four poor uses of land cover in the drylands: overcultivation, overgrazing, mismanagement of croplands and deforestation (Grainger, 1990).

4.1.2 Land cover change: a pressing global environmental issue

The causes of land cover change are many and varied. They may be anthropogenic or natural in origin. This chapter focuses on land cover change resulting from human actions, which have manifest themselves against the backdrop of natural change, transforming wildscapes into landscapes (Mannion, 2002). Humans live on the land, using its resources to provide basic needs of shelter and fuel, and farm the land for food. It is these functions that have driven major land cover changes, including deforestation, pasture and cropland expansion, urbanisation, land degradation and conversion of marginal lands, such as wetlands. That same land surface also controls a vast array of basic environmental properties, having an important influence on hydrology, climate, biodiversity and global biogeochemical cycles. As the surficial covering of the planet, land cover is one of the most significant variables in the understanding of local, regional and global environmental change. Moreover, land cover change can be both a cause and a consequence of environmental change, with an impact larger than that of climate change (Skole, 1994). The impacts of a land cover change are diverse and can be considerable. Land cover change is, for example, the single most important variable affecting ecological systems (Vitousek, 1994), the greatest threat to biodiversity (Chapin *et al.*, 2000) and is associated with some of the greatest uncertainties in the carbon cycle and so, ultimately, climate change

(Royal Society, 2001). The study of land cover change is, therefore, central to understanding environmental change (Riebsame, Meyer and Turner, 1994) and to disciplines such as geography (Goudie, 1993).

Land cover change, along with industrial metabolism, represents the two principal sources of global environmental change (Committee on Global Change, 1990). Thus, human interaction with land cover is important in scientific, social, economic and political arenas. How the needs and values of each of these arenas are reconciled in a contracting resource base is one of the most pressing environmental issues of the 21st century. Land cover has always been transformed by human action, with its change reflecting social and cultural developments throughout prehistory and history (Mannion, 2002). Unfortunately, the scale, type and rate of land cover change characteristic of the post-Second World War era is a hindrance to sustainable development (Table 4.1). It is of paramount importance, therefore, that land cover change is recognised as a central problem to the global environmental issues debate. As part of this debate it will be necessary to determine whether the alteration of a patch of land has led to its improvement or degradation and this requires discussion about environmental ethics, future resource demands, data quality and degree of scientific knowledge (Turner and Meyer, 1994). Although land cover is important and directly visible, it is also a highly complex variable in space and time. Further, it is also one with which there are great uncertainties that fundamentally limit the study of environmental change.

A key objective of this chapter is to raise awareness of the limitations of land cover data sets. The reason for this is not to criticise map producers, but rather to highlight the difficulties involved and dangers in using land cover maps unquestioningly. The latter is of particular importance within the context of global environmental change as, for example, many modellers seeking to predict future environmental scenarios utilise land cover maps without explicit recognition of the uncertainties involved which may fundamentally limit the resulting conclusions drawn. Furthermore, these future predictions have implications for policy-makers and social scientists who are concerned with the conditions within

Table 4.1 An example of the recent change of one land cover type, forests: Annual change in forest area, 1990–2000 (million ha) (Reproduced with permission from FAO, 2001; published by Food and Agriculture Organization of the United Nations)

	Loss			Gain	Net change
	Deforestation	**Conversion to plantation forests**	**Total loss**	**Natural expansion of forest**	
Tropical areas	−14.2	−1.0	−15.2	+1.0	−14.2
Non-tropical areas	−0.4	−0.5	−0.9	+2.6	+1.7
Global	−14.6	−1.5	−16.1	+3.6	−12.5

which humans live as they need the information with which to make policy and management decisions. All of the issues discussed in this chapter are of relevance to sustainable development, in particular, in the debate of how this is to be achieved. Land cover, and subsequently how it is used, is a principal determinant of environmental and human-life quality and thus requires consideration by **all** decision-makers.

4.2 Human causes of land cover change

Land cover and land use are connected; how humans use the land is translated into changed physical states of land cover. In order that land cover change be sustainably managed, an improved understanding of the drivers and causes of land cover change is required (Committee on Global Change Research, 1999). Achieving this enables the decision-maker(s) to move from an environmental concern, such as land cover change and its consequences, to the identification of human activities that are the principal drivers of that concern (Stern, Young and Druckman, 1992). What is particularly difficult in this regard, however, is the understanding of how the drivers operate at different spatial and temporal scales, with some acting slowly over centuries, and others visibly and rapidly (McNeill *et al.*, 1994). Moreover, it is unclear how the links between drivers of land cover change and the impacts of that change on environmental processes can be fully established when there is still uncertainty about how environmental systems operate. In an attempt to reconcile these difficulties, Rayner *et al.* (1994) suggested a wiring diagram approach which provides a schematic framework to understand how general human needs and wants are translated into land cover changes and, in turn, how these might impact on the environment (Figure 4.2).

4.2.1 Using the wiring diagram to understand the causes of land cover change

The wiring diagram presented in Figure 4.2 is multi-dimensional in its approach. It uses three kinds of analysis to explore the stimuli that contribute to land cover change and how these link with the environment. The first focuses on the understanding of the social drivers of land cover change, the second on human decision systems, such as public policy, and the third on the natural processes which are affected by and affect the first two. The three-pronged analysis is further organised into three levels of spatial scale: the micro-decision, the macro-decision and global-scale decision. The diagram is based on a number of assumptions: the first is that land use is the outcome of competition among potential uses; the second is that political institutions and other decision-making agents operating above the level of the decision-maker need to be included; the third is that both markets and

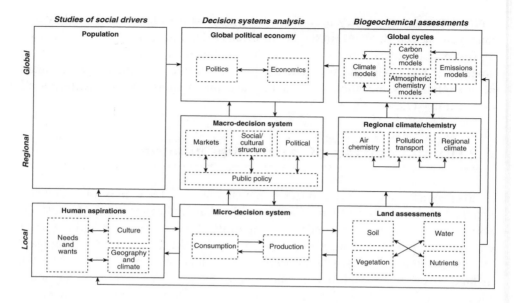

Figure 4.2 A wiring diagram illustrating how human needs and wants drive land cover change, which in turn impact on the environment at local to global scales. Boxes (modules and sub-modules) represent variables to consider and the arrows links between variables (information flows or processes). (Reproduced with permission from Rayner *et al.*, 1994; published by Cambridge University Press.)

administrative systems of decision-making are vital carriers of information throughout the system; and the last is that cultural filters for information flows are important (Rayner *et al.*, 1994). Despite these assumptions, the diagram is effective in providing a framework from which understanding and future direction of research into causes of land cover change can be generated.

4.2.1.1 Studies of social drivers of land cover change

Social drivers of land cover change are well documented. Population growth and human aspirations, both individual and collective, are key driving variables. Population growth is widely recognised as one of the most prominent driving forces behind increased human impact on the environment (Middleton, 1999a). As a driver of land cover change its effect can be seen at local, regional and global scales. Treated in isolation, there are three principal ways in which population impacts on land cover (Sage, 1994). The first is that population growth can lead to the expansion of cultivated land resulting in the alteration of the natural cover through clearance and resource needs; the second is that population growth can result in the intensification of production resulting, in the medium to long term, in unsustainable land use and subsequent expansion of cultivated land; and the third is that population growth can be scale neutral in local terms but have an impact on land cover elsewhere, as food and resources are imported or excess out-population migrates placing pressure elsewhere (Grigg, 1980; Bilsborrow, 1987). To study the relationship between population growth and land cover change requires basic demographic information (such as birth and death rates and migration) as well as details on how micro-decision-makers, macro-decision-makers and the global political economics influence decisions made by a population.

The driving forces of land cover change are often strongly interrelated and so should be considered together in a multivariate fashion (Stern, Young and Druckman, 1992). A popular approach in this regard has been to consider population growth together with affluence and technology in order to determine its impact on the environment.

Affluence is a component of the human aspirations module. Human needs and wants can be categorised into diet, shelter, prestige, security, recreation and spiritual satisfaction. The influence of each of these on land cover is determined on a local scale by factors such as health, age, gender and climate. Socio-cultural structures are also important at the local scale and link to the micro-decision system module that portrays the smallest units of decision-making whose actions directly or indirectly determine land cover change. Socio-cultural structures are important in determining an individual's goals, values and expectations, and these are linked to an individual's consumption patterns and characteristics. Consumption must be considered alongside production. The relationship between consumption and production varies spatially and temporally. Both of these are represented in the wiring diagram as part of the decision-making analysis. In some societies, consumption and production may be treated as one and the same, and in others they may be separate, even on different continents. Although consumption and production are determined by a society's goals, values and expectations, the impact of these on land cover change is also determined by local factors such as capital and income, labour and material availability.

With respect to land cover change, improvements in technology help to temper the impact of a growing population through the so-called 'technological fix'. Technological innovation arises when a growing population living within limited resources is forced to consider more productive ways of using the resources available (e.g. Boserüp, 1981). On the other hand, however, land cover change can be attributed directly to technology. New technologies also enable extraction and use of previously untapped resources (e.g. uranium) and encourage ever-increasing levels of affluence. Moreover, technological developments in agricultural production have allowed people to pursue other economic activities, enabling urbanisation, which, coupled with new transport technologies, separated consumption and production further so that a global society is now evident (Grübler, 1994). Through that global society, and recent developments in technology and communications, environmental awareness is being heightened with exchange of information on the environment, including land cover change, and good management practices more commonplace (Mannion, 2002). Technology, therefore, is a doubled-edged sword in terms of its impact on land cover; where and when technology is developed, used or abused are dictated by a society's organisation and values (Middleton, 1999a). Increasingly, it is evident that aspirations of humans are influenced by the stakeholders of the macro-decision-making system, which operates at both regional and global scales.

4.2.1.2 The decision systems analyses of the drivers of land cover change

Decision system analyses (Figure 4.2) can reveal how economic, socio-cultural and political structures and their interaction to drive public policy influence land cover change. The fusion of economic variables, such as prices and commodities, with governance and socio-cultural structures (e.g. access, accountability) to produce policy on issues such as the regional/national economy, land distribution, environmental issues, urban structure and infrastructure juxtaposes the behaviour and decisions of micro-decision-makers. In turn, policy formation is influenced by micro-decision-makers via the political voting system and the interaction of people operating with a common purpose (e.g. non-governmental organisations (NGO)).

There is inequality in the balance of the relationship between the macro- and micro-decision-makers across the globe. This has led to much conflict, with a number of land cover types emerging as highly contested spaces. Studies of such spaces reveal how the actions of decision systems operate on the ground. Conclusions drawn from research into the decline of forest area and quality in Sabah, Malaysia, by McMorrow and Talip (2001) were that decline had less to do with population pressure or misuse of common property resources by local people than with national and state economic and social policies. The study also surmises the impact of the global economy on forest loss in Sabah, acknowledging that the economic slowdown experienced since 1997 has led to calls to use the forest resource as an aid to economic recovery. This illustrates the growing role of globalisation as a cause of land cover change, a cause represented in the wiring diagram of Figure 4.2, within the global political economy module.

The growing power of global economic institutions, such as the World Trade Organisation, and other political alliances that control trade or economic subsidies (e.g. European Union (EU) and the General Agreement on Tariffs and Trade (GATT)) serves to impact on decision-making that causes land cover change at the macro-, or micro-decision-levels. The globalisation of decision-making raises concerns over democracy and accountability, as demonstrated by the growing number of anti-globalisation protests. Nevertheless, in order to address land cover change as a global environmental issue, there are many advantages to globalisation, particularly in terms of international environmental governance. The United Nations Environment Programme (UNEP) estimates that there are now in excess of 500 international treaties and other agreements concerned with environmental issues (French, 2002). Those specifically related to land cover change include the 1994 Convention to Combat Desertification, the 1992

Convention on Biological Diversity (CBD) (Box 5.1) and the 1992 UN Framework Convention on Climate Change (UNFCCC). The effectiveness of these treaties may be compromised by the vast complexity of the current global regime. Nevertheless, it is encouraging that, at the global level, the idea of low politics (focused on technological, equity and ecological concerns) appears to be replacing high politics (focused on market domination) (Rayner *et al.*, 1994). How these trends in the global eco-political system fall out through the dynamic interactive system that links humans to land cover change and, in turn to environmental change, remains to be seen.

4.2.1.3 Linking human drivers and decisions to their environmental impacts

The intersection of social drivers and human decision systems causes some land cover changes that are associated with some of the most important and pressing environmental concerns facing society today. Land cover exerts a large influence on many basic environmental processes and consequently a change in land cover can have a marked impact on the environment. The wiring diagram (Figure 4.2) usefully portrays the relationship between the human and the environmental dimensions in the study of land cover change. At the local scale, the land assessments module illustrates how the environmental impacts of land cover change on the key variables of soil, water, vegetation and nutrients are determined by the availability and suitability of land for alternative uses, its productivity and the sustainability of those uses. The alteration of the local environment is linked to regional scale climate and atmospheric chemistry through resultant trace gas emissions, evapotranspiration and surface albedo processes. In turn, regional atmospheric processes impact on the local scale environment via precipitation and pollution deposition. Regional scale climate and atmospheric chemistry are also linked to global scale systems, specifically to global biogeochemical cycles. This has implications for climate change. This, as stated earlier, requires the attention of the global political economy. The links between the environmental and human dimensions illustrated here reinforce that land cover and its use have profound implications for the environment at local to global scales, thus land cover change is a global environmental issue with significant implications for human habitability.

Despite the formulation of the links between the causes of land cover change and that of environmental change, the impact of a land cover change on the environment remains complex and difficult to generalise. Typically the impact of a land cover change varies as a function of the magnitude of the change (e.g. a conversion of cover typically has a greater impact than a modification), the area of land undergoing the change and its general spatio-temporal context (e.g. the nature of its surroundings and antecedent conditions). Depending on the nature of the change, numerous possible environmental impact scenarios, typically with complex feedback and knock-on effects, may be observed. The

Table 4.2 Impacts of urbanisation on the environment (Reproduced with permission from Douglas, 1994; published by Cambridge University Press)

Action	Impacts			
	Within city	**Around city**	**Downwind and downstream**	**Rural and long distance (including transfrontier)**
Airborne pollutant emissions	Smog particulates, heavy metal fallout (lead, etc.)	Contamination of agro-ecosystem and natural habitats	Pollutant plumes downwind, acid rain	Acid rain, change to global CO_2 budget
Waterborne pollutant emission	Loss of aquatic life in streams, contamination of aquifers and groundwater supplies	Contamination of local water supplies, productive and recreational fish stocks and irrigation water	Pollution of once good surface water supplies for downstream villages and towns	Deterioration of potential irrigation water, cumulative impact on major rivers (Ganges)
Solid wastes	Littering, illegal dumping, pollution of groundwater and streams, attraction of vermin, flies and disease vectors	Dumping of night soil, alienation of terrain for landfill operations, illegal disposal of rubbish	Siltation of streams, loss of channel capacity, in combination with increased urban storm run-off, greater frequency of flooding	Possible contamination of base flow to rivers, dumping of sewage sludge in oceans
Noxious chemicals in gaseous and liquid forms	Danger of Bhopal-type disasters	Illegal disposal of chemical wastes, high danger of Minimata-type incidents	Contamination of ecosystems, especially from mine tailings and industrial wastes	Damage to aquatic systems and food chains, especially through heavy metals
Growth, delivery, marketing and consumption of food	Traffic congestion around market areas, problems of food wastes and vermin	Intense competition for available market garden land, high use of pesticides, and possible contamination of groundwater	Wastes from food processing plant, attraction of birds and vermin to wastes	Impact of fertiliser residue on major rivers, pressure to grow crops on marginal land with consequent risks of soil erosion, siltation of reservoirs and streams
Building materials and construction activity	Severe soil erosion and ground disturbance during construction, urban flooding and siltation	Extraction of sand and gravel, leaving derelict areas unsuited for agriculture	Dust from urban area carried downwind, gravel workings affect stream channels and bank stability	Removal of forest trees to meet demands for wood, tropical forest removal affects genetic resources, water balance and global CO_2 budget
Fuel supplies	Air pollution from coal and wood burning, fire risk from paraffin stoves	Removal of growing timber, including windbreaks to meet fuelwood demands	Drift of pollutants downward, illegal dumping of ashes into streams	Reduction of forest cover to produce charcoal, competition with rural dwellers for fuelwood, impact of mining
Water supplies	Lowered aquifer and subsidence from pumping of groundwater	Alienation of land for water storages, modification of run-off by land	Reduced base flow in some streams, increased effluent disturbances	Competition between irrigation and urban uses of water

land cover change type of urbanisation can be used as an example to illustrate the inter-scale and complex effects of land cover change. Douglas (1994) describes how actions associated with urban areas impact on the environment (Table 4.2). This example highlights that the effects of a land cover change may be manifest at scales ranging from the local to global and over time periods from minutes to centuries. Thus, a local change can have marked impacts at global scales and *vice versa*. Inter-scale effects are common with, for example, local scale changes having implications for the broader environment and forming a global environmental issue since all environmental processes are linked through their functions (systemic in its global scale), and land cover change is worldwide in occurrence and is thus cumulative and global in scale (Turner *et al.*, 1990). Additionally, a global scale change may have local impacts. For example, climate change may necessitate the increased extraction of water by pumping from an aquifer which has its own consequential impacts (e.g. the lowering of the local water table causing subsidence and producing water shortages).

4.2.2 Prospect for the causes of land cover change

The wiring diagram presented by Rayner *et al.* (1994) in Figure 4.2 is a useful approach in isolating proximate and distal causes of land cover change. Each of the modules interact and function in different ways in space and time to produce land cover change of different types, in different locations and of different intensities. The nature, timing and location of these changes determine the environmental change consequences. All decision-making by stakeholders is linked and thus very complex, meaning that land cover change is inextricably woven with human decision-making to produce a dynamic interactive system. Thus far, this has produced a complex geographical pattern of the interaction between environment and humans.

To make better sense of the global patterns of land cover change, McNeill *et al.* (1994) suggested a prioritisation of the most important contemporary land cover types, which are subject to change (Box 4.1). In conducting such a prioritisation, it is hoped that a deeper understanding of the land cover change issue is reached and responses to address change are formulated effectively. This still remains an extremely difficult task, since certain land cover changes have extreme social costs, while others have significant environmental consequences. With every scientific, social, economic and political scenario, the prioritisation may alter. Land cover change and their causes are not static, thus future scenarios require attention. The World Summit on Sustainable Development which took place in Johannesburg, September 2002, not only provided the opportunity for all to look back on the ten years following the first Summit in Rio, but also to isolate important issues of the future. In terms of land cover change, there is one issue of growing importance that will require the attention of decision-makers. That is

---

Box 4.1: Important contemporary land cover types

Cover type	Rationale	Location
1. Tropical forest	• Large conversion extent • High rate of change • Wet regimes, high trace gas flux • Climate/water influence • Biodiversity • Difficult soil management • Sustainable development • Largest frontier • Developing countries	Amazon, West Africa, Southeast Asia, Central Africa, Central America
2. Tropical savanna and grassland	• Large extent of occupation • Large conversion extent • High rate of change • Frequent burning, non-CO_2 trace gases • Sustainable development • Dwindling frontier	Brazil, Sahel-Sudan, South Africa
3. Temperate and boreal forest	• Commercial timber harvest is dynamic • Agriculture unknown • Potential sink • Land intensive • Indirect effects	US, USSR, Europe, Canada, Scandanavia
4. Cropland	• Land intensive = trace gas • Land extensive = CO_2 • Rapidly increasing area • Tenure conflict • Sustainable development issues • Impact on frontier	Global
5. Temperate grasslands	• Unknown in key regions: central Asia • High soil carbon • Potential degradation	
6. Settled and built up	• Extent unimportant • Sphere of influence could be important: core/periphery	
7. Wetlands	• High loss rate, large area uncertainty	

Rationale based on the following criteria:

1. Importance for (a) the global physical climate system; (b) regional energy and water balances; (c) global biogeochemistry; (d) atmospheric chemistry; (e) biodiversity.
2. Importance for (a) land quality, soil fertility and biodiversity; (b) sustainable development, sustainable agriculture and resource development issues; (c) land tenure, land access and land use.
3. Current rate and magnitude of changes.

Reproduced with permission from McNeill *et al*., 1994; published by Cambridge University Press the issue of conflict as driven, principally, by resource competition. There is a growing awareness of the close links between illegal resource extraction, violent conflict, human rights violation and environmental destruction (Renner, 2002). On the one hand, the desire to control ever-diminishing, but increasingly lucrative, resources causes land cover change through natural resource pillage (e.g. illegal logging). On the other hand, increasing awareness of inequity and enforced property rights promotes conflict, for instance through terrorism and other acts of war, which has had a direct impact on land cover (McCarthy, 2000). Furthermore, conflicts cause large-scale displacement of people and loss of life and each of these have a consequence for land cover (Mannion, 2002).

Given the significance of land cover change, its associated issues and dynamism in space and time, it is imperative that accurate and up-to-date information on land cover is available to decision-makers to manage land cover change effectively and informatively. In particular, there is a desire for accurate land cover maps.

4.3 Land cover and land cover change information

Although land cover is something that appears to be simple and easy to observe, accurate data on land cover are surprisingly scarce. Rarely are appropriate mapped data available (Rhind and Hudson, 1980; Estes and Mooneyhan, 1994). As an example, in Britain the first complete survey of land cover was undertaken in the 1960s and not up-dated until the 1990s. The accuracy of each map was also lower than many would accept operationally. The accuracy of the Land Cover Map of Great Britain produced in the 1990s was estimated to be approximately 70%, only if problematical boundary regions were excluded (Fuller, Groom and Jones, 1994). Given that land cover is one of the most dynamic features usually depicted on a map (Belward, Estes and Kilne, 1999), it is apparent that even for a small nation, with a strong tradition of mapping, accurate and up-to-date land cover data may be unavailable. Even if a land cover map was available, the classes it depicts may also not be those of interest (Knox and Weatherfield, 1999). Those requiring land cover data may, therefore, be tempted to acquire the data themselves but will find that the acquisition of land cover data is far from trivial and the problems of mapping land cover tend to increase as the area to be mapped increases. It is not surprising, therefore, that maps representing land cover are associated with considerable uncertainties. Indeed, maps of the same region produced by different bodies may differ considerably. Moreover, without an authoritative body to oversee the production of land cover maps, different bodies may produce or choose an available map that matches a scientific or political point they wish to make.

Some important problems associated with land cover data sets were demonstrated by Townshend *et al.*'s (1991) comparison of 16 global land cover maps. Figure 4.3 summarises the areal extent of a set of major land cover classes depicted in each of the maps compared. Even a superficial analysis of Figure 4.3 reveals that the maps differ considerably: note, for instance, that the maps differ by a factor of approximately two in terms of their representation of forest cover. A further issue is that the maps also differ in their spatial representation, with relatively low levels of agreement in class labelling across the globe (DeFries and Townshend, 1994). As there is typically little information on the accuracy of the representation depicted in a land cover map, it is difficult to select a map for any particular application (Foody, 2002). This is an important issue, particularly as the integration of different maps into an environmental model (e.g. for predicting climate change scenarios) may result in dissimilar outputs. It is important therefore that the suitability (especially in terms of accuracy) of land cover maps be carefully addressed. For this reason, map producers are increasingly providing accompanying information on issues associated with the accuracy of land cover maps (e.g. Smith and Fuller, 2002).

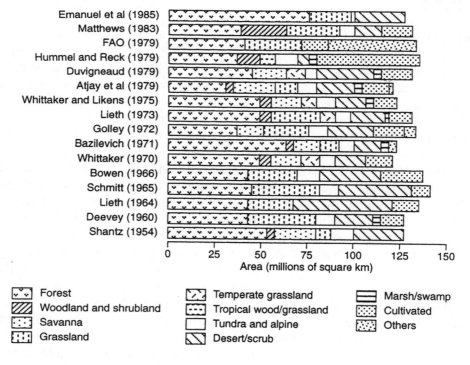

Figure 4.3 Comparison of the areal extent of major land cover classes represented in 16 global land cover maps. (Reproduced with permission from Townshend *et al.*, 1991; published by Elsevier.)

78

4.3.1 Problems in land cover mapping

Many reasons have been suggested to explain the difficulties encountered in mapping land cover. A vast array of concerns, ranging from the difficulties in defining classes through to technical problems in mapping land cover accurately, may be identified.

Land cover classes appear, superficially, to be simple to recognise and easy to characterise; however, they are often inherently vague and so difficult to map objectively (Rossum and Lavin, 2000; Bennett, 2001). Consider one important class, forest. Problems in mapping forested land cover abound. Before even encountering the technical problems of mapping forests, there is a substantial but basic definitional concern, what is a forest? This is a difficult question to which there are many answers (Bennett, 2001; Foody and Atkinson, 2002). In general, answers to this seemingly simple question will include some reference to a critical threshold of some measure of tree cover per unit area being exceeded. This raises the questions of what critical threshold and why? The threshold used by a map-maker may not match the requirements of a particular user. In practice, a variety of thresholds have been used for maps depicting forest. For example, the forest classes depicted in maps such as Figure 4.1 (DeFries *et al.*, 1998) and the International Geosphere Biosphere Programme (IGBP) global land cover map (Scepan, 1999) all require more than 60% tree canopy cover. However, some land cover data sets may be based on a different threshold. In its global Forest Resources Assessment (FRA) of 2001, the Food and Agricultural Organisation (FAO) uniformly adopted the 10% threshold. Moreover, the threshold used may vary between locations. For example, in the early assessments of global FRA, the FAO used thresholds of 10 and 20% canopy crown cover for developing and industrialised countries respectively. Changing the threshold used between assessments can result in significant, apparent changes in forest cover. The use of any threshold is problematic; with strict enforcement of the FAO's 10% threshold, a tropical region with 9.9% forest cover is mapped as non-forest while one with 10.0% cover is classed as forest in exactly the same way as a site with complete tree cover. The significance of this should not be underestimated. Not only does it make it difficult to accurately characterise basic land covers but it also means a change, such as a very large change in the percentage forest cover, may not result in a definitional change of land cover type and so need not be noted in some land cover change investigations (those focused on land cover conversion). All these concerns raised relate to just a simple definitional issue as to what constitutes a basic land cover class. Problems tend to magnify when more precise information is required: what type, age and diversity of forest?

Once a set of classes have been defined, the task of mapping land cover may begin. Acquiring land cover data by field observation is difficult, time consuming and expensive. At regional to global scales in particular, direct field observation is impractical. At such scales the only realistic means of acquiring land cover data is via remote sensing.

4.3.2 Potential of remote sensing as a source of land cover data

Remote sensing is extremely attractive as a source of land cover data (Lambin, 1997; Campbell, 2002). A typical remotely sensed image, such as an aerial photograph, is simply a representation of the reflectance properties of the earth's surface cover in visible light. For mapping large areas, imagery acquired by satellite remote sensing is generally used. The sensors are carried on board satellites that typically operate over a broad range of the electromagnetic spectrum, commonly acquiring imagery in visible and infrared wavelengths. The imagery derived from a typical remote sensing system is simply a representation of the way these different types of radiation interact with the land surface. The imagery is thus a representation of surface cover and so land cover properties.

As land cover types interact differently with radiation, they each have a distinctive remotely sensed response and so may be mapped from the remotely sensed imagery. Many factors control the detail of how radiation interacts with land cover. It is apparent that land cover type-specific spectral responses may be detected. Moreover, these spectral responses may be separable. Although inter-class separability varies, the spectral responses observed by a satellite sensor may be used to distinguish land covers, enabling their mapping, from remotely sensed imagery.

4.3.2.1 Advantages in using remote sensing for land cover mapping

The spectral separability of land cover types is only one reason for using remote sensing as a source of land cover data. Several other important features of remote sensing combine to give it an enormous potential as a source of land cover maps. First, as remotely sensed imagery are acquired from a sensor above the earth's surface looking down upon it (sometimes at an angle), the imagery have a map-like format. That is, the imagery shows the location of features on the earth's surface and their spatial context. Although not having the properties of a map, this map-like representation can be used as a map and provides an excellent base for mapping. Second, remotely sensed imagery are available at a range of spatial and temporal scales. This is particularly useful, given the range of scales over which land cover changes occur and on which their impacts may be observed. This widespread availability of imagery also enables the mapping of any location on the planet. Whether remote, inhospitable or in a region where access on the ground is difficult or perhaps even denied, it is possible to acquire a remotely sensed image of any site of interest from which a map can be derived. In general terms, there is a trade-off between spatial and temporal domains of sensor characteristics; relatively fine spatial resolution coverage can be obtained at a coarse temporal scale but if a site is to be monitored at a high temporal frequency coarse spatial resolution imagery may have to be used. Partly for this reason, coarse spatial resolution imagery is generally used to derive a range of regional to global scale maps that can be updated frequently (e.g. DeFries *et al.*, 1998; Scepan, 1999), while finer spatial resolution data may be used to map smaller areas with greater detail on a less frequent basis (e.g. Fuller, Groom and Jones, 1994; Smith and Fuller, 2002). Third, remotely sensed imagery are relatively highly consistent in space and time. A remotely sensed image is simply a depiction of the remotely sensed response of the earth's surface that is controlled directly by the land cover and is independent of human interpretations and definitions. There are, therefore, fewer problems in combining data from different sources and agencies that may, for example, utilise different class definitions. With a remotely sensed image, the map-maker can select the classes of interest to the particular investigation and seek to map those from the image without constraints imposed by others. Fourth, most satellite sensors used in mapping land cover acquire digital image data. This is important since the volume of data acquired by satellite remote sensing is very large. This is also helpful in manipulating the imagery and linking it with other spatial data sets, particularly within a Geographical Information System (GIS). Fifth, and finally, the remote sensing offers a very inexpensive means of surveying large areas rapidly. While the cost of image acquisition varies with time, and as a function of variables such as the level of pre-processing applied and user status, the cost, at the time of writing, of a full Landsat 7 Enhanced Thematic Mapper (ETM+) image was £450 and a NOAA Advanced Very High Resolution Radiometer (AVHRR) image £35. Moreover, some data sets are available freely through data archives (e.g. AVHRR data through the NOAA/NASA Pathfinder AVHRR Land (PAL) programme and a variety of data sets from Landsat sensors through the Global Land Cover Facility). These images cover large areas with the data acquired near-simultaneously. Thus while field survey of a region could take months or years, satellite remote sensing provides the potential for complete spatial coverage of the region to be acquired over a period of time measured in seconds.

Considerable advances in the use of remote sensing for mapping land cover have been made in the last three decades with the regular provision of high-quality satellite sensor data sets. Mapping of very large areas such as continents only became a realistic possibility in the 1980s when researchers began to utilise imagery acquired from the meteorological NOAA AVHRR satellite sensor (Tucker, Townshend and Goff, 1985). Indeed, the NOAA AVHRR has become the workhorse of large area mapping investigations. NOAA AVHRR data have, for example, been used to map the large vegetation regions such as the Amazon (Lucas *et al.*, 2000) and the globe (DeFries *et al.*, 1998). Other sensors have also been used, notably, for example, SPOT VEGETATION to map Canada (Cihlar,

Beaubien and Latifovic, 2001) and Landsat TM in mapping the UK (Smith and Fuller, 2002) and conterminous US (Vogelmann *et al.*, 2001). Recently launched sensors, such as the Terra Moderate Resolution Imaging Spectroradiometer (MODIS) (Friedl *et al.*, 2002) are likely to be used increasingly for mapping land cover over large areas.

4.3.2.2 Drawbacks in using remote sensing for land cover mapping

Despite the great potential of remote sensing as a source of land cover data, there are many problems encountered (Townshend *et al.*, 1991; Townshend, 1992; Wilkinson, 1996; Foody, 1999, 2002; Mather, 1999). A key fundamental concern is that land cover maps derived from remotely sensed data are often judged, sometimes incorrectly, as being of insufficient accuracy for operational use (Townshend, 1992; Wilkinson, 1996; Foody, 2002). This clearly limits the ability to both map and monitor land cover from remotely sensed data. It is, therefore, important that land cover maps be carefully and rigorously evaluated and accompanied by appropriate quality statements to facilitate their sensible use. Unfortunately, land cover maps are often used unquestioningly despite the considerable uncertainties associated with them (Woodcock and Gopal, 2000). For example, the IGBP global land cover map is estimated to have an area-weighted accuracy of 66.9%, considerably less than the map-maker's target of 85%, and the accuracy of individual classes varies from 40 to 100% (Scepan, 1999). The size and importance of the problems associated with uncertainties in the map will vary with the proposed application (DeFries and Los, 1999) but should not be ignored.

Many of the problems in mapping land cover from remotely sensed data arise due to technical problems relating to the calibration or inter-image normalisation required for multi-temporal analyses or spatial problems relating to the resolution of the imagery and ability to locate sites accurately within them as well as how to optimally extract land cover data from the imagery.

Considerable work has addressed these issues, and advances, particularly in satellite sensor technology and computer analysis of imagery, should reduce the problems in the near future. For example, common problems relate to cloud obscuring the land surface and can be reduced through an increase in the number of sensors and use of more intelligent sensors such as those that can be pointed away from cloud-covered regions or weather-independent radar systems. Other concerns such as the limits imposed by the spatial resolution can be reduced through the development of finer spatial resolution instruments and super-resolution techniques to extract more fully the information content of the data (e.g. Tatem *et al.*, 2002). Indeed much current research is addressing the many challenging issues that presently limit the ability to realise fully the vast potential of remote sensing as a source of land cover data. That said, at regional to global scales, remote sensing is the only practical source of land cover data. In the foreseeable future, data pertaining to land cover and land cover change will be, hopefully, more accessible. It is armed with these data and other ancillary information that land cover change can be managed in order that environmental consequences are sustainable and socio-economic benefits are reaped.

4.4 Human responses to land cover change

Humans respond to environmental change in a number of ways, namely by altering their social organisation, values or the environment itself. In the past, environmental change generally prompted a response of migration to another area. Contemporary responses, however, are more complex and can take place in several dimensions. For instance, responses may occur as environmental change is occurring or as it is anticipated. Actions may be deliberate in their responses. Responses may be co-ordinated, orchestrated by government or trade organisations, or unco-ordinated, for example, as conducted by individual households or small firms. Humans may choose to mitigate any environmental change, block the impacts of environmental change, adjust to the impacts or improve the robustness of social systems in anticipation of an environmental impact (Stern, Young and Druckman, 1992).

One concept that currently dominates the way in which humans respond to environmental change is that of sustainable development. Sustainable development requires that societies meet human needs by increasing productive potential and ensuring equitable opportunity for all. In terms of land cover change, focus must be on efficiency in resource management, resilience of ecological systems and equity. In determining the future of the world's land cover, each tract of land must be considered as part of a complex and interlinked system. Sustainable land management maintains the potential of the land to provide resources, and the ecosystem of which that tract of land is part, in a certain desired condition. Use of a tract of land should therefore be determined by its limits of exploitation, for instance, the rate of the use of forest cover for timber must be within the limits of regeneration and natural growth of that forest. Furthermore, all stakeholders must be considered, and the use of that tract of forest is determined by its ability to enhance both current and future potential to meet human needs. Equity encourages all stakeholders to have a common interest. This is of particular importance since many problems of resource depletion arise from disparities in political and economic power. Returning to the case of a tract of forest, it may be over-exploited by excessive logging because logging concessions have more power than the forest dwellers who have generally always used the forest sustainably.

Each of the Brundtland Commission Report's (1987) critical objectives has direct implications for land cover and requires the attention of every individual, nation and international body. To this end, the concept of sustainable development has initiated both discussion and action that will have impact on the future of the world's land cover. Whether these achieve sustainable development is a matter for the crystal ball. However, execution of these initiatives and policies will assist decision-makers to decide which are not sustainable and to chart new ones. What makes the pathway(s) to sustainable development even more challenging is that different land cover types cannot be considered in isolation since the use of one land cover type has knock-on effects on other types. For example, the sustainable management of forest cover in the tropics relies heavily on the sustainable use and maximum productivity of agricultural land (Poore, 1993). Furthermore, much of the world's population lives without day-to-day intimate contact with the terrestrial biosphere (Trudgill, 2001). Yet, the area of the terrestrial biosphere required to meet the consumption of this population, also known as its ecological footprint, is growing rapidly (Wackernagel and Rees, 1996). This augments the concern for the status of land cover which forms an important part of the global ecosystem of which people are a part (Wayerhaeuser, 1998). Therefore, it is evident that sustainable land cover change is a complex concept which requires integrated solutions. Here, we consider one land cover type to outline challenges faced and progress made.

4.4.1 Changing forest cover: a challenge for sustainable development

Forests are a land cover type vital to the sustainable well-being of local communities, national economies and global biospheric functioning (Myers, 1996; Mannion, 2002). Despite this, they are currently over-exploited and under-utilised (Myers, 1983). There has been much discourse on the future of forests and the role of management and policy in securing their future and hence the habitability of the earth (Woodwell, 1995; Boyce and Haney, 1997; Noss, 1999). For example, the International Tropical Timber Agreement (ITTA) aims to achieve the sustainable utilisation and conservation of tropical forests, and the International Union for Conservation of Nature and Natural Resources (IUCN) emphasises that conservation depends on development and that lasting development is impossible without conservation. Poore (1993) considered these assertions for tropical forests stating that each nation is responsible for the management of its natural resources and that a balance needs to be reached between using these resources to maintain and improve the standard of living of its population and conserving them to ensure the survival of the resources for future use.

There are many stakeholders with an interest in forest management, and their needs vary across the globe and also within nations at regional and local scales (section 5.4). In order that the balance of use is achieved, Myers (1996) proposed a policy programme featuring a range of proposals to be considered by all decision-makers. One proposal suggests enhancing the institutional status of forests, so that the management of forests is driven by those who know the actual and potential forest outputs. In order that this be achieved effectively, there urgently needs to be a full scientific evaluation of the world's forests so that our economic understanding of forests be improved. Economic analyses of forests that evaluate the entire range of goods and services from forests need to be undertaken. Moreover, the analyses should be geared to social equity as well as economic efficiency. Care needs to be taken to ensure that the pursuit of social equity and economic efficiency really does balance the needs of all sectors of society, rather than favouring policy- and decision-makers at the expense of the poorest. Another proposal focuses on the detrimental impact of subsidies that exert adverse effects not only on the forest themselves but on the economy in the long run. Such subsidies may seem perverse to some, but could suit decision-making elites, or those who pursue economic gain at the expense of environmental sustainability. Myers (1996) also suggests that the key policy question of the relationship between sustainable development and forests be broached in two mutually supportive ways; that forests themselves should develop sustainably and that forests be used to support sustainable development in general. Myers (1996) acknowledges that the policy measures required will be difficult, but warns that they will not be as difficult as a world bereft of many of its forests.

4.4.2 Trends in forest management, conservation and sustainable development of forest resources

The publication of the UN FAOs 'State of the World's Forests, 2001' report provides an excellent insight into how far forest policies have developed in recent years. The report highlights some of the policy, technical and institutional measures taken to improve forest management and conservation, reflecting the move to balance environmental, social and economic objectives.

Efforts of note include an increase in the use of plantations to provide industrial woods thus reducing timber harvesting in natural forests. Where natural forests are used there has been some improvement in harvesting practices (e.g. Reduced Impact Logging (RIL)), though these are still not widely practised. Moreover, many countries have recently imposed bans or restrictions on timber harvesting. Increasing attention has been focused on institutional and governance issues in the drive for sustainable forest management. These issues include the fight against illegal forest activities, including illegal logging and corruption. Community participation in forest management is also noted as a significant feature of national policies and programmes the world over.

The FAO report also highlights the difficulty in reconciling the conflicting demands of different stakeholders for the many goods and services provided by forests. Particular attention in the past decade has been on the role of forests in addressing the climate change issue. The Kyoto Protocol on Climate Change (1997) created a new, legally binding treaty for industrialised nations to meet the voluntary emission targets set at the 1992 UN Conference on Environment and Development (UNCED). This meeting led to an emphasis on national reporting of carbon budgets as the focus of national contributions to global climate change and has initiated the idea of international trading of carbon credits (Pfaff *et al.*, 2000). The ability to trade carbon credits arises when transferable allowances on the amount of carbon every nation is able to emit on an annual basis are set (see Box 2.2, section 7.8.1). Those nations exceeding their targets could sell their excess to other nations which could use them to increase their emissions by an equivalent amount. There are many advantages of this approach to reducing greenhouse gas emissions, not least economic (Manne and Rutherford, 1994). Explicit reference to land cover change practices as a way of reducing emissions was made in the Protocol since the slowing of slash and burn deforestation and encouragement of regrowth of forest cover are mechanisms recommended for achieving set emission targets (Potter, 1999). Indeed, the use of forest cover has proved a popular strategy for terrestrial carbon management. Table 4.3 presents four main strategies and case studies that have been used with the aim of climate change mitigation. Such practices are likely to become more common if the Kyoto Protocol is ratified, having a profound effect on forest cover. This depends, though, on forestry management practices which are included as eligible measures to achieve set targets. Biological diversity conservation is another major issue that has received considerable attention in the last decade. Major developments that have occurred include efforts to integrate conservation and development needs, community-based conservation, a greater emphasis on ecosystem management and the adoption of a bioregional approach, whereby protected areas are considered within a greater geographical and land use context (FAO, 2001).

Most countries now have a national forest programme: since 1985, 128 countries have developed or upgraded their national forest programme, and these programmes are supported by the Intergovernmental Forum on Forests (IFF) and the Intergovernmental Panel on Forests (IPF) (Vajpeyi, 2001). In many cases, this has initiated a revision of forest policies and legislation and wider stakeholder participation in decision-making processes. The case of the United States (US) provides an excellent example of this. The World Resources Institute (WRI) has published its vision of a sustainable US forestry sector, in which a more integrated approach to using and managing forest resources is advocated (Dower *et al.*, 1997). The WRI outlines ten steps towards sustaining the diverse, valuable and irreplaceable resources of the US forest (Box 4.2). It is envisaged that these steps, which include elements of innovation, experimentation, and leadership by government, communities, resource managers, manufacturers and consumers, will assist the US forestry sector to meet its sustainability challenges (Johnson and Ditz, 1997).

Table 4.3 Overview of the main forest activities for terrestrial carbon management (Reproduced with permission from Townshend *et al.* (1991); published by Elsevier.)

Strategy type	Explanation and approach	Case study
• Reduce carbon sources	The protection and conservation of the carbon pools in existing forest cover (i.e. wood and leaf biomass) would prevent the emission of carbon to the atmosphere through burning and/or decomposition of forest biomass. Dixon *et al.* (1994) estimated that carbon emissions from land would be reduced by 1.2 Gt of carbon per year if deforestation were eliminated.	The Rio Bravo Conservation and Management Area in Belize protects 14 000 ha of forest cover from being sold off to farmers. The project is managed by Programme for Belize (a Belizean NGO). Stuart and Costa (1998) estimated that without protection the total forest cover would be lost within five years, and by protecting the forests 2.5 million tonnes of carbon would be conserved over the 40-year life of the project.
• Enhance carbon sinks	Large areas of secondary forests exist, particularly in tropical countries. These could be protected from and/or better managed in instances such as fire, logging and shifting cultivation. In doing so, Houghton (1996) estimated that 1–2 Gt of carbon per year may be accumulated in forests.	Reduced Impact Logging techniques in Sabah, Malaysia were implemented to reduce the impact of logging on forest cover. A project implemented by the Innoprise Corporation in Malaysia for 1400 ha of dipterocarp forests in Sabah aimed to reduce logging damage by 50% and conserved an estimated 40 t of carbon per hectare in comparison with that of conventional logging techniques (Costa, 1996).

Table 4.3 (Continued)

Strategy type	Explanation and approach	Case study
• Expand carbon sinks	By establishing new areas of forest cover or encouraging regeneration of forest in previously cleared areas, carbon is sequestered from the atmosphere and stored in terrestrial carbon pools. The amount of carbon sequestered in this way depends on species type, site and management. Brown *et al.* (1996) estimated that 38 Gt of carbon could be sequestered over the next 50 years in this way.	A Forests of the Future campaign organised by The Sunday Times, a UK newspaper, and Future Forests, a UK-based environmental business, planted 132 000 saplings at six sites across the UK where industry, progress or agriculture had altered the land cover. Five further sites have been identified for phase two of the campaign. The target is to plant one million trees to offset some of the carbon emitted by UK citizens (Futureforests website).
• Carbon substitution	The use of sustainably produced biofuels to provide energy results in a reduction of carbon emissions to the atmosphere compared with that when using fossil fuels since any carbon emitted using biofuels is offset by the regrowing biofuel. Sampson *et al.* (1993) estimated that biofuels could offset fossil fuel emissions by up to 4 Gt of carbon per year by 2050. A further reduction in carbon emissions may be achieved through the use of wood products in place of other materials that are associated with the emission of carbon (e.g. cement).	British BioGen, a trade association to the British Bioenergy Industry, has initiated a number of projects aimed to use wood to heat domestic and commercial properties. One project, at the EcoTech Rural Business Centre in East Anglia, is now heated by a 250 kWt automatic wood chip boiler. EcoTech is on a 4.5 ha site including a visitors centre, offices, conference room and training facilities. The project was supported by Breckland Council, the Rural Development Commission and the European Union (British Biogen website).

Box 4.2: Ten steps towards a sustainable US forestry sector

Step 1. Develop and implement regional or State sustainable forest-sector plans: Move away from an *ad hoc* planning process. The planning process should have breadth so that all major forest products and services are addressed; have the participation of all who use, manage or benefit from the State's forests; be based on the best available ecological, social and economic information on the State's forest sector; and culminate in specific binding commitments by government agencies, large corporations, NGOs and others.

Step 2. Establish a national network of demonstrating sustainable forests: To be managed by partnerships of federal, State, industrial and NGOs to generate realistic and public information on the comparative costs and benefits of conventional and more sustainable forest regimes. The network should be used to test new ideas and document, stimulate and live with real-world constraints to provide findings useful and credible to resource managers.

Step 3. Slow fragmentation and enhance stewardship of private forest lands through tax returns: By reform of federal estate tax law so that heirs to forest lands do not need to sell or sub-divide their land thus avoiding the fragmentation of larger, more environmentally valuable forest habitats into smaller, less valuable parcels. Tax benefits also provided for land owners who demonstrate sustainable forestry practices.

Step 4. Restore and enhance timber productivity on degraded lands through innovative financing mechanisms: Provide low-interest loans, cost-share agreements, profit-sharing and other financing from public and private sources to encourage land owners.

Step 5. Protect and restore critically endangered forest ecosystems through targeted incentive programmes, land acquisitions and land swaps: Use of a wide range of technical and scientific, legal and economic tools and strategies to integrate conservation objectives into forest management. The choice of conservation options would depend on who owns the land.

Step 6. Encourage forestry efforts within the US to sequester carbon, increase fibre supplies and enhance rural development: Takes advantage of addressing more than one sustainability challenge at one time. For example, several US utility companies have financed forestry activities in at least six countries to offset emission of CO_2 from new fossil fuel plants. The valuable contribution forests make to carbon storage and should encourage better forest management, leading to a more sustainable US forest sector in the long term.

Continued on page 90

Continued from page 89

Step 7. Make the environmental performance of forest companies and their products more open to scrutiny: Apply standards such as the ISO 14001 and EMAS to companies to allow business decision-makers and stakeholders to benchmark and choose companies based on their environmental performance.

Step 8. Integrate sustainability into corporate goals, planning and operations: To stay competitive, forest sector firms need to take trends in timber productivity, biodiversity, climate change and clean production into account. Those firms with the vision and commitment to do so will create new opportunities for improved and diversified production.

Step 9. Cultivate a more robust concept of sustainability in US forestry education: Educating tomorrow's professionals in sustainability issues by the integration of such issues into the mainstream curriculum of relevant timber and business/management courses.

Step 10. Bolster US international leadership to improve the sustainability of forest management worldwide: All countries have a stake in sustainable forest management; the US could provide international leadership in technical know-how, legal and policy development, and private-sector management and marketing skills. The US government could work with other governments to build up their capacity to design, monitor and enforce sound environmental practices, including a forest data and monitoring system and indicators of sustainable forest management.

Source: Dower et al., 1997

Forestry is increasingly affected by globalisation. How forests are being used and managed is being influenced by freer flows of labour, capital, goods and information between countries (FAO, 2001). In the international domain, there has been an increase in co-operation between organisations such as the International Tropical Timber Organisation (ITTO), the World Trade Organisation (WTO), the United Nations Commission on Trade and Development (UNCTAD) and NGOs across the globe (Vajpeyi, 2001). Indeed, at the international level, forests have been under particular attention since the UNCED in 1992 at which the 'non-legally binding authoritative statement of principles for a conservation and sustainable development of all types of forests' and Chapter 11 of Agenda 21 combating deforestation were adopted. Two institutions in particular that have been established to meet the challenges posed by forest cover change and enable forests to play a part in sustainable development are the IFF/IPF (1995–1997 and 1998–2000 respectively) and the World Commission on Forests and Sustainable Development (WCFSD) (Myers, 1996). October 2000 saw an agreement on an international arrangement on forests, including the establishment of the UN Forum on Forests

(UNFF) (FAO, 2001). There has also been a strengthening of intergovernmental co-operation on forests at the regional scale. For example, at the 1993 Ministerial Conference on the protection of forests in Europe, the ITTOs Year 2000 Objective was adopted with the goal of putting internationally traded tropical timber on a sustainable basis. At the local level, private companies have voluntarily or pro-actively adopted environmentally and socially sound practices, and increasingly are collaborating with other companies, communities and environmental groups on issues relating to sustainable forest management. In all cases of co-operation, NGOs and other civil groups have become increasingly active in forest-related advocacy, legal action and natural reserve management (FAO, 2001).

4.4.3 Future challenges in responding to land cover change

> The 1990s have been instrumental in terms of defining a common global vision for the future of forests and their relation to people's lives... The groundwork has been laid, but realising the vision of sustainable management, conservation and development of the World's forests will depend on a number of factors. These include the ability to finance and share equitably the costs and benefits of sustainable forest management, continued and strengthened political commitment, and the translation of political commitment into effective action on the ground.
>
> *FAO, 2001, p. xiv*

Indeed, the next decade is one of many challenges if progress in sustainable development, with respect to forest cover change in particular and land cover change in general, is to be realised. Global demand for use of land is likely to continue to accelerate (Turner and Meyer, 1994). It is imperative therefore that any policy or initiative is periodically evaluated and modified. Forests have been used as an example of what can be achieved in the management of land cover change. However, each land cover type within a nation's border cannot be con-sidered in isolation and furthermore, national policies will have worldwide effects and need to fit into a worldwide picture of land cover change management.

Effective management requires that the impact of the development, in the sustainable development ideal, on land cover is accommodated. In particular, how development may vary with technical, socio-economic and political changes (Bongaarts, 1994). This, coupled with future environmental change, particularly climate change (Repetto and Austin, 1997), makes land cover change manage-ment a complicated task, fraught with problems. Policies must aim to follow a moving target. Forecast modelling is useful here; however, their already difficult task is compounded by the need to link social and natural science models. The basis of any projection of land cover change is information on the driving forces of land cover change (Stéphenne and Lambin, 2001), yet models of these driving forces have not been designed to link with a natural science model. For example, the data to be used in such models may not suit the requirements of a natural science model. A good example of this is in population data which have been collected according to political boundaries rather than natural boundaries. Furthermore, the patterns of environmental change across the earth's surface do not mirror the patterns exhibited by one driving force of land cover. In reality, as a result of inter-scale effects discussed earlier in this chapter, land cover changes are increasingly experienced far from their driving forces (Turner and Meyer, 1994). Yet there is prevalence amongst those who can influence land cover change policy that the cause of land cover change can be explained using simplistic models of the role of population, poverty and infrastructure (Mather, Needle and Fairburn, 1998). Popular myths preside over the causes of land cover change, though it has been shown that the findings of meso- and micro-scale studies and data are at odds with global scale assessments, because they are time- and place-specific. Such myths include that 'population growth and poverty causes deforestation' and 'urbanisation is unimportant in global land cover change' (Lambin *et al.*, 2001).

The natural science models which are used to forecast environmental changes require data on the very nature of land cover and its change, since environmental change at the local scale through to the global scale and land cover are inextricably linked (Bonan, 1997; Chase *et al.*, 2000). Faced with these complexities a new generation of models that are able to deal with spatial, temporal and causal dimensions at once are required (Irwin and Geoghegan, 2001). However, this leads to more complex models that incorporate increasingly sophisticated mechanisms compromising the general operationality of the models and their potential for being extrapolated to large and heterogeneous areas (Lambin, 1994).

A further complexity in the management of land cover change is that pertaining to the uncertainties associated with our knowledge of land cover and its change. Uncertainty is a problem associated with science in general and environmental science specifically (O'Riordan, 2000); however, in terms of land cover change it is exacerbated by a lack of reliable and accurate data on land cover and its change. Lack of understanding and agreement on particular aspects of land cover change prevails. For instance, there is poor understanding of the factors underlying and determining the transition from shrinking to expanding forest cover as observed in developed countries and whether a similar transition will occur in developing countries. Moreover, the environmental impacts of reforestation are under debate (Mather, 1993). Another area of uncertainty surrounds the concept of desertification as a land cover change (Mainguet and Da Silva, 1998). Those interested in the use of marginal environments such as the Sahara region of Africa face contradictory and opposing viewpoints concerning the occurrence and impact of environmental degradation (Agnew, 2002). Agnew (1995) identifies four sources of confusion, namely paucity of reliable data, spatial and temporal variability of dryland systems, ignoring resilience, and the institutionalisation of environmental problems. Slaymaker and Spencer (1998) attribute the confusion to the superficial grasp of the physical processes underlying desertification. Such uncertainty compromises the success of policy relating to this land cover change type.

Uncertainty in land cover change science is further exacerbated by the multi-dimensionality of the issue (in both space and time). Environmental changes may have feedback effects on land covers and human driving forces of land cover change. These effects, real or perceived, have a further set of human dimensions to the extent that they generate societal responses intending to manage or ameliorate harmful changes (Turner and Meyer, 1994). One example of this is the practice of managing forests to mitigate the rising atmospheric CO_2 forcing climate change as outlined in Table 4.3. Uncertainty and debate surrounds the relative strengths and weaknesses of different forest types to store and sequester carbon and, further, whether carbon storage and sequestration processes may be measured with accuracy and with appropriate packaging for policy-makers' perusal. Moreover, there is little known about how offset proposals impact on social and environmental systems (Smith *et al.*, 2000). Policy formulation at any level needs to respond to the dynamics of the complexities and uncertainties they currently face in the management of land cover and its change.

4.5 Conclusion

It may be inevitable that the 21st century will be characterised by major land cover change. Since land cover change has both certain and uncertain implications for the global environment, and humans who are a part of that environment, there is a clear case for informed decision-making to mitigate and manage the impacts of land cover change. In order to do this effectively, land cover data are important. Unfortunately, presently, these have varying degrees of accuracy and utility.

Decisions made by each one of us, as individuals, a society or as part of a global network lead to land cover change through the dynamic and interactive relationship between people and environment. Land cover change is a complex problem which requires integrated solutions. Thus, as a global environmental issue, land cover change is important to all, involving all, either directly or indirectly. As such, it is important that those in political, economic, social and scientific arenas work together to address this global environmental issue, as encouraged by the concept of sustainable development. However, the best framework to achieve sustainable development, and thus address land cover change, is yet to be determined.

Further reading

Mannion, A.M. (2002) *Dynamic World: Land-Cover and Land-Use Change*, London: Arnold.
An undergraduate textbook on land cover and land use change written from a geographical perspective with reference to past change and natural change as well as recent and human-induced change.

Global Environmental Issues

Meyer and Turner II, B.L. (eds) (1994) *Changes in Land Use and Land Cover: A Global Perspective*, Cambridge: Cambridge University Press.
A detailed book which discusses the links among human activity, land cover change, and environmental change and suggests a research agenda to tackle the land cover change issue.

Lambin, E.F. *et al.* (2001) The causes of land-use and land-cover change: moving beyond the myths, *Global Environmental Change*, **11**, 261–69.
A review paper outlining the main causes of land cover change, dispelling the many myths surrounding them.

Campbell, J.B. (2002) *Introduction to Remote Sensing*, 3rd edn, London: Taylor and Francis.
A student textbook on remote sensing for earth observation outlining the general steps required for land cover mapping from remotely sensed data and characteristics of the main sensors used for land cover mapping.

Food and Agricultural Organisation (2001) *State of the World's Forests*, FAO statistics, www.fao.org (accessed 15 June 2002).
The 2001 edition of a series of reports produced every two years on the status of forests. Recent major policy and institutional developments and key issues concerning the forest sector.

Chapter 5
Conserving Biodiversity Resources

Frances Harris

5.1 Introduction

Biodiversity is the range of life forms on our planet, as seen in the variety of living organisms and the range of ecological communities. The most obvious interpretation of biodiversity is that it concerns the different types of animals and plants we see. However, biodiversity is not just about having animals and plants. Biodiversity is also concerned with the genetic variability within a species. Genetic biodiversity broadens the range of attributes exhibited by a species such as disease resistance, or drought tolerance in the case of plants, and so enhances the likelihood of some individuals surviving through environmentally challenging times. Biodiversity also goes beyond concern for individual plants and species to be concerned with ecosystems. The variety of ecosystems in the world is vast and includes rainforests, boreal forests, swamps, savannas, deserts, tundra, corral reefs, alpine meadows and urban environments. Such ecosystems are important as mutually interdependent communities of plants and animals living within a distinct environment (temperature, rainfall, soil type, altitude). Furthermore, the variability of habitats and ecosystems in our world plays a role in regulating global environmental processes. Each form of biodiversity is important.

5.2 Why is biodiversity important?

Biodiversity plays many roles in our lives, both directly and indirectly. Before considering why we should conserve biodiversity, it is important to identify the

Global Environmental Issues.   Edited by Frances Harris
© 2004 John Wiley & Sons, Ltd   ISBNs: 0-470-84560-0 (HB); 0-470-84561-9 (PB)

ways in which biodiversity has value for us (Table 5.1). Biodiversity can be of economic value to many societies. At the local level, this can be through the subsistence value of wild foods and gathered products (firewood, fodder, game) which provide household needs for free. This can play a substantial role in the livelihoods of the poor (Koziell, 2001), as shown in India (Jodha, 1986) and Africa (Harris, 2003). At a local, regional or national level, timber and non-timber forest products may produce commodities to be traded and exported. At the international level, biodiversity is a tradeable commodity providing the

Table 5.1 The many ways in which biodiversity is valued (Developed from Koziell, 2001; Blench, 1998; Grimble and Laidlaw, 2002.)

	Direct	**Indirect**	**Non-use**	**Future**
Economic	Subsistence: firewood, fodder, game meat, building materials, medicines, dyes, gums, resins	Coping strategies (wild foods)		Medicinal plants
	Tradeable: timber, fish, meat, ivory, medicinal plants, skins	Role in production of marketable commodities – biological control, ecotourism		
Biological	Genetic diversity for plant breeding Evolutionary value Agents of biological control	Ecosystem resilience and sustainability Watershed protection Regulating global ecosystems	Photosynthesis and carbon fixation	Evolutionary value (e.g. crop genetic diversity) Key role in maintaining ecosystems
Societal	Recreation	Religious Spiritual values Aesthetic value	Existence value	Adaptation to change/coping strategies
Ethical	Stewards of earth's resources Precautionary principle Intergenerational equity Protection of threatened species/ecosystems		Satisfaction of knowing it exists and will remain	

raw materials (compounds or genes) for the pharmaceutical, cosmetic and agricultural industries. Almost a quarter of medical prescriptions are for drugs extracted from plants or micro-organisms (Attfield, 1999). Genetic diversity in crop varieties is also becoming increasingly valuable to the agricultural sector (Chapter 6, Box 6.1).

Biodiversity can also be of economic value through its role in ecotourism. Increasingly, countries are recognising the value of their environment in attracting tourists, who are willing to pay considerable sums (and in hard currency) to experience environments or see wildlife. Whether it be whale watching on the seas, polar bear spotting in northern Canada, tracking gorillas in the Great Lakes district of Africa or bird watching in rainforests, this can be a way in which biodiversity can bring economic benefits to a local and national economy. With roles from famine foods for remote communities in marginal environments to the global ecotourism market, many livelihoods rely on biodiversity.

In addition to products of economic value, biodiversity serves to sustain ecosystems. At the genetic level, biodiversity produces the means for continuing evolution, and contains genetic material for plant breeding, which is particularly important for the world's agricultural systems. Plants and animals play specific roles in ecosystems, especially biological control of disease. At the ecosystem level, the range of natural systems provide 'life-support systems' for the planet such as regulation of climate and global ecosystems. Biodiversity provides important ecosystem functions such as watershed protection, carbon fixation via photosynthesis and coastal protection (e.g. mangrove swamps). It is difficult, if not impossible, to attribute an economic value to these environmental values of biodiversity; however, they are central to the planet's ability to regulate, adapt or evolve its environment for long-term sustainability.

Biodiversity also provides an aesthetically pleasing environment for recreation, religious and spiritual benefits. In some cultures, biodiversity may have a religious or cultural value, particularly for those whose belief or religion relies on particular environments, for example sacred forests. Many societies rely on their local environment for amenity value or for traditional products. Some societies value biodiversity for the sake of knowing that something 'exists'. It has been argued that biodiversity contributes to general well-being: that the richness and diversity of life forms are in themselves valuable and enhance our lives.

Overall, many people feel we have a moral duty to protect threatened species and ecosystems, and to conserve biodiversity for future generations. The underlying philosophies vary from the belief that all organisms (including humans) are morally considerable beings in themselves and therefore should be conserved (biocentric individualism), feeling that we should preserve biodiversity as a failure to do so might result in harm to humanity (anthropocentrism), to philosophies which emphasise the role of species and ecosystems in maintaining the biosphere (ecocentric holism) (Oksanen, 1997). Finally, in some religions there is a belief that we are stewards of the earth's resources, and we should conserve what we have been given for future generations (Judeo-Christian). The precautionary principle warns us against damaging or prejudicing biodiversity without knowing the full implications of our actions.

We do not know what will happen in the future. Undoubtedly the world environment is changing, and biodiversity may be crucial in enabling us to adapt to that change. It is impossible to screen all existing biodiversity for its potential future use, and then identify what to keep and what to let go. Attfield (1999) argues that as we cannot be sure where useful biodiversity will be found, we should preserve all habitats, possibly in proportion to the genetic diversity found in each of them. It is anticipated that the future value of biodiversity will concern medicinal plants, evolutionary diversity (especially crops), adapting to changing environments, and ecosystem management.

5.3 The global distribution of biodiversity

Species richness is the most widely used measure of biodiversity (Gould, 2000) and when this is mapped across the globe, it is clear that there are areas of high species diversity. The neotropics, indo-tropics and afro-tropics contain approximately two-thirds or more of all terrestrial species, with the neotropics containing the greatest overall biodiversity. Biodiversity hotspots are areas particularly rich in species or containing rare or threatened species. 'Up to 44% of all species of vascular plants and 35% of all species in four vertebrate groups are confined to 25 hotspots comprising only 1.4% of the land surface of the earth' (Myers *et al.*, 2000). Oceanic biodiversity is also unevenly distributed: at least one quarter of all marine species live in coral reef ecosystems. In the last 20 years, hydrothermal vent communities were discovered which exhibited a surprising range of organisms. Roberts *et al.* (2002) analysed the geographic ranges of 3235 species of reef fish, corals, snails and lobsters. Those taxa that were found to have restricted ranges were more vulnerable to extinction. They found that the ten richest centres of endemism covered 15.8% of the world's coral reefs (0.012% of oceans).

From a political and management point of view, this uneven distribution of biodiversity across the globe means a few countries contain 50–80% of the world's biodiversity. These include Madagascar, Indonesia, Brazil, Australia, Malaysia, Columbia, China, Thailand, Ecuador, India, Mexico and Peru. Research on biodiversity hotspots has suggested that these should be seen as the focal point for conservation efforts. There is some logic in concentrating biodiversity conservation efforts on the small percentage of the earth's surface which will yield the most biodiversity. However, this means the burden of conservation falls disproportionately on those countries containing biodiversity hotspots. As Swanson has succinctly commented, 'Diversity and development appear to be inversely correlated' (Swanson, 1999, p. 312). As biodiversity is not directly correlated with wealth, many of the countries which are well endowed with biodiversity are not necessarily wealthy enough to engage in conservation activities, be it proactive conservation measures or just conservation through foregone economic opportunities. Furthermore, these countries may also need to build up their cadre of ecologists and conservationists to enable them to assess and manage their biodiversity without recourse to foreign experts for assistance.

5.4 Stakeholders in biodiversity conservation

Given the many ways in which biodiversity has value, it is clear that people and communities use and rely on biodiversity in different ways to provide many things. Thus there are a range of people and groups who have an interest in the biodiversity conservation and opinions on what should be the priorities for conservation. All are stakeholders in biodiversity (Figure 5.1).

At the global level, society seeks to preserve biodiversity because of its role in ecosystem functioning and regulation (particularly its role with respect to climate control), its existence value (for future generations, for spiritual reasons, for recreation and tourism) and also for its future value in helping us to adapt to unforeseen problems such as new diseases. In contrast, indigenous groups value biodiversity for its role in their day-to-day livelihoods, for example, genetic biodiversity provides a broad range of landraces of crops, and natural vegetation can supplement diets based on crops with wild foods. In extreme circumstances, these wild foods are coping strategies when crops fail completely. Biodiverse environments also provide a range of products which can either be

Figure 5.1 Stakeholders in biodiversity.

used in the household or collected and sold, generating income at the household level (e.g. bush meat, grasses to make mats, fruits and nuts). National governments see biodiversity as a national resource. Like indigenous groups, national governments also view biodiversity as an income generating resource but on a larger scale: tropical forests are logged for international sale, and tropical habitats can be developed to attract international tourism. Businesses also see biodiversity as an income generating resource, in particular the pharmaceutical industry, which sees in biodiversity potential new chemicals and drugs to treat diseases, and the agrobiotechnology industry, which is interested in genetic biodiversity and the potential discovery of genes bearing particular attractive traits for new crop varieties. International aid and development organisations are also stakeholders in biodiversity. This is a diverse group of stakeholders. Their aims vary from biodiversity conservation to economic development (or a mixture of the two), and can be quite specifically targeted (e.g. conservation of a particular species). Aid and development organisations are also accountable to their supporters: members or donors who can influence priorities. Although a broad and varied group, aid and development organisations are important in financing biodiversity conservation.

Given each stakeholder group's view of biodiversity, it is not surprising that their priorities for biodiversity conservation differ, and it can be the start of conflict between stakeholder groups. A particular resource may be the subject of competing demands from different stakeholders. For example, a forest may provide wild foods and raw materials for locals, small-scale trade activities (e.g. medicinal plants or bush meat collectors and vendors), logging or selective timber export for larger traders, a source of national revenue via ecotourism, and internationally, carbon sequestration, watershed protection for a region (possibly transboundary) and regulation of the global climate and ecosystem. It may also be a habitat of a rare species, and be valued by society for its role in the landscape, and in addition may have historical and cultural meaning. Any sort of management for biodiversity conservation would need to take into consideration, and balance, the competing needs of stakeholders and the many values exhibited. Conservation of the forest to meet global needs for ecosystem regulation could mean prohibiting logging by national governments and potentially limit the activities of indigenous groups as well, leaving both of these latter stakeholders without a means of income generation or livelihood. In contrast, allowing countries to intensively log forests would ultimately degrade the environment and resources on which indigenous groups rely for their livelihoods, as well as diminishing resources needed for future climate regulation, and development by pharmaceutical and agrobiotechnology companies.

Hence agreement about the need to conserve biodiversity does not necessarily mean agreement on the best way to achieve this, or who the beneficiaries of biodiversity conservation should be. Membership of these stakeholder groups is not distinct. An individual with family roots in a forest dwelling community may see the value of forest biodiversity to rural livelihoods and also to the national economy. Hence the individual could represent several stakeholder groups. Furthermore, stakeholder groups can combine their efforts to achieve their goals, as indigenous groups work with aid organisations, or pharmaceutical business teams up with national government to conserve rainforests and identify potential new drugs, thus also meeting some of the concerns of global society (see Box 5.3). Stakeholder analysis (Grimble *et al.*, 1995) has been identified as an important method for identifying differing user groups, their composition (among which there may be overlap) and their views on access to and use of resources. Dialogue and negotiation between such groups may be necessary to enable sustainable management of biodiversity resources to take place.

5.5 Threats to biodiversity

Biodiversity is threatened in many ways. Immediate causes are often related to man's activities. Cincotta, Wisnewski and Engelman (2000) established that in 1995, 20% of the world population was living in hotspots covering 12% of the earth's surface. Furthermore, these authors estimated that population growth rates in hotspots (1.8%/yr) were significantly higher than the rate of world population growth (1.3%/yr). As populations grow, develop and modify their environment, there are many ways in which their activities may adversely affect biodiversity. Populations exploit wild living resources, be they plant or animal. This may be through harvesting and consuming wild produce, or more planned landscape change through the expansion of agriculture, forestry and aquaculture. The fear is that such land cover change will result in biodiversity loss, particularly when it results in habitat fragmentation or degradation (Burger, 2000).

Tourists are often attracted to areas of high biodiversity, particularly coastal areas, yet their presence can put tremendous pressure on the environment they are so keen to see and enjoy (Burger, 2000). Movements of people and animals can inadvertently transport species to new areas. Species introductions alter local niches and food webs, and may introduce new predators or diseases which affect the viability of species established already. Successful competition by introduced species can reduce or completely eradicate pre-existing species. People also cause pollution, be it of soil, water or the atmosphere, and this too can impact on local biodiversity. Globally, climate change affects the distribution of species.

Coastal areas are particularly pressured environments. In general, coastal zones provide 'home to more than 60% of the World's population, support the Worlds' densest human populations, and are the site of two thirds of the World's large cities' (Holdgate, 1993). Shallow coastal waters are important in fisheries and coastal ecosystems such as mangrove swamps, and coral reefs are important in the life cycles of commercial fishing. High population densities in coastal areas bring about land cover change (from agricultural, urban development) and pollution (sedimentation of water courses, sewage, erosion).

5.6 Trends in conservation

Many strategies have been used over the years to try to conserve biodiversity. They can broadly be divided into *in-situ* and *ex-situ* approaches. *Ex-situ* approaches involve collecting and storing things away from the environment in which they were found. Thus seed collections, gardens, orchards and zoos all seek to preserve biodiversity in a safe and controlled environment away from where they were found naturally. While such an approach removes biodiversity from its threatened environment to a safe area, it also removes it from the dynamic environment in which it was growing and evolving. Recreating the 'ideal' conditions is an exacting task, and reproduction to propagate the species can be difficult or impossible. This form of conservation preserves species, but not necessarily habitats or ecosystems.

In-situ approaches concentrate on preserving and conserving biodiversity within the environment in which it is found. *In-situ* conservation is:

> the conservation of ecosystems and natural habitats and the maintenance and recovery of viable populations of species in their natural surroundings and, in the case of domesticated or cultivated species, in the surroundings where they have developed their distinctive properties.
>
> *CBD, 1992*

In-situ conservation allows the ecosystem and species to continue to evolve within their environment, adapting and changing in response to changing environmental conditions. Thus the focus is on the habitat and ecosystem, as well as the species.

Increasingly, *in-situ* conservation is seen as a more viable method of biodiversity conservation, and it is recommended as the preferred method in the Convention on Biological Diversity (CBD) (Box 5.2). The CBD defines a protected area as 'a geographically defined area which is designated or regulated and managed to achieve specific conservation objectives'. The rules regulating the protected areas vary from preservation to controlled resource harvesting. The International Union for the Conservation of Nature (IUCN, 1994) defines seven types of protected area, ranging from wilderness area or strict nature reserve to managed resource-protected areas. The most obvious form of protected area is a national park. Worldwide, 1.5% of the earth's surface is enclosed in marine or terrestrial protected areas.

The designation of a protected area can be controversial, as there are few, if any, areas in the world which are not already home to an existing human population. If the aim is to exclude people from an area, alternative land must be found for people, and they must be persuaded to move. The park or protected area then requires some sort of management or policing to ensure that the area fulfils its aims. This can include management of habitats, and patrolling to keep people (particularly poachers, plant collectors or fishermen) out of the area. Where a region has provided a resource which is important to local livelihoods, this can create conflict with locals. The more valuable the resource, the more lucrative the market for it, the harder it will be to keep people out. Management and patrolling of protected areas requires an infrastructure of trained personnel (e.g. ecologists), institutions (parks authorities, guards, managers) and long-term funding. Yet finding funding to support such programmes for the long term is difficult. Donors usually want to make a targeted financial input towards achieving a set objective, and then hand it over to local managers. However, biodiversity conservation is a long process (arguably endless) which does not necessarily provide short-term goals with which donors can be associated.

Traditionally, *in-situ* conservation has involved separating people from biodiversity, sometimes through enforced exclusion (relocation) and fencing. Such 'fortress conservation' (Adams and Hulme, 2001) is now seen to have several disadvantages, most noticeably cost and lack of local support. Protected areas which aim to exclude people must be patrolled and possibly fenced. In remote areas where population density is low, interference from people may not be deemed significant enough to warrant patrolling (e.g. remote national parks in North America). However, it has already been noted that biodiversity hotspots are often also densely populated areas, with high rates of population growth. In many developing countries, protected areas have displaced local residents, who may have lost lands and livelihoods. They may return to the area to hunt wildlife or gather traditional goods (for subsistence or small-scale marketing). Having been removed from the area, they no longer have the same incentive to conserve the environment. Such protected areas need considerable patrolling if they are to be effective.

Managing protected areas is very expensive. In developed countries, where charities and companies are keen to support biodiversity conservation, funding is available. However, in developing countries, it is difficult to find the long-term funding needed to maintain conservation initiatives. The lack of funds to implement conservation plans can result in 'paper parks', protected areas which have been planned and recorded in government offices, but which are practically non-existent at the ground level. Exclusion of people from protected areas is not always beneficial for biodiversity conservation. Ethnobotanical research shows the role of local populations in holding environmental knowledge, and managing their environments. This can include maintaining habitats, and deliberately conserving diversity in species. Detailed research in West Africa (Fairhead and Leach, 1996) showed how farmers modified their landscape to create the environment they wanted. During rainforest clearance farmers select what plants to leave to grow, allowing them to shape the regrowth and development of secondary forest. Local populations can be involved in conserving and protecting specific plants to ensure that they are available in times of hardship.

People know where to find wild foods and medicinal plants, and may even manage their environment to favour the successful growth of naturally occurring specimens of these plants, without actually cultivating them (Harris, 2003). Farmers select and maintain a range of crop varieties to ensure sufficient bio-diversity to meet changing needs for crop characteristics in variable environments (Kandeh and Richards, 1996; Mortimore and Adams, 1999). These examples show that rather than being seen as threats to biodiversity, in some cases people, and in particular their local knowledge and environmental management, are import to the conservation of biodiversity, especially communities and assemblages of plants making up habitats, rather than just isolated species.

New models of *in-situ* conservation are being developed which seek to respect the livelihood needs of local populations, their role in environmental management, and gain their support for conservation. Such a participatory and development-focused conservation is new, but initial hopes are that it will be better for local people and biodiversity conservation in the long term. The aim is to increase awareness of the value of the conservation of biodiversity by ensuring that locals benefit from it. This can be via negotiated access to and offtake of natural resources, such as buffer zones surrounding protected areas (e.g. Korup national park, Cameroon), by involving local people as park wardens and guides, so providing local jobs (e.g. in Costa Rica, Vaughan, 2000), or through granting more of the management and planning of biodiversity conservation to locals (e.g. CAMPFIRE programme, Zimbabwe, Box 5.1).

Involving local people in conservation achieves several goals. Locals feel involved in, rather than alienated from, the conservation process. The conserva-tion activities gain from local peoples' knowledge of their local environment. Local people can remain *in-situ* as custodians of the environment, rather than being displaced, and this avoids the need to resettle people. However, such participatory conservation also takes time, as such a change in attitudes requires retraining of staff for conservation, and building up of trust and mutual respect between local people and professional conservationists.

One of the significant things seen in CAMPFIRE is the linking of con-servation and development. Indeed, this is now seen as a greater priority in conservation work: that conservation of nature must not be to the detriment of locals, but should be linked to poverty alleviation and development initiatives to benefit local people. This has proved a challenge for organisations which are specialised in conservation, and now must become experts in community develop-ment as well (Adams and Hulme, 2001). There is also recognition of the potential role of markets in biodiversity conservation. Market forces can be a driving force for biodiversity conservation when the sustained presence of biodiversity can generate income. Thus ecotourists who wish to see wildlife provide the financial incentive for conservation, and beautiful environments (particularly coastal and island regions) can become tourist resorts (section 5.8). Increasingly, it is believed that nature must pay its way, as exemplified by the slogan 'use it or lose it'.

Box 5.1: The CAMPFIRE programme

The Communal Areas Management Programme for Indigenous Resources (CAMPFIRE) in Zimbabwe has been praised for engaging local people in biodiversity conservation in a productive and mutually beneficial way. This programme links biodiversity conservation with development initiatives, so that biodiversity conservation also contributes to local peoples' needs for poverty alleviation, income generation and capacity building (providing training in skills and employing people as conservation staff) (Murphree, 1993).

Rather than excluding people from land to permit fortress-style conservation of wildlife, local residents are encouraged to take on the role of wildlife wardens themselves through a range of incentives. Based on the premise that conservation will be popular when it brings benefits to locals, the CAMPFIRE programme encourages local people to see wildlife as a resource, part of their natural capital, which can be exploited sustainably. If wildlife is managed sustainably, some culling (licensed hunting) and some viewing (safari) are possible. If permits for hunting are issued by locals, and the revenue is returned to locals, they then have a financial incentive for managing wildlife. Of course, the concern for safety of crops and families from wildlife remains, but fencing now surrounds the villages and farms, giving wildlife larger areas to roam. The income generated provides jobs for locals as wardens and gamekeepers, and is also used to fund development projects such as schools, roads and hospitals. This provides the incentive for the whole community to support the project. Initial evaluation of CAMPFIRE was positive, although as time passed some criticisms have arisen (e.g. Bond, 2001). In addition to being a model for more participatory conservation, the programme is also seen as a more successful form of nature conservation, as biodiversity is no longer conserved in small 'islands'. Instead of refuges, possibly linked by corridors, biodiversity is conserved across wider regions, providing a wider territory for individual animals to roam.

5.7 The Convention on Biological Diversity

Alongside the movement towards more participatory conservation, involving people at the local level, there is also a co-ordinated global effort to conserve biodiversity. At the United Nations Conference on Environment and Development (UNCED) in 1992 in Rio, the CBD (Box 5.2) was developed to address issues of biodiversity conservation.

This convention had three main objectives:

1. to conserve biological diversity,
2. to promote the sustainable use of its components, and
3. to encourage the equitable sharing of the benefits arising out of the utilisation of genetic resources.

Box 5.2: The Convention on Biological Diversity

The Convention on Biological Diversity (CBD) was developed at the United Nations Conference on Environment and Development (UNCED), held at Rio de Janeiro in 1992. It built on a series of conventions and agreements relating to the conservation of birds, fisheries and endangered species, such as the Ramsar Convention on wetlands and the Convention on International Trade in Endangered Species (CITES). The broad aims of the convention, as outlined in Article 1, were:

> the conservation of biological diversity, the sustainable use of its components and the fair and equitable sharing of the benefits arising out of the utilization of genetic resources, including by appropriate access to genetic resources and by appropriate transfer of relevant technologies, taking into account all rights over those resources and to technologies, and by appropriate funding.

In order to achieve these aims, Articles 5–9 of the convention encourage each nation to 'develop national strategies, plans or programmes for the conservation and sustainable use of biodiversity'. Further, nations should identify key components of biodiversity for conservation, and monitor biodiversity over time, especially in relation to activities which may have adverse impacts on biodiversity conservation. Such monitoring activities will also provide useful data on changes in biodiversity. The CBD is in favour of *in-situ* conservation of biodiversity wherever possible, and hence encourages the establishment of protected areas, and management of resources, ecosystems and habitats within protected areas and in neighbouring areas. *In-situ* conservation measures are to be complemented by *ex-situ* conservation methods, where appropriate. The CBD also encourages the rehabilitation and restoration of degraded areas. Developing nations may require training in conservation, park management, *ex-situ* conservation techniques, public education and awareness raising.

The CBD recognises the need for balancing the competing demands of existing resource uses and users (particularly indigenous and local communities) and the need to conserve biodiversity for the longer term, and recommends a mixture of market forces and participatory methods to

Continued on page 107

Continued from page 106

encourage biodiversity conservation. In addition to focusing on the conservation of existing biodiversity, the CBD considers management of the risks associated with modified organisms and the introduction of new species which may threaten existing ecosystems, habitats or species. Following the CBD, the Cartagena Protocol was developed dealing specifically with this. Its objectives are:

> In accordance with the precautionary approach contained in Principle 15 of the Rio Declaration on Environment and Development... to contribute to ensuring an adequate level of protection in the field of the safe transfer, handling and use of living modified organisms resulting from modern biotechnology that may have adverse effects on the conservation and sustainable use of biological diversity, taking also into account risks to human health, and specifically focusing on transboundary movements.

The third main aim of the convention is the equitable sharing of benefits arising from the use of biodiversity. These benefits may be via access to genetic resources (e.g. plant breeding) or biochemical products (e.g. medicinal plants). Once again, the transfer of technology and skills may be necessary to enable developing countries to develop these resources in the country. Alternatively, arrangements can be made with organisations in other countries. These arrangements are often governed by laws concerning intellectual property rights and trade rules, and have subsequently been the subject of much debate (see section 5.9, Box 6.1).

The CBD made provision for funding of its recommended activities through the Global Environment Facility. This recognised the need for developed countries to contribute to the financial costs of monitoring, conservation, and technology and skills transfer to developed countries, particularly the poorest countries and small island developing states.

The CBD has provided an indication of the strength of political will to conserve biodiversity. By 2001, 177 countries had ratified the CBD (Koziell, 2001), making it one of the most broadly supported international environmental agreements. However, the political will to conserve biodiversity needs to be matched by action at the ground level. Practical implementation of the CBD has raised new challenges. Biodiversity is valued by different people in different ways, and some of those values cannot be equated in economic terms. The international will to conserve biodiversity is global. However, the many stakeholders seeking to conserve biodiversity have different strategies to achieve this goal. Poorer countries in the south need financial, technical and logistical assistance to enable them to carry out initial biodiversity monitoring, and set up biodiversity conservation programmes. Even if countries are keen to conserve biodiversity, they may not have the technical staff necessary to carry out inventories of biodiversity, and plan conservation measures. There is considerable debate over the best way to conserve biological diversity and over what constitutes sustainable use of bio-diversity. In some cases, the needs of biodiversity conservation may be in direct contrast to national development plans (particularly where economics are reliant on natural resource harvesting (e.g. timber) or natural resource-based (e.g. ecotourism)). Countries may seek financial compensation for development opportunities foregone for the sake of the global good. Perhaps the most challenging aspect of the CBD is ensuring the equitable sharing of benefits arising from biodiversity (see Boxes 5.1, 5.3 and 6.1 for three examples).

5.8 Linking conservation and tourism

Many countries which are fortunate enough to have highly biodiverse environments have found that this opens possibilities for marketing that environment for tourism. Nature tourism concerns people who want to see wildlife or plants. Ecotourism, less clearly defined, usually involves tourism involving small- to medium-sized operators, low environmental impact, interaction with local culture and a willingness to give up some creature comforts (Vaughan, 2000). There are many ways in which people can achieve this: wildlife safari holidays in East Africa, specialist nature watching trips (gorillas, birds, whales, polar bears), snorkelling and diving off coastal areas.

The challenge is to use the development of ecotourism to conserve habitats so that biodiversity can flourish. This could mean limiting tourist numbers, and ensuring that financial revenue from ecotourism is reinvested in management to control land cover change and biodiversity loss. However, countries which seek to develop biodiversity-related tourism are also faced with the fact that this is a highly competitive industry: there are many competing destinations, and the market is affected by trends, as well as by people's perception of the host country (political stability, accommodation, accessibility, perceptions of personal safety). National parks do not always cover their running costs from gate fees (Vaughan, 2000). Furthermore, the development of tourism requires investment in infrastructure, such as airports, roads, sewage and accommodation. Although the goal is to provide an incentive to save biodiversity, ecotourism will still have some impact on biodiversity, as access to wildlife or protected areas will require roads and trails, and may introduce exotic species or pathogens (Vaughan, 2000).

5.9 Biodiversity and business

Biodiversity can 'pay its way' at the genetic level as well as at the ecosystem level. This is seen with respect to the 'future value' of biodiversity as a potential source of new medicines for the pharmaceutical industry, cosmetic products, and genes or biological control agents in the agricultural industry.

Currently, people consume approximately 7000 species of plant; however, only 150 are commercially important, and 103 species account for 90% of the world's food crops. Rice, wheat and maize make up significant proportions of the calories and proteins people derive from plants (Thrupp, 2000). As plant breeding refines and develops crop cultivars, the genetic diversity on which our food supply is based becomes smaller and smaller. A well-targeted pest or disease could jeopardise food supplies. Hence it is important to collect and document landraces of crops: 'geographically or ecologically distinctive populations (of plants and animals) which are conspicuously diverse in their genetic composition' (Thrupp, 2000). These landraces may, at a future date, provide the genetic diversity vital to maintaining the viability of our main food crops. Debates about sourcing these landraces and the way in which benefits are transferred from local farmers through collectors to agrobiotechnology companies are discussed in Box 6.1.

The pharmaceutical and agrobiotechnology industries have several ways of accessing new resources for product development. Existing *ex-situ* collections of seeds and plants, often with associated catalogues describing attributes, provide one route to new product identification. Ethnobotanical studies can identify promising plant products. Studies of indigenous agricultural systems can identify successful intercropping arrangements, allelopathic interactions between crops and biological control agents. Therefore, there are alternatives to broad-scale random screening of plant products.

The pharmaceutical industry has considerable amounts of money to invest in product development, and if harnessed appropriately, this could be coupled with biodiversity conservation. In terms of the CBD, this 'biodiversity for business' has implications for the interpretation of the 'sustainable use' clause of the convention, as well as the interpretation of the 'equitable sharing of benefits arising from biodiversity'.

Many biodiversity-rich countries do not have the scientific expertise and technical capacity to develop pharmaceutical products themselves, and therefore seek to engage in partnership with companies to enhance exploitation of these resources. The companies collect raw material from biodiversity-rich countries (possibly with the assistance of knowledgeable locals), and then take the material away for screening, processing and possibly eventual product development. A new crop, or a new drug, may eventually be developed and sold, usually under patent, providing the potential for profit for the company, without much return to the initial source of the raw material (see Box 6.1). Although considerable research and development costs are involved, such bioprospecting can be highly lucrative if successful. One factor which can aid success is when people with local knowledge (e.g. traditional healers) assist in the initial product identification. Yet often these people do not receive a share in the benefits, leading to accusations of biopiracy.

The CBD seeks to redress this with the development of intellectual property rights for products and acknowledgement of the role of local knowledge in developing products. This has increased the hopes of income generation from bioprospecting among developing countries. Arrangements based on royalties can result in waiting for up to 20 years for product development to be successful, which is off-putting to biodiversity-rich countries. Increasingly, deals focus on a range of benefits such as fees for each sample provided, or milestone payments, as well as non-monetary benefits such as training, capacity building, technology transfers and support for conservation projects.

However, the complexity and vagaries surrounding the legal and institutional frameworks for benefit sharing are putting companies off developing collaborative programmes for bioprospecting (Ten Kate and Laird, 2000). Many companies prefer to acquire material from *ex-situ* collections, when it is harder to allocate benefit sharing as the collection could have been made decades ago. There are some examples of successful collaborations. The Merck/INBio agreement in Costa Rica is seen by some as a model for replication elsewhere (Box 5.3). Shaman Pharmaceuticals, in California, is another example of a company which seeks local help in identifying potential medicinal products. Shaman Pharmaceuticals recompenses local, traditional healers through immediate payments, as well as paying for local collecting or growing fees.

Box 5.3: Mixing biodiversity and business for mutual benefit: the Merck/INBio agreement

Among the many concerns about how to use market forces to encourage biodiversity conservation while equitably sharing the benefits of market profits, one particular case stands out as a success story. In 1991, the largest pharmaceutical company, Merck & Co., and the Instituto Nacional de Biodiversidad in Costa Rica (INBio) joined together in an agreement which would provide the pharmaceutical company with plant, insect and soil samples from Costa Rica, and INBio and Costa Rica with initial financial returns, plus a share in any royalties deriving from any pharmaceutical products subsequently developed. In this win–win situation, 'the contract provides Costa Ricans with an economically beneficial alternative to deforestation and concurrently advances the research efforts of Merck' (Blum, 1993).

Costa Rica, a small country in Central America, is well endowed with biodiversity. Costa Rica has diverse habitats and climates and contains ecosystems representative of South America, the West Indies and tropical North America. More than one quarter of Costa Rica's land is protected in some type of national park or preserve. It is estimated that Costa Rica is

Continued on page 111

Continued from page 110

home to between 5 and 7% of the world's species, containing more biodiversity/acre than any other nation, and that this biodiversity contains 12 000 plant species, 80% of which have been described, and 300 000 insect species, 20% of which have been described.

However, it is also a poor country, with a natural resource-based economy (coffee and bananas). The options for development include cutting and selling the forest, or finding an alternative way to use the forest to achieve long-term income so that Costa Rica could earn money from its biodiversity without destroying it. The aim of the agreement was to provide a way of earning money from Costa Rica's biodiversity which was an alternative to deforestation.

From the pharmaceutical industry's point of view, rainforests, which hold a wide diversity of species, are the potential source of many future drugs and products. However, if these rainforests are lost, then that potential future product is also lost. Therefore, it is in the pharmaceutical companies' interests to find ways of persuading countries who have rainforest not to chop them down, but to maintain them sustainably while the companies slowly inventory, catalogue, and examine all plants, animals and micro-organisms for potential products.

The Merck/INBio agreement is:

> a collaborative research agreement under which Merck agreed to pay INBio a sum of US $1 million for all of the plant, insect and soil samples the institute could collect in addition to a percentage of the royalties from any drugs that Merck develops from samples provided by INBio.
>
> *Merck & Co., as cited in Blum, 1993*

Thus INBio would supply approximately 10 000 identified and classified samples to Merck, which would have first right of refusal to evaluate the samples. Merck can then test the samples for chemical activity. In this agreement, Merck acquires a guaranteed number of samples, pre-identified and classified, benefiting from the locally adapted knowledge of Costa Rican scientists, and Costa Rica retains sovereignty over its resources, INBio receives payment, technology transfer (equipment, two Merck scientists providing training to local staff), and the Costa Rican Ministry of Natural Resources gains a 10% of the fee plus 50% of any future royalties. Should drugs be developed from this bioprospecting, the money raised from royalties could be very significant in relation to the income from its prime exports of coffee and bananas. Costa Ricans benefit as the development of their biotechnology industry provides more jobs (as part of the programme INBio trained many new parataxonomists to collect samples) and improves the general economy.

Continued on page 112

Continued from page 111

Thus the agreement meets many of the goals of the CBD: international co-operation, using market forces to encourage biodiversity conservation, transfer of technology between developed and developing nations, apparently equitable, sharing of benefits arising from the use of biodiversity. However, the big question is whether this agreement is unique or whether it provides a model for replication by other biodiversity-rich but economically poor countries.

Factors which contributed to the success of the agreement include (Blum, 1993):

- Technical expertise in Costa Rica to process samples prior to delivery to Merck.
- Transportation system to transfer samples to Merck laboratories in Spain and USA quickly.
- Scientific collection system already in place in Costa Rica.
- Merck already established in Costa Rica.
- Costa Rica's dedication to preserving biodiversity.
- Costa Rica's unique natural resource endowment.
- Costa Rican government is stable and democratic.
- Costa Rican population is well educated (98% literacy).
- INBio is a non-profit, private scientific organisation which inventories and catalogues Costa Rica's biodiversity. INBio has 'a computer database of species names, conservation status, distribution, abundance, way of life, and potential uses in medicine, agriculture, and industry'.
- All funds above operating costs are placed by INBio in a special fund managed by the government to ensure money is directed to conservation activities.
- Transparency.

Not all of these conditions are present in every country seeking to develop a similar agreement. Costa Rica as a nation is well aware of the importance of its natural resources, and with so much land in parks and protected areas, it is possible for the country to develop such a programme. In countries where the biodiversity resources are held under different land tenure systems, alternative agreements may need to be arranged. Costa Rica's political stability encourages external investment and long-term collaboration. Many other developing countries suffer from political instability which prejudices their ability to manage long-term environmental conservation projects and longer-term economic agreements. Finally, Costa Rica's relatively well-educated population provides a resource which can work on the project: training needs are less than in other countries.

5.10 Conclusion

Biodiversity has been described as the 'common concern of mankind' (CBD 1992: Preamble). What exactly is our common goal with respect to biodiversity? Is the aim **preservation**, defined as: 'to maintain or restore a current or earlier state of affairs for the foreseeable future' (Attfield, 1999, p. 139), or do we 'seek to protect resources with a view to their eventual use' (Attfield, 1999, p. 139), which is **conservation**? The first values biodiversity for what it is at present, and does not countenance any change in view of changing environments, circumstances and resources. The second recognises that biodiversity has potential values in the future, which we do not wish to lose, but does not deny that some change may occur. Preservation is particularly difficult to enforce when it is recognised that biodiversity exists in a dynamic state: natural ecosystems involve succession, fluctuations in populations, feedback mechanisms, major periodic events such as fires or floods and evolution. What is the specific state of affairs which is to be conserved? Furthermore, some habitats, such as agricultural landscapes, are only maintained through regular human activities (section 6.4). The role of people in creating and maintaining biodiversity is not to be underestimated. While it has long been recognised that hedgerows, grazing or regular ploughing have contributed to landscapes and biodiversity in temperate farming systems, recent interest in hunter–gatherer and indigenous agricultural systems has shown that these populations also modify their environments in ways which affect biodiversity. Are the management practices carried out by humans now part of the ecosystem which must be conserved?

Biodiversity is important to many people; however, they do not all have the same goals with respect to biodiversity. Some people have short-term needs with respect to biodiversity (e.g. subsistence products), whereas others see its 'future value' in enabling adaptation to a changing environment (climate, disease, agriculture...) as being most significant. The distribution of global biodiversity is skewed towards the countries which can least afford to invest in its conservation, yet conservation of biodiversity is as important to those nationals (for subsistence reasons) as it is to the wider international community (for future value and global ecosystem function). The CBD seeks to develop an international biodiversity management regime without alienating sovereign rights from individual countries.

Although biodiversity is seen as a common concern of mankind, its global distribution raises questions as to who 'owns' biodiversity (Horta, 2000). In this market-driven world, there are cultures which find it unpleasant to think of anyone 'owning' or having rights to biodiversity when it is part of a God-given environment. As biodiversity diminishes, and more people realise its potential importance, the scramble for biodiversity, via bioprospecting, raises questions as to who owns biodiversity from a property rights, or selling, point of view.

Already, the equitable sharing of benefits arising from the use of biodiversity is proving complicated. Business and market forces are ahead of the development of regulatory institutions and agreements, and further legal analysis is required.

In the introduction to this book, the distinction was made between technocentrics, who believe that alternatives can be found or developed to resolve environmental problems, and ecocentrics. Technology will not save species or habitats themselves, although technological ingenuity may produce alternatives to reliance on habitats or species. As put by Pimm *et al.* (1995, p. 347) 'Ingenuity can replace a whale-oil lamp with an electric light bulb, but not the whales we may hunt to extinction.' Once a species is gone, it is irretrievable.

Further reading

CBD (1992) *The Convention on Biological Diversity.* Secretariat for the Convention on Biological Diversity, United Nations Environment Programme.
http://www.biodiv.org/convention.articles
This is the full text of the convention. This is a useful reference concerning priorities of biodiversity conservation.

Gaston, K.J. and Spicer, J.I. (1998) *Biodiversity. An Introduction,* Oxford: Blackwell Science. This is a useful introductory text for students new to studies of biodiversity, and provides useful introductions to further literature.

Hulme, D. and Murphree, M. (eds) (2001) *African Wildlife and Livelihoods. The Promise and Performance of Community Conservation,* Oxford: James Currey.
This book is the result of several years of research on wildlife conservation in Africa. It begins with a theoretical discussion of wildlife conservation issues, but also contains chapters from many authors concerning biodiversity conservation in east and southern African countries.

Koziell, I. (2001) *Diversity Not Adversity. Sustaining Livelihoods with Biodiversity,* London: International Institute for Environment and Development and Department for International Development.
This publication from IIED focuses on the role biodiversity can play in poverty alleviation. It provides clear information on the many values of biodiversity to rural livelihoods, and suggests ways policy-makers can support biodiversity conservation for poverty alleviation.

Swanson, T.M. (ed.) (1998) *The Economics and Ecology of Biodiversity Decline: The Forces Driving Global Change,* Cambridge: Cambridge University Press.
An interesting discussion of the challenges surrounding biodiversity conservation.

Ward, H. (ed.) (2000) International Affairs, Special Biodiversity Issue. *International Affairs* 76.
This special issue of International Affairs focuses on biodiversity, with articles concerning biodiversity and business, the biotechnology trade, the importance of agrobiodiversity to food security, tourism and biodiversity, and valuing nature.

Part Three

Meeting our Needs

Chapter 6
Food Production and Supply

Guy M. Robinson and Frances Harris

6.1 Introduction

Agriculture is the most widespread modification to the natural environment associated with human activity. A complex variety of agricultural systems exist worldwide depending upon biophysical and human inputs, which combine to transform natural ecosystems into simpler forms designed to produce food and fibre for human consumption. The transformations have long been associated with certain environmental problems. These are examined in this chapter in terms of contemporary debates on the severity of agriculture's environmental footprint, especially in the context of the globalisation of agri-food production and consumption systems.

The complexity of the relationship between agriculture and environment rises as the degree of alteration of natural ecosystems is increased. Hence, subsistence-farming systems tend to modify rather than substantially alter the natural system, placing relatively little stress upon the environment, though it can involve forest clearance. In contrast, the 'industrial-style' farming typical of large parts of the developed world not only involves profound alteration of ecosystems, the use of agro-chemicals, irrigation and application of biotechnology, but also produces various environmental disbenefits in the form of habitat destruction, soil erosion, pollution of watercourses, and human health problems related to industrialised food processing methods. Nevertheless, it is the industrial model that is advancing in many parts of the world in response to the growing pressure of population increases and the absorption of farming into globalised agri-food production systems. In an increasingly globalised world food market, the pathway of food from a field to our plates is mediated by markets, transportation and food distribution systems, as well as fashions and tastes around the world.

Global Environmental Issues. Edited by Frances Harris
© 2004 John Wiley & Sons, Ltd   ISBNs: 0-470-84560-0 (HB); 0-470-84561-9 (PB)

Blaikie (1985) emphasised the importance of the social relations of production and their determination of the nature of land use. Following his approach, the farm household can be seen as being part of two kinds of social relations: the local and the global, with the impress of the latter increasingly assuming greater prominence. Hence a full understanding of the environmental impacts of agricultural production requires a consideration of the political and structural frameworks within which agriculture rests (Le Heron, 1988). In particular, it is necessary to acknowledge the absorption of the production sector (i.e. the farm) into the larger food supply (or agri-food) system, which includes off-farm inputs such as suppliers of seeds, fertilisers and machinery (sometimes termed 'backward linkages') and the roles of food wholesalers, retailers, processors, distributors and consumers (sometimes termed 'forward linkages' or 'downstream linkages'). It is the critical influences and components of the overall agri-food system, especially the increasing involvement of large trans-national corporations (TNC), which often have the most adverse environmental consequences. This is demonstrated throughout this chapter.

The potential for environmental problems associated with agriculture has grown considerably in recent decades, and this provides numerous issues to be addressed in this chapter. In particular, there is the opportunity to draw marked contrasts between the growing recognition of problems associated with industrial agriculture (and the multi-faceted but chaotic set of controls being established on production processes by governments and other agencies) and the perceived virtues of alternative models of agriculture that place a high emphasis upon sustainability and maintenance of environmental values. After considering the basic character of the agri-ecosystem, this chapter discusses key environmental issues associated with these systems, both in the developed and in the developing worlds, concentrating on impacts that occur on farms as opposed to the upstream and downstream parts of the system. It then addresses contrasting approaches to the development of agriculture: the 'technical fixes' of the Green Revolution and genetically modified (GM) foods, and the ecocentric approach linked to discourses on sustainability and extensive methods as opposed to intensive ones. These discussions are set within the context of the growing impress of the globalised economic system, which increasingly is shaping the character of agricultural production worldwide.

6.2 The agri-ecosystem

Within agricultural production there is an integral link between land-based activity and accompanying modification of the natural environment. This modification produces an agri-ecosystem in which an ecological system is overlain by socio-economic elements and processes. This forms 'an ecological and socio-economic system, comprising domesticated plants and/or animals and the people who husband them, intended for the purpose of producing food, fibre or other agricultural products' (Conway, 1997, p. 166). The agri-ecosystem can be viewed as part of a nested hierarchy that extends from an individual plant or animal and its cultivator, tender or manager (farmer), through crop or animal populations, fields and ranges, to farms, villages, watersheds, regions, countries and the world as a whole (Figure 6.1). Hence the interactions between the component parts of the system can be studied at a variety of spatial scales.

The primary agricultural management practice is the cultivation of the soil, which acts as the reservoir of the water, minerals and nutrients that are needed for plant growth (Figure 6.2). Cultivation involves selecting plants likely to produce a satisfactory yield; propagation, in which tillage of the soil ensures suitable conditions for planting or sowing and for feeding the crop; and protection from competition for the primary resources by weeds and from direct or indirect reduction in yield potential by animals, pests and pathogenic organisms (Tivy, 1990).

There is a delicate physical and biological balance that renders agriculture possible, but which also restricts production in various ways. This helps to explain why less than 15% of the earth's land mass has been cultivated and why current

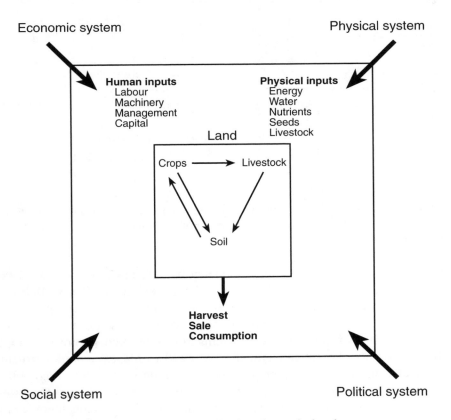

Figure 6.1 Simple conceptualisation of an agricultural system.

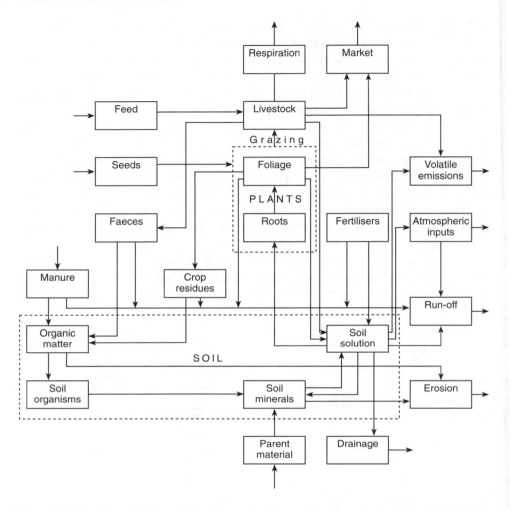

Figure 6.2 Nutrient cycles in agricultural systems. (Extract from *Farming Systems of the World* by A.N. Duckham and G.M. Masefield; published by Chatto & Windus. Used by permission of the Random House Group Limited.)

estimates claim that this proportion can only be extended to 25% with huge invest-ment in either or both irrigation and other technological inputs (Pierce, 1990).

In agri-ecosystems the farmer is the essential human component that influences or determines the composition, functioning and stability of the system. This system differs from natural ecosystems in that agri-ecosystems are simpler, with less diversity of plant and animal species and with a less complex structure. In particular, the long history of plant domestication has produced agricultural crops with less genetic diversity than their wild ancestors. In agri-ecosystems the biomass of the large herbivores, such as cattle and sheep, is much greater than that of the ecologically equivalent animals normally supported by unmanaged terrestrial ecosystems. Cultivation means that a higher proportion of available light energy reaches crops and, because of crop harvesting or consumption of crops by domestic livestock, less energy is supplied to the soil from dead and decaying organic matter and humus than is usually the case in unmanaged eco-systems in similar environments. Agri-ecosystems are more open systems than their natural counterparts, with a greater number and a larger volume of inputs and outputs (Figure 6.2). Additional inputs are provided in the form of direct energy from human and animal labour and fuel, and also in indirect forms from seeds, fertilisers, herbicides, pesticides, machinery and water. The dominant physical or natural resource inputs to the farming system are climate and soils.

Within the agri-ecosystem there is a reciprocal relationship between environmental factors and agricultural activity, with numerous linkages and feed-backs affecting both human-created agricultural systems and the environmental systems upon which farming activity depends. In agricultural systems, natural ecosystems are modified to increase productivity through control of soil fertility, vegetation, fauna and microclimate. This is intended to generate a greater biomass than that of natural systems in similar environments, though this may also gener-ate undesirable environmental consequences. In particular, farming alters the character of the soil and, through run-off, effects can be extended to neighbouring areas, for example nitrate pollution of watercourses and groundwater (Box 9.1), and to wildlife.

6.3 Agricultural intensification

The simplest way to increase agricultural output is to increase the area of land that is farmed (whether in crops or pasture for livestock). This has been the process whereby, over millennia, large proportions of the land surface have been brought under cultivation. As a result, farming today tends to be focused on the best land, in terms of climate, soil quality and topography. Pressure to produce more food to supply growing populations has also led to exploitation of more marginal lands, giving lower yields and a higher risk of crop failure, and greater risk of damage to the underlying biophysical base through erosion, seasonal flooding, drought and biological infestations. However, this has often given rise to the development of new farming techniques, such as terracing slopes to control erosion, irrigation in drier areas and development of crops better suited to particular environments.

An alternative to expansion onto new land has been intensification of production on the land already cultivated, that is increasing the output per unit area of land. To achieve this, farmers can add more nutrients to the soils, through fertilisation, and also breed higher yielding varieties of crops. Farmers must concentrate on maintaining soil fertility, while avoiding the problems associated with long-term intensive cultivation such as pests, diseases, soil erosion and salinisation. Farmers select crop varieties especially suited to the climate and soils so that they are higher yielding. Livestock rearing may also be intensified through improved livestock housing, supplementary feeding and more controlled breeding programmes. Mechanisation may increase the timeliness of agricultural operations as well as enabling farmers to work a larger area of land more effectively.

Both the adoption of more intensive methods and the spread of farming activity into virgin territory give rise to modifications to natural ecosystems and development of distinctive agri-ecosystems. However, the changes may also give rise to negative feedbacks to these ecosystems, sometimes causing harm to the human population and being labelled as 'environmental problems' with their own attendant 'solutions'. The following are examples of common problems.

6.3.1 Soil erosion

In terms of modification to natural ecosystems one of agriculture's greatest impacts is upon soil resources, producing worldwide denudation rates that are over one thousand times the estimated natural rates of erosion, increasing from 20 million to 54 000 million tonnes per annum. In comparison, it may take between 100 and 500 years to generate 10 mm of topsoil under natural conditions of vegetation (Parry, 1992). Soil erosion occurs incrementally, as a result of many small rainfall or windblow events, and more dramatically, as a result of large but relatively rare events.

Overall, 15% of the earth's ice-free land surface is afflicted with some form of land degradation. Of this, accelerated soil erosion by water is responsible for 56% (affecting 11 million km^2) and wind erosion 28% (affecting 5.5 million km^2) (Pimental, 1995). Worldwide, the United Nations' Food and Agriculture Organisation (FAO) estimates a loss of between 5 and 7 million ha of productive land each year through soil degradation and loss. Hence one of the key problems facing world agriculture is how to optimise the use of the soil resource whilst preserving its quality. In industrial agriculture the latter may be enhanced by addition of artificial fertilisers, but the main processes of tillage, particularly using mechanised vehicles, can cause major changes in soil structure. Particularly in North America, this has been countered by use of the so-called minimal tillage, conservation tillage or 'no till' farming systems, for example direct drilling of seed in untilled or little-tilled soil. Other practices that restrict erosion include crop rotations, winter cover crops, contouring and terracing (Canter, 1986).

Although the United States has the world's largest conservation service, it is estimated that over 1% of the farmed area is seriously degraded each year by soil loss. Most severe loss of topsoil occurs in the intensively farmed cotton and soybean region east of the lower Mississippi, the wheat belt of the Great Plains and the Appalachia foothills of Georgia and the Carolinas. The on-site economic costs in the USA are in excess of $27 billion, of which $20 billion is accounted for by the cost of replacing soil nutrients lost to erosion and $7 billion for lost water and soil depth (http://soilerosion.net). In South Africa between 300 and 400 million tonnes of soil are lost each year, or nearly $3\,t\,ha^{-1}$ of land. In effect, for every tonne of agricultural crop produced in the country there is a loss of 20 t of soil. The cost of replacing the soil nutrients lost is Rand1000 million (http://www.botany.uwc.ac.za/envfacts).

In semi-arid areas mismanagement of the soil can produce permanent devastation, widely referred to as desertification. This covers a variety of conditions, including deterioration of rangelands, forest depletion, dune encroachment, wind erosion of topsoil and deterioration of irrigation systems. It is difficult to separate human-induced processes of desertification from 'natural' processes, but some estimates suggest that one-eighth of the world's land surface is actively experiencing desertification or is at high risk (Pimental, 1995). In addition to semi-arid areas in the developing world, such as the Sahelian Zone in Africa, parts of southern Africa and north-east Brazil, there are also several semi-arid areas in the developed world, for example around the Mediterranean and south-west USA.

Light sandy soils are most susceptible to wind erosion and hence in the UK the light peats of the fens and the sands under arable cultivation in East Anglia, Lincolnshire and Yorkshire experience the highest losses of soil. These losses have increased in recent years, with considerable discussion of the roles of climate, soils and cultivation practice as determinants of the location and severity of this erosion (Boardman, Foster and Dearing, 1990; Davidson and Harrison, 1995; Evans, 1996). Localised rates of erosion as high as $250\,t\,ha^{-1}$ have been recorded (Robinson and Boardman, 1988). The incidence of severe erosion is often associated with periods of exceptionally heavy rainfall during particularly intense storms, although much erosion also occurs under periods of prolonged lower-intensity rainfall, especially on soil compacted by agricultural machinery (Robinson, 1999). Speculation on the future impact of climatic change suggests that higher winter rainfalls, with an increased probability of high rainfall intensities, raise the prospect of increased erosion in Western Europe, particularly in areas of cereal cultivation.

6.3.2 Irrigation and salinisation

In 1990, irrigated agriculture consumed 2500 km^3 of water on 18% of the world's cultivated land (Pierce, 1990). This represented a sevenfold increase in irrigated area during the 20th century, with 20% of the world's cropland being irrigated. This accounts for one-third of the world's food output. The greatest contribution of irrigation to national food output occurs in countries where padi rice is a significant crop and/or where semi-arid climates occur, for example Pakistan (65% of the cultivated area is irrigated), China (50%), Indonesia (40%), Chile and Peru (35%), India and Mexico (30%) (Rangeley, 1987; Figure 6.3).

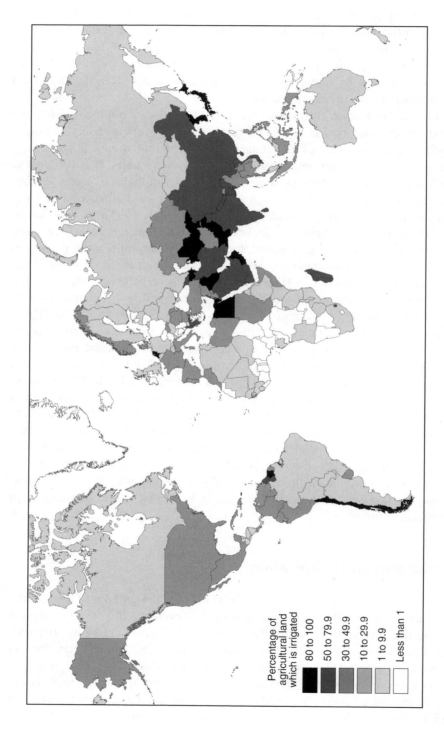

Figure 6.3 Proportion of agricultural land which is irrigated. (Reproduced with permission from Pierce, 1990; published by Pearson Education Limited.)

Percentage of
agricultural land
which is irrigated

80 to 100
50 to 79.9
30 to 49.9
10 to 29.9
1 to 9.9
Less than 1

The majority of the irrigation development in Asia is for the expansion of rice cultivation, which already accounts for three-quarters of food grain consumption there.

Pierce, 1990, p. 131

Hence it is not surprising that in key Asian countries, e.g. India, Pakistan, and the Philippines, and also Mexico, much of the capital assistance for the so-called 'Green Revolution' package to revolutionise agricultural output in the 1960s and 1970s was dominated by expenditure on the extension, upgrading and maintenance of irrigation systems. In the same decades the majority of Middle Eastern countries allocated between 60 and 80% of their agricultural investment to irrigation (UN Water Conference, 1977). Many of these extensions to the irrigated area, as typified by the opening of the Aswan High Dam in Egypt in 1969, have been costly, large-scale projects reflecting the fact that most of the sites with plentiful supplies of water for irrigation have been utilised already. Moreover, there is little evidence that economies of scale are present in the larger schemes and hence costs are very high, contributing in several cases to the high indebtedness of developing countries (Kreuger, Schiff and Valdès, 1992).

Substantial areas of this irrigation are based on withdrawing groundwater from aquifers that receive little annual replenishment from rainfall. Moreover, even where recharge does occur, irrigation usage often exceeds recharge. Hence even giant aquifers, such as the Ogallala in the United States, face diminishing well yields and rising pumping costs (Parry, 1992). In some cases, for example central California, these problems have led to recent reductions in the amount of irrigation. Yet there is no doubt that irrigation water effectively applied can significantly improve the returns from land in arid and semi-arid regions. Irrigation brings increased yields of between three and five times over rain-fed farming. But the returns to soil water in high input rain-fed farming systems in temperate regions far exceed those of irrigated systems in tropical regions. Indeed, some argue that it is rain-fed systems that should be given priority everywhere (Berkoff, 2001). An additional important feature of soil water is that it supports the production of the major internationally traded crop, wheat. In 2000, soil water provided the 200 km^3 of water associated with the production of the 200 million tonnes of grain entering world trade. Although not a high proportion of the 7600 km^3 of soil water (c. 4000) and fresh water (c. 3600) used annually at the global level, this is equivalent to the total usable water in the Middle East and North African region. It is also equivalent to three times the usable fresh water in the Nile system, twice the fresh water available in the Central Asian river systems and dwarfs the water available in famous rivers such as the Colorado (c.18 km^3/year) (Allen, 1997).

Whilst irrigation has had dramatic impacts upon crop productivity and in extensions to the cultivated area, especially in dry climate regions, there have been some negative impacts associated with irrigations alterations to the natural water–salt balance, increasing the extent and risk of saline and alkaline soils.

Secondary salinisation and alkalisation occur when the natural drainage system is unable to accommodate the additional water input. This causes a rise in groundwater levels, and capillary action can transport dissolved salts to the active root-zone and surface areas. The extent of this process depends on the depth of the groundwater, but generally the higher the salt content of the groundwater, the greater the depth through which this saline solution can damage crops.

Following the expansion of irrigation in the 1970s, one estimate claimed that nearly 70% of the 30 million ha of irrigated land in Egypt, Iran, Iraq and Pakistan were suffering from moderate to severe salinity problems (Schaffer, 1980). A further 7 million ha in India were also being adversely affected following extensions of irrigation in the central and western portions of the Indo-Gangetic plain, Gujarat and Rajasthan. More recently there have been more disturbing reports about the loss of good-quality soils through salinisation of soils in the vast alluvial river basins of the tropics in South and Southeast Asia and China, especially in the tributary basins of the Ganges and the Indus (Postel, 1999). Such problems also occur outside the developing world, with significant salinisation occurring in Australia, especially in conjunction with irrigation in the main river basin, the Murray–Darling (Robinson, Tranter and Loughran, 2000). Similar problems have been recorded in several parts of former Soviet Central Asia where there has been increased extraction of river water for growing cotton. In addition to exacerbating salinity problems, the high levels of water consumption have also contributed hugely to the diminution of the Aral sea, which has dramatically shrunk in size in recent years.

Despite the many large-scale projects aimed at improving water supply for agricultural purposes, there are stark predictions relating to future failures of food supplies due to diminished supplies of water. Such failures are projected for the Sahelian region of Africa, South Asia and large parts of Latin America as a consequence of shifting rainfall belts. Approximately one-third of the world's population (1.7 billion) already live in countries that periodically experience significant deficits in water supplies, and the population affected may rise to 5 billion by 2025. Moreover, in central Asia, north and southern Africa, because of a combination of higher temperatures and pollutant run-off, decreases in rainfall will be associated with declining quality of water that is available. Against this portrayal of impending disaster, it is possible that some regions may benefit from predicted warming, which may enable new crops to be grown, for example extending the cultivated area northwards in parts of Canada.

Recent estimates by Rosegrant and Cai (2002) suggest that irrigation can increase crop yields by a factor of between 3 and 5 when compared with yields from rain-fed farming systems. However, irrigation in many parts of the tropics returns relatively limited amounts of water to the soil when compared with rain-fed systems in the temperate zone; that is, the efficiency of irrigation in tropical agriculture tends to be limited. Another key factor to consider is that estimates predict consumption will account for as much as 87% of renewable fresh water within the Middle East and North Africa by 2025. The huge pressure of demand upon available freshwater supplies in this region can be contrasted with the much lower proportional demand upon water resources in Latin America. In China and India the corresponding figure is predicted as 33%, though this represents an over-all three- to fourfold increase in water consumption since 1950. This expansion of both industrial and agricultural water usage is giving rise to annual shortages in drier parts of both these countries.

For the future, the already significant differences between the developed and developing worlds with respect to irrigation are likely to continue. The former will largely consolidate its current area of irrigated farming, seeking gains from technological advances. Meanwhile, the area under irrigation will be expanded in developing countries, but at the expense of decreased reliability of freshwater supplies. The competition for fresh water between irrigators and those who want both to stop further withdrawals from the environment and to restore fresh water to the environment is likely to become more intense.

6.3.3 Artificial fertilisers and waste from livestock

Maintaining soil fertility is crucial to enabling repetitive cultivation of soils, whether for food or for livestock fodder and pasture. The key soil nutrients are nitrogen, phosphorus and potassium, although many other nutrients are also important in maintaining soil fertility. Traditionally, farmers all over the world used organic materials such as household waste and livestock manure to fertilise fields. Inorganic fertilisers, developed since the Second World War, provide soil nutrients as pellets or in solutions of known composition that can easily be spread onto fields. Inorganic fertilisers are attractive to farmers because of the ease with which they can be used, and the fact that their composition is known. In contrast, manure and waste material are of variable quality, bulky, and also carry risks of disease. However, the development of inorganic fertilisers has been both a blessing and a curse for industrialised farming systems in developed coun-tries. Their ease of use has actually resulted in overuse, leading to environmental problems.

Inorganic fertiliser use is much greater in the industrialised farming systems of developed countries than in developing countries. Since their inception, many farmers have adopted inorganic fertilisers to use alongside animal manures, but sometimes they have replaced manure use completely. The use of inorganic fertilisers in developed countries increased dramatically in the second half of the 20th century, with average rates of use between 120 and 550 kg N ha^{-1} (Conway and Pretty, 1991). Such high rates of use are based partly on recommendations for crop application, but have been exacerbated by fertiliser subsidies. Crop recom-mendations vary according to soil type, rainfall and climate; therefore a range of values is given to accommodate seasonal and field variability. Crop uptake of nutrients also varies with the stage of growth of the crop. Ideally, inorganic fertilisers are applied in just the right amount, at the right time, so that crop uptake is high and efficiency of fertiliser use is high. However, in practice, fertiliser use efficiency varies from 20 to 70% (Conway and Pretty, 1991). The remaining fertiliser may be leached from the soil, or removed from the fields via run-off after rainfall events. Estimates of such losses from arable crops range from 35 to 155 kg N ha^{-1} (Conway and Pretty, 1991). This overuse has generated new environmental problems such as eutrophication of waterways and the contamination of groundwater and wells (see Box 9.1).

In contrast to the situation in industrialised agriculture in developed countries, the low levels of use of inorganic fertiliser in developing countries is causing concern for the future of soil fertility and agriculture in these regions. Subsistence farmers cannot usually afford to buy much, if any, inorganic fertiliser. In Africa, average inorganic fertiliser use is 10 kg ha^{-1}, although this varies from place to place and crop to crop. The ability of smallholder farmers to purchase and use inorganic fertilisers is limited partly by cost, but also by problems associated with fertiliser availability at the appropriate time of year.

Studies of nutrient balances have been carried out to assess whether farmers are maintaining soil fertility. These studies assess the nutrients added to farmland (via inorganic fertilisers, manure, atmospheric deposition, sedimentation) and the nutrients removed from farmland (predominantly in harvested biomass such as crops, but also through erosion, leaching and volatilisation) to determine whether there is a net flow of nutrients into, or out of, the soil. National-level studies suggest that, overall, soils in countries in Africa are losing substantial amounts of nutrients. Stoorvogel, Smaling and Janssen's (1993) study of 38 countries estimated that annual losses of nutrients in 1983 were 10 kg N ha^{-1}, 4 kg P$_2$O$_5$ ha^{-1} and 10 kg K$_2$O ha^{-1}, and these were predicted to rise by 2000. A more recent study by Henao and Baanante (1999) reinforced these conclusions, claiming that 86% of sub-Saharan African countries are losing more than 30 kg N ha^{-1} each year. These studies, based on FAO production statistics and national estimates of fertiliser use and crop production, are, of necessity, fairly crude estimates of what is happening at the field level. Indeed, the value of such figures has been assessed critically by other researchers on African farming (Scoones and Toulmin, 1998; Mortimore and Harris, in press). However, the broadly accepted conclusion is that farmers are 'mining' their soils for nutrients without taking sufficient precautions to return nutrients to the soil, therefore jeopardising the long-term fertility and productivity of their soils. Of course, Africa is a diverse continent, with a broad range of climates, soils and farming systems, and within this there are examples of farming systems in which soil fertility is being built up (Tiffen, Mortimore and Gichuki, 1994; Harris, 1996). The debate is how such examples of more sustainable farming systems can be replicated throughout Africa, and there is a drive to develop policies and aid programmes which will enable farmers to manage their soils more sustainably (Scoones and Toulmin, 1999).

The trend towards intensification, specialisation and greater economies of scale in agriculture is well illustrated by livestock production in the United States where small numbers of very large producers account for significant proportions of the country's livestock products (Royer and Rogers, 1998). Livestock farmers have intensified traditional extensive activities by applying new systems of stock rearing and fattening. Hart and Mayda (1998) note that in 1949 three of every four farms in the USA had a flock of barnyard hens and a few cattle, two of every three had a milk cow or two, and more than half slaughtered a few pigs. Today less than one farm in ten has any pigs, dairy cows or chickens, though over half keep beef cattle, which have become the favoured livestock of small farmers because of their low labour, feed and land demands.

The trend towards large-scale production has been dictated not only by farm-based economics but also by the demands of consumers and the processing sector. Consumers increasingly desire 'leaner cuts of meat in convenient ready-to-use packages of predictably uniform quality' (Hart and Mayda, 1998, p. 60). In order to meet this demand, the processors require a steady supply of animals of nearly identical size, shape and quality. They also prefer to have dealings with a limited number of producers capable of regularly delivering large numbers of 'standard' animals rather than a larger number of small producers whose supply is of more variable quality. In addition, at the processors' behest, animal geneticists have developed more prolific breeding stock which will grow faster and produce leaner meat on less feed. Such animals tend to be better suited to large holdings because they require a greater capital investment and more skilful management.

In feedlot production of beef, originating in the western USA, single holdings have capacities in the tens of thousands of head of cattle, with the attendant problems of feeding and waste disposal from such large numbers. The development of centre-pivot sprinkler irrigation systems fed from deep wells has also enabled the farmers to increase their own production of feed crops such as maize, sorghum and alfalfa. The rapid development of feedlots in the area has also attracted the meatpacking industry, as the costs of shipping feed grains and cold-processed beef have shifted in favour of the latter. So new plants, designed to process one type of meat highly efficiently, were established near the feedlots, again generating major problems regarding waste disposal and odour. ConAgra has the largest meatpacking plant in the United States – in Greeley, Colorado – which is supplied from two feedlots, each of which can hold up to one hundred thousand head of grain-fed cattle (Schlosser, 2002). A similar process of reliance on feedlots and large new processing plants has occurred in Canada, primarily in the province of Alberta (Broadway, 1997, 2000), and there have been related developments in the processing of milk in New Zealand and Ireland.

6.3.4 Landscape change and loss of biodiversity

As discussed in Chapter 4, one of the prime causes of land cover change is agriculture. Agricultural expansion changes vast areas of natural vegetation to farmland by forest clearance. Forest clearance in the tropics may be temporary, as in slash-and-burn systems, leaving scope for regeneration of secondary forest, or it may be permanent, enabling longer-term cultivation of land. In temperate areas, increased agricultural production is now achieved through intensification, rather than expansion.

In Europe the landscape of the countryside has been formed by agricultural practice during millennia, and influenced by current changes in agricultural policies. These intensive farming methods have been strongly supported by government policies. For example, in the European Union (EU) the Common Agricultural Policy (CAP) has encouraged farmers to intensify production to obtain maximum advantage from the available price supports. In particular, the returns on cereals, whose prices have been artificially supported, have encouraged farmers to plough land that has either never been ploughed before or has only been ploughed when cereal prices were extremely high. This has impacted sharply upon landscape features in lowland areas. Long-term destruction of hedgerows, woodland, rough grazing, downland, moors and wetlands has been one readily identifiable outcome. These changes have also been closely linked to the development of mechanisation and to the economics of its use. Uplands have also been adversely affected. For example, there has been a reduction in rough grazing related to the increased numbers of sheep brought about by the headage payments available under the Less Favoured Areas (LFAs) scheme introduced in 1975 (Wathern et al., 1986, 1988). In designated LFAs farmers have raised stocking rates to increase their headage payments, exceeding the carrying capacity of the semi-natural vegetation of the rough grazings. They have then sought to make grant-aided improvements to their rough grazing under grassland conversion projects, at the expense of dwarf shrub vegetation and open moorland. Loss of rough grazing has been especially prominent in the UK's national parks (Robinson, 1994).

The removal of field boundaries has been well documented in the UK, especially for hedgerows. Shoard (1980) quotes a figure of about 192 000 km in all, or about 7200 km a year of hedgerows removed in England and Wales between 1946 and 1974. A loss of around 6400 km per annum in England was reported in surveys in the mid-1980s (Barr et al., 1986). However, this has not been a simple one-way process, and measures designed to promote hedgerow replanting have had some success. One estimate reports 25 600 km of hedgerow replanted in England between 1984 and 1990, though around 85 000 km were grubbed up during the same period (Bryson, 1993) and there was a further net loss of more than 16 000 km in the first four years of the 1990s. Measures such as the UK's Farm and Conservation Grant Scheme (1989–1993) and the Agriculture

Improvement Scheme (1985–1989) encouraged replanting and other desirable conservation and landscape improvements (Ghaffar and Robinson, 1997). These have been succeeded by the Hedgerow Incentive Scheme, launched in 1992, and intended to restore c. 900 km of hedgerow per annum. This is now part of the Countryside Stewardship Scheme, offering ten-year agreements to farmers to restore damaged or neglected hedges in return for financial incentives.

Agriculture has also become a central factor in the loss and decline of biodiversity. Where distinctive semi-natural areas survive, it tends to be despite agricultural practices rather than as a result of them (as it once was). One example of this loss of biodiversity through intensified farming activity occurs throughout the temperate zone, where some bird species have been especially badly affected by the use of agricultural chemicals and changes in habitat caused by modern farming practices (Gregory and Baillie, 1998). For example, in the UK the number of skylarks has declined by three-quarters in the last quarter century, largely because of the change from spring-sown to autumn-sown cereals (Gregory *et al.*, 1999; Figure 6.4; Table 6.1). Spring-sown cereals allow the stubble of the previous crop to be left unploughed over the winter, providing food and cover for the birds.

This loss of biodiversity is currently extremely rapid in the world's tropical forests, with rapid rates of forest clearance apparent in the forests of Amazonia and Southeast Asia. Although some of this is associated with commercial timber operations, in many cases it represents the advance of subsistence cultivators. In India, for example, less than half of the forested lands remain under closed canopy forests and many of the remaining forestlands are in various stages of

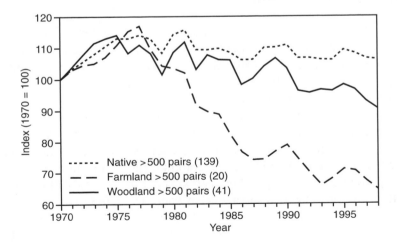

Figure 6.4 Headline indicators for populations of wild birds in the UK. The indicator is produced jointly by the Royal Society for the Protection of Birds/The British Trust for Ornithology/ Department of the Environment, Food and Rural Affairs. (Reproduced with permission from Gregory *et al.*, 1999; published by RSPB).

Table 6.1 The decline in bird species in the UK, 1972–1996. The indicator is produced jointly by the Royal Society for the Protection of Birds/The British Trust for Ornithology/Department of the Environment, Food and Rural Affairs. (Reproduced with permission from Gregory *et al.*, 1999; published by RSPB)

Species	% UK population decline 1972–1996	Species	% UK population decline 1972–1996
Turtle dove	85	Song thrush	66
Grey partridge	78	Reed bunting	64
Tree sparrow	76	Yellowhammer	60
Skylark	75	Lapwing	46
Corn bunting	74	Linnet	40

deterioration. However, from the 1970s there have been imaginative protection policies implemented involving local communities. Local forest protection committees have been established. These have been given the responsibility of protecting degraded land and granted rights that allow the use of a range of timber and non-timber forest produce. Such schemes have had some spectacular successes in terms of biological regeneration and increased income flows (Pretty, 1995). Yet, despite the spread of such schemes, problems over who owns the forest and the rights of local communities have led to conflict and misuse of resources. In other countries, the lack of control over forest settlement by subsistence farmers continues to be a major problem (MacMillan, 1994).

6.3.5 Pollution and health scares

Industrial farming is characterised by economies of scale: large areas of monoculture enable mechanisation and timeliness of agricultural operations (soil preparation, sowing, spraying with fertilisers or pesticides and herbicides, mechanised weeding and harvesting). In livestock production, large units again benefit from economies of scale and make the most of specialist knowledge concerning animal production. However, the nature of these systems means that they are more prone to disease outbreaks than traditional mixed farming systems (where crop and livestock production are mixed, and a broader range of crops and animals is produced, so that crop rotation is more common and rapid). In monocultures it is more important to utilise agri-chemicals to control disease within the system: generally using agri-chemicals in a precautionary way prior to outbreak of disease (whether in crops or livestock). Some industrial livestock systems also use heavy metals or antibiotics and drugs routinely to enhance production levels. In a traditional mixed farming system (whether a temperate organic farming system or smallholder farm in the tropics), the by-products of crop production are useful for animal production (e.g. crop residues used as animal fodder) and the by-products of animal production (e.g. manure) are useful for crop production. In more specialised industrialised farming systems, which tend to focus on either crop or livestock production, the by-products of the system become a burdensome waste which must be disposed of, rather than a useful product which can be recycled as an input into another part of the farming system (Figure 6.2).

In reviewing the environmental problems associated with agriculture, Conway and Pretty (1991) describe the three categories of pollution arising from agriculture as being contamination of water, contamination of the farm and natural environment, and contamination of food and fodder. The first is through pesticides, nitrates and phosphates, organic livestock wastes, silage effluents and other wastes arising from the processing of crops. For example, the residues of persistent pesticides, collectively known as 'drins' (aldrin, dieldrin, endrin and isodrin), can be washed into streams and freshwater lakes. The second is, in part, through pesticides and nitrates, but also from ammonia emissions from livestock in padi rice fields, and from metals or pathogens resulting from livestock wastes. The third is the result of pesticides and nitrates. In addition, gaseous wastes from agricultural production can pose a danger in enclosed environments (e.g. N_2O from silage in silos, ammonia and H_2S in intensive animal production units) and also contaminate the wider atmosphere (NH_3 and methane from livestock manures and padi rice fields, burning of cereal straw or even forests and savannas to clear land for agriculture).

The presence of these various environmental disbenefits associated with industrial farming was first brought to wider attention in the 1960s following the publication of Carson's (1962) book, *Silent Spring*. Subsequently there has been a huge surge in media and public awareness of these 'ills' of modern farming. This has undoubtedly led to a major shift in the public perception of farmers: no longer are they automatically regarded as the custodians or stewards of the countryside, but instead more negative views are reported, associated with concerns over food quality, destruction of valued landscapes, pollution of watercourses and mistreatment of farm animals.

This increasingly negative view has been reinforced periodically by health 'scares' directly related to the downstream consequences of farming activity. In particular, there has been concern over such issues as the presence of Salmonella in chicken and certain types of cheese and the spread of bovine spongiform encephalopathy (BSE or 'mad cow' disease) in the late 1980s, which literally brought the environmental consequences of intensive farming home to the consumer. Such food scares provoked calls for a different type of evaluation of agriculture, extending beyond the narrow economic viewpoint that generally underscored government policy (Dwivedi, 1986; Blakemore, 1989; North and Gorman, 1990; Lang, 1998). The continued occurrence of such problems, as exemplified in the recent foot-and-mouth outbreak in the UK (Ilbery, 2002) and global concern for the effects of GM crops (section 6.4.2), partly illustrates continuing deficiencies in the nature of environmental regulation. Nevertheless, there have been some significant new regulatory changes, in some cases associated with non-governmental agents such as supermarkets and consumer associations. These regulations reflect a growing concern in the developed world for the quality of food and food safety issues (Lang, Barling and Caracher, 2001).

6.4 Technocentric solutions: the Green Revolution and GM crops

With reference to the types of agricultural system proposed for the future, there are two distinctive and competing approaches. One focuses primarily upon increasing yields in order to tackle rising demand for food, and the other places greater weight upon minimising agriculture's environmental impacts. These two approaches are at opposite poles of a spectrum that ranges from a technocentric viewpoint to an ecocentric one. These different visions appear in many different guises, including a prominent role in current debates about the nature of sustainable development, in which some equate sustainability largely with terms such as 'environmentally friendly' and 'environmentally sensitive', whilst others refer to 'economic efficiency' and 'technical solutions' (Robinson, 2001; Chapter 10).

With regard to the impacts of agriculture on the environment, the different concerns and outcomes of these contrasting views can be seen clearly in the examples selected here. These are the technocentric approaches, exemplified in the Green Revolution and the development of GM foods, and the ecocentric approach of organic farming.

6.4.1 Green Revolution 'solutions'

In the 1960s, considerable efforts were made through the so-called Green Revolution to increase food yields to meet rising population numbers. Agricultural research focused on breeding high yielding varieties (HYVs) of staple crops such as wheat, rice and maize. These varieties were most successful when used in conjunction with a 'package' of inputs: fertiliser, irrigation, pesticides and agronomic techniques. When everything was used together, the package generally resulted in higher yields. However, when farmers were unable to purchase the whole package, the HYVs were not as successful. The technologies and crops developed were suited to environments where water was available for crops, and farmers could purchase inputs. Thus the Green Revolution was successful in increasing irrigated (padi) rice production in Southeast Asia, but had more limited effects in more marginal farming environments, and was not pursued substantially in Africa.

The Green Revolution undoubtedly increased agricultural production, particularly in irrigated rice-based farming systems in Asia. However, longer-term analysis has shown that this technological solution to food production concerns has had some lasting socio-economic and environmental effects that have not been so positive. Resource-poor farmers were unable to maintain the Green Revolution technology, and were out-competed by their slightly better-off rivals. Many farmers had to sell their land after incurring debts and were left as landless labourers, working on larger neighbouring farms or joining the stream of migrants to the cities. Adoption of Green Revolution varieties of rice displaced traditional varieties, resulting in a monoculture that was more prone to disease and pests, and also resulting in the loss of biodiversity within these crops, as farmers stopped using and saving their traditional crop varieties (Box 6.1).

Box 6.1: The development of biotechnology and the equitable sharing of benefits derived from agricultural biodiversity

One of the aims of the Convention on Biological Diversity (CBD) (1992) is:

> the fair and equitable sharing of the benefits arising out of the utilization of genetic resources, including by appropriate access to genetic resources and by appropriate transfer of relevant technologies, taking into account all rights over those resources and to technologies.

This aim has proved difficult to achieve when genetic biodiversity is harnessed for agricultural development.

Local farmers keep and maintain many landraces of crops to provide a portfolio of varieties to suit the variable conditions in which they farm. By growing many varieties (e.g. drought tolerant, pest resistant) in the same field, farmers increase the likelihood of a successful crop whatever the environmental conditions in a single growing season. With the increasing awareness of the importance of biodiversity, organisations such as the International Union for Plant Genetic Resources and International Agricultural Research Centres have made collecting such genetic material, with the accompanying local knowledge, a priority. These publicly funded international organisations have become the 'librarians' of crop genetic biodiversity. The majority of plant breeding for commercial purposes is carried out by private plant breeding firms: six large corporations dominate the commercial food and farming sector, and ten seed companies control 30% of the seed trade (Pretty, 2001). For plant breeders, the publicly funded collections of local landraces of crops can be a gold mine of genetic diversity, which may contain the desired trait to be inserted into a new commercial variety of crop. Farmers have donated material to the collectors extensively, and these international repositories of genetic material freely pass on material to plant breeding companies. Biotechnology companies

Continued on page 136

Global Environmental Issues

Continued from page 135

invest their considerable research expertise (knowledge, equipment, tech-niques, funds) in using the raw genetic materials to develop new plant varieties that are likely to be commercially successful for a broad range of farmers. The final product, a commercially viable and tested crop variety ready for release, is then sold to farmers by the biotech company (or under license to subsidiary companies). Thus the chain of knowledge is from local farmers to collectors to the collections of international institutions, and then on to private companies (Table 6.2).

Such a system suffers from several problems, which affect the aim of 'equitable sharing of benefits' in the CBD. Each stage of crop development has a value. The initial stages involve transfer of knowledge and raw material without charge, and later stages involve selling of a commercial product. However, farmers are not financially rewarded for reproducing the genetic material containing the desired trait. The publicly funded international research institutions are not rewarded either. Thus the current system is accused of resembling colonialism: scientists harvest the raw materials they need from the developing countries, develop a new product and then sell it back to the original countries at much higher prices. Some call this 'stealing of genetic material and knowledge from gene-rich developing countries' 'biopiracy' (Merson, 2000; Moran, King and Carlson, 2001). Developing countries lack the technical expertise and financial resources to develop

Table 6.2 Transfer of knowledge from farmers to biotechnology companies

	Farmers	International public sector research organisations	Private agro-biotech companies
Role	Custodians of landraces	Librarians of genetic biodiversity	Holders of technologies and financial resources for development of new crop varieties
Expertise	Local knowledge relating genetic characteristics to environment	Increasingly limited funding for research	Expertise in research and technology
Driving force	Landraces conserved to broaden stability of farming system in the face of environmental uncertainty	Potential to follow altruistic rather than profit-oriented research goals	Driven by market forces

Continued on page 137

Continued from page 136

genetic material themselves: training and transfer of technology are required to rectify the imbalance between developed and developing countries, and to enable developing countries to take charge of agri-biotech research to meet their own needs (Zerda-Sarmiento and Forero-Pineda, 2002).

As the economic value of biological resources and knowledge related to genetic resources are identified, there is increasing interest in the allocation of property rights to biodiversity management (Cullet, 2001). While the biotech industries have seen much potential in development of genetically modified organisms (GMO) which they are keen to capitalise on, the ethical and regulatory issues relating to GMO development and the sharing of benefits arising from the use of biodiversity are lagging behind. Biotech companies have developed technically enforced intellectual property rights in the form of F1 hybrids, and the 'terminator technology' (genetic use restriction technologies) (Eaton et al., 2002). These have proved extremely unpopular with farmers and NGOs working in developing countries, as in these systems, the end developer in the chain of knowledge development takes all the benefits. The idea of claiming intellectual property rights on a 'new' plant is less culturally acceptable in some countries than in the West. It is also extremely hard to enforce. There is considerable (unresolved) debate about how to ensure that farmers' rights are respected in the light of new agreements regarding international trade in property rights (Shiva, 1997; Ghijsen, 1998; Cullet, 2001; Feyt, 2001; Moran, King and Carlson, 2001; Gaisford et al., 2002). There are fears that the 'corporatization of property rights in plant materials may threaten farmers' lives and international and national public agricultural research' (Lipton, 2001, p. 839).

6.4.2 Genetically modified crops

Genetically modified (GM) crops were developed in the 1990s, and have been grown on a commercial scale in large areas of the USA and Canada as part of the deployment of biotechnology to increase crop yields. By 2000, roughly 28 million ha of GM crops were being grown worldwide, a fifteenfold increase in just five years (Middleton, 1999b). The five principal GM crops, soybeans, maize, cotton, rapeseed and potatoes, are grown commercially in nine countries of which France, Spain and South Africa are the most recent adopters, having grown GM crops for the first time in 1998. The USA has nearly three-quarters of the total global area devoted to GM crops. In 1999, half of the 29 million ha devoted to soybeans in the USA were planted with GM herbicide-resistant seeds, intended to give easier control of weeds, less tillage and reductions in soil erosion. However, the cultivation of GM crops remains contentious as the scientific community and general public continue to debate their safety.

Genetic engineering technology involves the insertion of genes with known characteristics and/or products into a strain of plant or animal previously lacking the desired trait. Amongst the key developments of genetic modification have been successful introduction of genes that create tolerance for herbicides, the creation of inactive plant genes thereby removing undesirable characteristics in crops, and the introduction of transforming agents such as plant viruses to create a desirable product, for example making the plant more palatable, nutritious or combative against disease. The technology has the potential to confer resistance to diseases, pests or pathogens, alter the nutritional quality of foods (e.g. Vitamin A-enhanced rice), prolong food shelf-life (e.g. 'Flavr Savr' tomatoes), enable crops to be grown in marginal environments (e.g. saline tolerant crops), as well as change crop characteristics to complement agronomic practices to increase yields. So, proponents of GM crops feel the technology has the potential to increase agricultural yields, and so reduce threats of famine or food shortages.

Despite the claims that GM crops will solve problems of food shortage in developing countries, it is noticeable that to date, most developments of GM crops have been for crops in temperate farming systems, rather than focusing on subsistence crops in the tropics. There has tended to be a dominance of new transgenic crops engineered for herbicide tolerance rather than for any intrinsic improvement in crop food quality or pest resistance (Lappé and Bailey, 1999). Indeed, two-thirds of all transgenic food crops to date have been engineered for herbicide tolerance. This is in part due to the fact that the private companies funding research and development in GM crops are looking for financial return for their efforts. This means concentrating research and development on crops for which there is a recognisable market, with likely adoption by farmers who can afford to purchase seeds year after year, and a food production system that will provide a market for the GM crop once grown. Industrial farming systems provide such a suitable market. Furthermore, linking herbicide tolerance with a company's own brand of herbicide also increases profits of the companies, for example Monsanto's Roundup-Ready™ Rice is suitable for use with their herbicide glyphosate, American Cyamid's IMI™ Rice seed works with their imidazolinone herbicide, and AgroEvo's Liberty Link™ Rice works with their Liberty™ herbicide (http://www.grain.org/publications/reports/rice.htm).

While some research has been carried out towards the development of GM crops for developing countries, two major problems have set back research and development for these beneficiaries. First, private sector companies are unwilling to invest large sums of time and money when they feel an ultimate market is not present. They are more interested in developing crops for industrialised farming systems. This has renewed calls for funding for public sector research bodies, such as the Consultative Group on International Agricultural Research (CGIAR), a system of international agricultural research stations, to take on this role. Secondly, there has been considerable bad press for biotechnology companies concerning their appropriation of genetic material conserved by smallholder farmers in indigenous communities for patented new crop varieties (Box 6.1). Many farmers traditionally hold back some seed from harvest to sow the next year's crop. The 'terminator gene' used by biotechnology companies ensures that this is not possible with GM crops, forcing farmers to buy new GM seed every year.

In terms of recent biotechnical developments it has been this emergence of GM foods that has offered the most divisive views of the future of the marriage between farming and technological advances. The rapidly expanding biotechnology corporations such as Monsanto, Hoescht and Calgene claim that GM crops have the potential to eradicate global food shortages, thereby providing global food security for the poorest parts of the world. Yet, strong opposition to these claims has been mounted in several developing countries and amongst environmental lobby groups in the developed world, especially in the UK.

Some observers argue that humankind has been breeding new varieties of crops (and animals) for centuries, and GM crops are just a new generation of crop breeding, using more accurate biotechnology. However, a broader concern relates to the fundamental difference between bio-engineered and conventional crops. Both the old and the new methods can alter the genetic composition of crops, but only 'traditional' plant breeding methods ensure a degree of uniformity from generation to generation. In contrast, there is no assurance of this occurring with genetic engineering technology. With GM technology comes the ability to transfer genes from unrelated species, and this may either seem inherently unnatural or downright frightening. Although proponents of GM crops argue that many tests are carried out prior to their general use to ensure safety both to the environment and to the consumer, it is difficult to gauge the long-term effects of release of this new GM material into the world's ecosystems. Opposition to GM crops in the UK has focused on the probability that the integrity of plant species may be compromised as engineered crop acreage increases and pollen-mediated gene flow affects native plants which are congenes for the transgenic crops.

Since 1999 there have been attempts to organise concerted opposition to GM crops in developing countries who fear 'biodevastation' as a threat posed by GM crops. There is concern regarding the potential reduction of biodiversity as farming becomes more reliant upon heavy use of herbicides. For example, a field of cotton with genes resistant to a particular toxin may enable only resistant plants to survive to reproduce. This type of highly undesirable potential outcome is one of the central arguments made by those concerned that there has been insufficient consideration of the long-term consequences of this technology. In India, 1500 organisations have campaigned against Monsanto's Bollgard cotton on the grounds that it can destroy beneficial species and can create superior pests (Power, 2001). The International Convention on Biodiversity includes reference to the need for an international protocol on biosafety to ensure safe handling and transfer of GM organisms between countries.

People's perceptions of risk, view of the environment (Chapter 1) and the precautionary principle all come into play with respect to the 'GM debate'. Ultimately, consumer confidence in GM foods will dictate whether they are a success or failure. Many organisations (from regulatory bodies of organic farming standards to major supermarket chains) are taking a GM-free stance as a way of reassuring customers and enhancing their own popularity and hence market share.

There are some interesting parallels between the debate on GM crops and those in the past on the Green Revolution. Both were championed as technological solutions to the global need for more foods. However, both have fallen short (so far, in the case of GM) of meeting the needs of poorer farmers in marginal environments, in particular with respect to the staple crops of African farmers, such as sorghum and cassava. The nature of the seeds (whether F1 hybrids or those containing terminator genes) ensures that farmers are tied to purchasing new seed each year, and in each case, the successful growth of the new crop is linked to purchasing supplementary agro-chemical products (in the case of herbicide-tolerant GM crops). The Green Revolution increased food yields, but also resulted in considerable social change as resource-poor farmers were forced out of farming, and environmental problems associated with monoculture, pesticide and fertiliser use. However, GM crops are still too new to judge whether, like the Green Revolution, this technological fix will bring extensive environmental or social problems.

6.5 Toward sustainable agriculture

6.5.1 Defining sustainable agriculture

In contrast to 'technology-fix' solutions to food production problems, ecocentric approaches emphasise different outcomes, typically stressing the need to consider the environmental costs of development. For over a decade such approaches have been part of the broad discourse on sustainability, in which an increasing amount of attention is being devoted to the notion of sustainable agriculture. The term 'sustainable agriculture' has been popularised during the last 20 years (Altieri, 1987), but its practice is as old as the origins of agriculture. Yet, its definition remains problematic (Bowler, 2002b). There has been extensive consideration of what constitutes a sustainable agricultural system, and what are the implications of the demands that such a system may have for farming in a particular area. However, it is difficult to derive a satisfactory definition that is not too simplistic and which has some practical characteristics within it that can be recognised within farming practice.

A comparison between the views of academics and farmers in the United States on what constitutes sustainable agriculture and how it might be achieved in practice revealed several statistically significant differences (Dunlap *et al.*, 1992).

Farmers tended to focus on economic factors rather than environmental ones and did not recognise any need for reducing purchased inputs or energy use. This contrasted with academic views that stressed the core ecological aspects of sustainable agriculture, for example, 'the ability of the agro-ecosystem to maintain productivity when subject to a stress or shock' (Conway, 1997, p. 177). The academic views also varied between academic disciplines, reflecting the need to formulate future policy on sustainability that encompasses a broad interpretation of the term. Sustainable agriculture is primarily an approach to agriculture that balances agronomic, environmental, economic and social optima, as implied in the definition provided by Francis and Younghusband (1990, p. 1):

> Sustainable agriculture is a philosophy based on human goals and on understanding the long-term impact of our activities on the environment and on other species. Use of this philosophy guides our application of prior experience and the latest scientific advances to create integrated, resource-conserving, equitable farming systems. These systems reduce environmental degradation, maintain agricultural productivity, promote economic viability in both the short- and long-term, and maintain stable rural communities and quality of life.

Doering (1992) suggests that sustainable agriculture has four key aspects that differentiate it from most current commercial farming, with sustainability emphasising limited inputs, specific practices (e.g. organic farming) and management perspectives based on ecological and social considerations (e.g. biodynamics and permaculture) (Bowler, 2002a,b): it implies less specialised farming, which often requires mixed crop and livestock farming for reduced dependence upon purchased inputs; it implies that off-farm inputs should not be subsidised and that products contributing to adverse environmental impacts should not receive government price support; it implies that farm-level decision-making should consider disadvantageous off-farm impacts of farm-based production, for example contamination of groundwater, removal of valued landscape features; and it may require different types of management structure, for example family farms as opposed to corporate 'factory' farms.

Whatever definition of 'sustainable agriculture' is chosen, it is widely agreed that there is an urgent need for a greater input of ecological and environmental information in agricultural policy. At present much of the research on sustainable agriculture focuses on individual plots or fields or is performed at farm level. Yet, most of the environmental issues associated with agriculture are manifested at larger scales, and hence the need to consider measurable changes in water quality, soil quality, biodiversity and other environmental indicators at these scales (Allen *et al.*, 1991).

A major problem has been to identify policies that can deliver sustainability in an environmental sense rather than just the economic sense largely preferred by governments. This applies throughout the world despite the very different economic and social conditions that prevail. In developing countries, farmers are often engaged in non-farming economic activities to supplement meagre farm incomes, such as harvesting of natural products, trading and practicing rural crafts. Seasonal migration (section 8.4.1) or remittances from urban-based family members provide funds to support farming or invest in improving land quality (e.g. Tiffen, Mortimore and Gichuki, 1994). In developed countries, farmers are not only responsible for food production, but also environmental management and provision of a recreational environment for urban dwellers and holidaymakers. As food prices diminish, farmers are encouraged to keep producing through area payments as well as receiving small payments to provide attractive landscapes from agri-environment schemes. They are also encouraged to diversify into non-traditional farm-related enterprises (e.g. holiday homes, horse livery, bed-and-breakfast, nature trails). This wider view of farming may move policies away from the promotion of farming as an industry in itself to a wider view of farming as a component of a sustainable rural economy.

6.5.2 Policies for sustainability

In some sectors of agricultural production the industrial model of farming has been challenged directly by environmental regulation that refers to growing concerns over food quality, environmental pollution and sustainability. Indeed, the achievement of sustainable (or 'alternative') forms of agricultural production is now widely recognised by governments as a long-term policy objective (Billing, 1996), and many countries are developing what they term 'sustainable agricultural strategies' as part of their national environmental and agricultural plans (DoE/MAFF, 1995, for England and Wales).

The shift in agriculture from productivism to more environmentally friendly farming is being achieved through a range of measures aimed to encourage, and ultimately enforce, the adoption of improved farming practices (Table 6.3). Such a carrot and stick approach enables environmentally aware farmers to make the transition to less productivist systems within a market-dominated farming system, and also ensures that there is a minimum level of environmental standards which are met by all farmers. This is achieved through self-regulation whereby

Table 6.3 The promotion of environmentally friendly farming in the UK

Approach	Examples
Voluntary	Certification schemes such as Freedom Foods, LEAF, Organic farming, Fair Trade
	Agri-environment schemes supported by government/EU funding
	Integrated crop management
Codes of practice	MAFF guidelines
Market-led incentives	Farm Assured Scheme
Legislation	For example, guidance on manure use within nitrate vulnerable zones

farmers are aware that poor environmental management will reduce productivity in the long term, and guidance from the government as codes of practice, a range of kite-mark schemes aimed at reassuring consumers about the standards by which food was produced, government- or EU-subsidised grants for environmental measures (e.g. hedgerow planting), and legislation. Many of the kite-mark schemes concern variations on low-external input agriculture: farming systems which are less dependent on purchased fertilisers, pesticides and herbicides, and instead focus on recycling nutrients within the farming system to build up soil fertility. The ultimate extreme is organic farming (Box 6.2), but integrated nutrient management, integrated crop management, and integrated pest management are all terms pertaining to farming systems that aim to limit inputs to some extent.

In the UK, two common kite-mark schemes are the Soil Association symbol for organic food, and the Red Tractor indicating compliance with the government's Farm Assured scheme, but recently new schemes are also coming onto the market: the RSPCA's Freedom Foods mark, indicating that animal products are

---

Box 6.2: Organic farming

The definition of organic production given by the United States Department of Agriculture (USDA) (quoted in Foster and Lampkin, 2000) is commonly used and has been widely adopted:

> A production system which avoids or largely excludes the use of synthetic compounded fertilizers, pesticides, growth regulators, and livestock feed additives. To the maximum extent feasible, organic farming systems rely upon crop rotations, crop residues, animal manures, legumes, green manures, off-farm organic wastes and aspects of biological pest control to maintain soil productivity and tilth, to supply plant nutrients, and to control insects, weeds, and other pests.

The number of farms in the developed world on which production meets this definition is small, but increasing rapidly (Table 6.4). There are a number of overlapping contemporary trends that are contributing to the dynamics of organic production (McKenna, Le Heron and Roche, 2001). These include:

(a) increased consumer concern over food safety;
(b) 'greening' of the images of corporate food producers and production practices;
(c) formalised and wide ranging guidelines prescribing food qualities at various organisational scales;
(d) movement of large-scale capital-intensive food producers;
(e) continued use of discourses incorporating 'sustainability' to define trends in production and consumption.

Continued on page 144

Continued from page 143

Table 6.4 Characteristics of organic farming (Source: Robinson, 2003)

	US	UK	EU
No. of certified organic growers	5021 in 1997	1000	104 000 in 1998
Rate of growth of number of growers	20% per annum	Quadrupled between 1993 and 1998	Tripled between 1993 and 1998
Cropland	344 050 ha = <.2% in 1997	540 000 ha in 2000 (including land in conversion) i.e. 0.5% increasing to 3%	2.7 million ha in 1998 = 2.1%
Sales	$2.3 billion per annum	>£350 million	Nd
% of food sales	<2%	<2%	Nd

In 1997 nearly half of the certified growers in the United States were in California. About 2% of US top speciality crops (lettuce, carrots, grapes and apples) are grown today under certified organic farming systems (Greene, 2000). The largest certification scheme in the UK for licensing organic food production is the Soil Association Symbol Scheme, complying with EU regulations but imposing additional standards. In 2000 there were 540 000 ha in organic production or in conversion in the UK, more than twice the area in 1999 (Soil Association, 2000), and approaching 3% of all farmland.

In the EU, organic accreditation is underpinned by European Council Regulation 2092/91, providing specific rules for the production, inspection and labelling of products, and with implications for farming systems and the environment. There is considerable variation between countries as to the extent of certified organic land, with the highest proportions occurring in Austria (7%), Finland (5%) and Italy (5%). Organic aid schemes in the EU have also increased considerably through the 1990s, with 25 000 agreements in Italy, 18 800 in Austria, 13 500 in Sweden and over 10 000 in Germany (Foster and Lampkin, 2000). Governments in many parts of the developed world have helped to promote this advance of organic farming. For example, the UK's Organic Farming Scheme was increased by £24 million in 1999, to encourage an additional 75 000 ha of new conversion (MAFF, 2000). Between 2000 and 2006 an additional £140 million is to be allocated to organic conversion, providing payments for conversion.

In Europe and the United States, retailing of organic produce has often been associated with small-scale, 'local' outlets rather than chain stores such as supermarkets. In contrast, in the UK, supermarkets have played a large role in the organic sector, starting with sales of 'organic' produce by

Continued on page 145

Continued from page 144

Safeway in 1981. This has provoked concern from organic food organisations, some of which have contended that the supermarkets' approach to organic food is inimical to the underlying ethos of organic production (Clunies-Ross, 1990). Although supermarkets have initially been obliged to accept the principles of supplier standards for organic produce based on internationally accepted criteria, some are now introducing their own criteria which can be applied to domestic suppliers. In so doing they may champion cheaper alternatives to fully 'organic' production, such as produce grown using integrated crop management (ICM), which promotes natural predators and crop rotation but does not eliminate the use of pesticides and fertilizers. In 2002, nearly 90% of Sainsbury's UK fresh produce supplies were being grown to ICM standards, largely through farmers in the LEAF (Linking Environment and Farming) network, which promotes 'green' farming.

As these trends have evolved in recent years, organic farming has been presented as a growing arena for the intersection of the interests of capitalist modes of production and family farming (Whatmore, 1995; Campbell and Coombes, 1999). For example, TNCs have gradually increased their emphasis on food safety, 'freshness' and 'green' values as a means of enhancing their economic advantage. This can lead to economic goals taking over from more ecological values even in the context of 'organic' production. Hence organic farming may be subsumed into the industrial model of consumption and production.

produced in a humane fashion, and the LEAF mark, developed by the Linking Environment and Farming Network. All of these schemes are voluntary, in that farmers choose to apply for certification. However, each results in financial benefits to the farmer, as kite-marked foods are often sold at premium prices. The Farm Assured scheme is increasingly becoming an economic necessity, as middlemen in the food chain insist on farmers having the 'Farm Assured' status before buying their produce. Thus, although voluntary, market forces are making it a necessity.

The Fair Trade kite mark is concerned with ethical trading of products to ensure that producers receive a fair price for goods supplied. This has applied to tropical cash crop products such as tea, coffee and bananas. Concerns have grown over ethical aspects of production and trade, and the treatment of workers and producers within farming systems in developing countries. 'Ethical trade' has included fair trade agreements, safe working conditions for disadvantaged producers and employees, and sustainable and environmentally safe natural resource management. In part the concerns have been both consumer-driven and trade-driven. For example, literature on 'consumer theory' refers to ethical consumerism as the 'fourth wave' of consumption, seeking to 'reaffirm the moral dimension of consumer choice' (Gabriel and Lang, 1995). It is a consumer response that links the global and the local, but with global concerns as a key component (Bell and Valentine, 1997). There is much in common between this consumerist concern and the growing debate about the morals and ethics of international trade (Brown, 1993). For example, the creation of the World Trade Organisation (WTO) has highlighted the absence from the trade agenda of 'issues of sustainable resource management, the regulation of commodity markets, and poverty reduction strategies' (Watkins, 1997, p. 110).

The essence of ethical concern over production and trade is encapsulated in the following quote from Browne *et al*. (2000, p. 71):

> The improvement in trading relationships through ethical trading, enforced by organic concepts of production, contributes to the accumulation of both natural and social capital, through greater sustainability of natural resources and increased access by producer groups to networks of production and trade.

In developing these ethical arguments, Whatmore and Thorne (1997) have shown how traditional commercial networks for some 'plantation' products now exist alongside new networks associated with concerns for rural social justice. They focused upon trade in coffee. The long-established commercial network has been based on commercial imperatives in which an unequal power relationship exists between numerous small-scale producers of coffee and relatively few dealers. These dealers sell to processors who mass-produce coffee for sale as globally recognised 'brand' coffees. In this system profits heavily favour the end producer and the retailer. The alternative system that has emerged in the last two decades has a different arrangement based on partnership, alliance, responsibility and fairness. This gives greater power to small-scale growers working within locally based co-operatives. In this 'fair trade' coffee network, growers are paid a guaranteed minimum premium price whilst maintaining critical parameters of quality control and marketing deadlines.

Non-governmental organisations (NGO) such as Oxfam have been crucial to the development of fair trade networks, encouraging the growth of trading companies like Cafédirect to emerge as the key link between local producers in the developing world and consumers in the developed world. However, the changing nature of the market in the developed world has also been significant, with increasing numbers of consumers prepared to pay a higher price for a fair-trade product, such as Cafédirect's ground and freeze-dried coffee as opposed to a conventionally marketed product.

Although governments have been slow to introduce legislation directly promoting sustainable agriculture, they do rely on legislation to ensure a bare minimum of environmental standards is maintained. In Europe, a recent area of legislation enforcement concerns the designation of nitrate vulnerable zones, in which rules regarding the use of nitrogen fertilisers and manures are strictly set out (Box 9.1). In addition, throughout the developed world there are now a range of agri-environment policies (AEP) consisting of any policy implemented by farm agencies or ministries utilising funding from agricultural support budgets, and aimed primarily at encouraging or enforcing production of environmental goods. The latter may include something quite explicitly targeted by the policy, such as hedgerow restoration, or it may refer to a more loosely stated concept, such as the generation of a desirable type of countryside. These targets may or may not be a joint product with the traditional farm outputs of food and fibre. Within the EU, just under 3% of the CAP is allocated to AEPs intended to promote 'environmentally friendly' farming. These have been part of EU policy for over 15 years, following the creation of Environmentally Sensitive Areas (ESA) in 1986 (Robinson, 1991) and have been taken up on a voluntary basis. In particular, AEP has been part of agricultural reforms in Western Europe, in which output-related support has gradually been changed in favour of area-based payments and payments for the supply of environmental goods. Further moves in this direction have been made under the European Commission's *Agenda 2000*, which advocates more reductions in price support for arable crops, beef and sheep meat.

One example of an agri-environment scheme in England is the Countryside Stewardship Scheme (CSS), introduced in 1991, giving farmers outside ESAs opportunities to follow environmentally friendly practices, on a voluntary basis, if they have particular types of land deemed worthy of special protection and management (Simpson and Robinson, 2001). The broad objective of the scheme is to conserve and create landscape and wildlife features within specified landscape types and to enhance public access to the countryside. It has been described as a market approach to conservation (Bishop and Phillips, 1993) because of its reliance upon landowners competing to offer environmental services and only receiving payment if they are deemed to be offering value for money. A sister scheme, Tir Cymen, was introduced in 1993 in Wales (Banks and Marsden, 2000) and more recently a similar one, the Countryside Premium Scheme, has been launched in Scotland (Egdell, 2000). In all of these schemes, five-year management agreements entitle farmers and landowners to claim financial payments that are attached to particular items of work. Within the management agreements in CSS there are four major elements:

1. Land management measures supported by annual maintenance payments: These specify ways in which current land management practices, such as ploughing, grazing and mowing, are to be continued or adopted to produce conservation objectives.
2. Supplements for additional work: These enable land management measures to be achieved, for example cutting back docks and thistles prior to introduction of grazing animals.
3. Public/educational access: Under which agreements for new access provision may be negotiated.
4. Capital items: One-off payments, for instance to reinstate a length of hedge-bank or dry stone wall or fence against livestock, can be negotiated (Morris and Young, 1997).

Types of land eligible are chalk and limestone grassland, lowland heath, waterside, coast, uplands, historic landscapes, old orchards, unimproved areas of old meadow and pasture on neutral and acid soils, community forests, peri-urban countryside, and traditional boundaries including damaged and neglected hedges, walls, banks and ditches. Emphasis is placed upon applications that are deemed to offer positive changes and where land has a special conservation interest or adjoins such areas.

In terms of take-up, at the end of 1998, 143 055 ha had been enrolled in 8614 management agreements (an average of 16.61 ha per agreement). As shown in Figure 6.5, the largest numbers of agreements have tended to be in the uplands, with concentrations in the south-west. The highest densities of agreement per unit

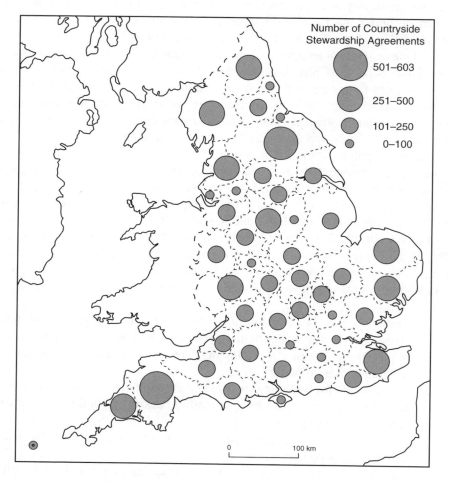

Figure 6.5 Take-up rates for the CSS in England (to Summer 2000). (After Simpson and Robinson, 2001.)

area of farmland have also been in upland areas and the west, notably the Peak District, parts of Devon, the Isle of Wight, Cumbria and west Cornwall (Evans and Morris, 1997).

The range of agri-environmental schemes, related legislation and codes of practice, coupled with consumer demand, all indicate that there is a movement within agriculture to move away from maximising production at any cost to realising that the environmental effects of farming must be taken into consideration, either at the producer level or later, as governments are faced with 'cleaning up' after farming. Hence the driving forces behind the move to more environmentally friendly farming are a growing section of consumers and government policy. Yet, one of the closest approximations to an environmentally sustainable farming system, organic farming, is not widespread in the developed world. As Table 6.4 shows, organic farming still only contributes to a small amount of the food market in Europe and America, despite its high profile in the media. For many consumers, the desire to promote more sustainable farming conflicts with the need to save money and buy the cheapest food, whatever the environmental burden associated with its production. Support for AEPs and organic farming is still a small percentage of the overall CAP in Europe. Despite the environmental concerns associated with farming systems, as yet, the effort and investment made to support a transition to more environmental farming systems is small.

6.6 Conclusion: the environment and the wider agri-food system

The above discussion has concentrated on some of the immediate impacts of agriculture upon the environment. Stress has been placed upon the types of influence exerted by the industrial model of modern farming, especially in the developed world, which is often driven by farmers' responses to government policy as well as reactions to prevailing economic imperatives. Increasingly, the latter have been part of globalising tendencies in which both food production and distribution have been radically restructured in favour of a more global scope and character, with TNCs playing an increasingly important role, especially in activities 'upstream' and 'downstream' from farms. This 'globalisation' is shaping our lives in profound cultural, ideological and economic ways (Goodman and Watts, 1997). Indeed, the concept of globalisation has become part of the standard vocabulary within the social sciences, with a general acceptance of the notion that we are experiencing a new and qualitatively different phase of capitalist development.

Food production and supply has always involved some element of international trade (Sylva, 1665; section 1.4). During the colonial period cash cropping in the tropics provided large amounts of imported goods (food, fibre, rubber) for the colonial powers. However, globalisation involves more than just ever-wider trading networks. At the heart of globalisation is the way in which the geographical outcomes of economic processes have increasingly become the function of links and dependencies that extend well beyond local, regional or even national environs. Instead they incorporate diverse, multi-faceted interactions between people and locations in many different parts of the world. This can perhaps be seen most readily in the sphere of consumption, where Coca Cola, McDonalds, Levi jeans or Microsoft computer operating systems, for example, offer an essentially uniform experience wherever they are consumed in the world (Ritzer, 1996).

Nevertheless, this uniformity is challenged in various ways, for example, the growth of a new culture of food consumption that centres on certain consumers' desires for a healthier diet and the rediscovery of traditional cuisines (Marsden, 1999). This cultural dimension to food consumption patterns represents a counter-current to the 'delocalisation' of the agri-food system associated with globalisation and the increasing similarity of lifestyles and habits in different parts of the world (Kuznesof, Tregear and Moxey, 1997). The presence of a local dimension to consumption patterns within the globalised trend that is represented in the presence of the same chains of restaurants in all major centres (such as McDonalds, KFC, Pizza Hut and Burger King) has been recognised as a process of 'relocalisation' (Murdoch and Miele, 1999) in which mass consumption patterns are mediated by local specificities. Amongst the characteristics of relocalisation are concerns for the place or region of origin of food as part of a desire for authenticity, greater variety and concerns over the standards of mass production and processing practices in the wake of various food 'scares', such as those relating to BSE, *E. coli* and GM organisms. It has also increased the amount of research that looks at the nature of the agri-food chain, generally treating it as a network of linkages from farmer to consumer (Figure 6.6) and acknowledging the importance of retailers and consumers in shaping the nature of farm-based production.

Throughout the discussion the hidden hand of globalisation has been implicit when considering many of the environmental impacts that occur on individual farms. However, as demonstrated when considering ecocentric approaches to agricultural development, there are some countervailing tendencies that are also helping to shape agriculture, primarily in terms of increasing recognition of the need to restrict the environmental externalities of any agricultural system by moving away from industrialised, globalised farming systems, and promoting environmental values and an alternative economy based on principles of sustainability. The latter has championed the setting aside of farmland from intensive cultivation (e.g. the payment-in-kind (PIK) program in the United States, set-aside schemes in both the EU and the USA, and Australia's Land Care scheme), promotion of environmentally friendly farming (e.g. ESA in the EU), and the adoption of sustainable agriculture practices (e.g. the growth of organic farming).

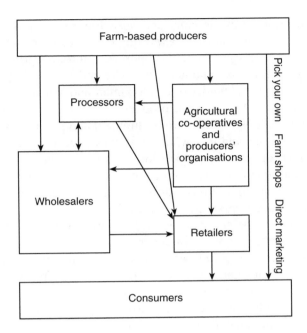

Figure 6.6 Simplified network of linkages in an agri-food system. (Reproduced with permission from Kaldis, 2002.)

However, in terms of government support there is no contest: encouragement of production remains overwhelmingly the main priority in the EU and the USA. Hence, assertions that farming in the developed world has entered a post-productivist era (or a third food regime) in which environmental concerns have assumed more significance, seem overstated (Robinson, 1997; Wilson, 2001; Evans, Morris and Winter, 2002).

One of the chief aspects of globalisation has been the increased role of TNCs in both the agri-input and the food processing industries, as part of vertical integration across the whole agri-food 'chain', and hence the need to examine the totality of the chain rather than just farm-based production, for example linkages between food retailers and processors, and their influence on the production sector. Furthermore, the closer relationships between changing consumption patterns of food and other sectors in the agri-food system requires more attention, for example the impacts on farming of major developments such as convenience frozen foods, contract production for supermarkets and media advertising. These have all contributed to the exposure of family-run farms worldwide to a wider range of external pressures, and have promoted new forms of production. As a result, certain parts of agriculture have been absorbed within more general urban-industrial structures and processes, and new industrial-style complexes have emerged, sometimes referred to as agribusinesses.

A good example of the latter are those TNCs who have expanded high-value food production in major agri-exporting countries in the developing world, utilising low-cost labour, government support (e.g. through structural adjustment programmes), globalised communications links and knowledge of the growing niche markets of the developed world (Goodman and Watts, 1997). This development is underpinned by intensive production techniques, with strong quality control from planting to storage, to trans-shipment and point of sale. This control has to satisfy corporate retailers who themselves need to meet the consumers' demand with respect to size, shape, colour and content. In establishing and implementing quality conditions a process of social and economic differentiation is reinforced in producing regions whereby smaller producers may be excluded from globalised food networks (Marsden, 1997).

A recent development in developing countries is the growth of high-value but perishable products for Western supermarkets. With the advent of low-cost air travel, farmers are also able to grow more perishable products as cash crops, such as fruit, vegetables and cut flowers for the European market. Often these cash cropping contracts are negotiated directly with supermarkets from the developed world. Such market-oriented farming brings benefits, but also makes farmers vulnerable to fluctuations in global commodity prices and changes in relative values of international currencies. Farmers producing products for distant markets may find they have little interest in consuming the product themselves, and therefore, should the market fail they will be left with nothing to eat. Globalisation of world trade and cheaper international transportation means that food can be transported around the world from wherever it is produced most cheaply to the market which consumes it. Is it sensible to ship food over such long distances (incurring large amounts of the so-called 'food miles') if it can be produced locally? Have the full environmental implications (e.g. transport, pollution tax) been taken into consideration when calculating prices? On the other hand, is it sensible to subsidise farmers in Europe and North America (arguably, in part, to maintain the appearance of the landscape and countryside) if food can be produced more cheaply in other parts of the world? Is it more environmentally friendly to consume locally (industrially) produced foodstuffs, or imported organically produced foods? Markets drive food production and the direction of food flows. Although global food shortages are cited as a justification for further research on increasing farming productivity, food shortages have usually more to do with unequal food distribution rather than global food shortage. Food consumption in the West is much higher than in developing countries. The ecological footprint of our food consumption patterns is high. Increasingly, TNCs and supermarkets are playing major roles in determining food production patterns. Are there standards (organic production, ethical trade, labour conditions, environmental effects) which should be adhered to in food production, at whatever cost to the consumer, or is low-cost food the priority, at whatever environmental or social cost?

Further reading

Tivy, J. (1990) *Agricultural Ecology*, Harlow: Longman.
This book provides a general introduction to the physical characteristics of the agri-ecosystem. It focuses on the way in which human modification of natural ecosystems has occurred via domestication of plants and animals, yielding less complex but highly specialised production systems.

Pretty, J. (1995) *Regenerating Agriculture: Politics and Practice for Sustainability and Self-reliance*, London: Earthscan.
This book examines ways in which less environmentally destructive farming systems can be obtained. It deals with the attempt to create more sustainable production systems through the introduction of non-industrial methods such as organic farming and the application of environmentally friendly practices.

Scoones, I. and Toulmin, C. (1999) *Policies for Soil Fertility Management in Africa*, London: Department for International Development.
This book provides examples of sustainable agriculture in an African context, focusing on management systems that maintain soil fertility – the basis for sustainability.

Bowler, I.R., Bryant, C.R. and Cocklin, C. (eds) (2002) *The Sustainability of Rural Systems: Geographical Interpretations*, Dordrecht: Kluwer.
A series of essays covering broader concerns about the sustainability of rural systems linking environmental considerations to the wider social, economic and political contexts.

Lappé, M. and Bailey, R. (1999) *Against the Grain: The Genetic Transformation of Global Agriculture*, London: Earthscan.
This book provides a critical assessment of the latest element in the 'industrial' model of farming–GM crops. This extends beyond environmental considerations to deal with the political and economic contexts within which the crops have been developed.

Goodman, D. and Watts, M.J. (eds) (1997) *Globalising Food: Agrarian Questions and Global Restructuring*, London and New York: Routledge.
The changing context of agricultural production is examined in more detail in a series of essays, with a focus on the processes of globalisation and the mass consumption of indus-trially produced foods.

Schlosser, E. (2002) *Fast Food Nation: The Dark Side of the All-American Meal*, New York: Perennial.
The role of American-based TNCs in globalisation processes is analysed.

Chapter 7
Meeting Society's Demand for Energy

Nick Petford

The major cause of the continued deterioration of the world environment is the unsustainable pattern of production and consumption.

Earth Summit, Rio de Janeiro, 1992

7.1 Introduction

In any society, access to energy is central to economic and social development and improved quality of life. However, most commentators accept that global patterns of energy use cannot be sustained for the current world population of c. 6 billion people. Given that energy consumption is expected to double in the next 50 years, meeting this level of global demand without unsustainable long-term damage to the environment represents a considerable challenge. Presently, three depletable resources – oil, coal and natural gas (fossil fuels) – provide approximately 90% of the industrialised world's energy needs. In contrast, biomass (fuel wood, crop and animal wastes) and physical labour remain the main energy sources in the developing world. The combined effects of burning fossil fuels and deforestation have potentially severe medium to long-term environmental consequences. Many of the more fundamental problems about how best to alter patterns of energy consumption amongst consumers (especially in the industrialised nations) are political in the sense that the rich nations must decide how far they are prepared to help. For example, these decisions may affect jobs in the developed world, as well as how we protect the environment and use natural resources in a sustainable way. It is a simple fact that the world is slowly but surely running out of its depletable energy reserves. Thus, irrespective of the environmental consequences of burning fossil fuels, sooner or later we will have to rely on other, non-depletable sources to meet the ever-growing

Global Environmental Issues.   Edited by Frances Harris
© 2004 John Wiley & Sons, Ltd   ISBNs: 0-470-84560-0 (HB); 0-470-84561-9 (PB)

demand for energy. Candidate sources include hydroelectricity and geothermal, wind and solar energy (Box 7.1 provides a discussion of nuclear power). These energy resources are renewable in the sense that continued use has little or no impact on their future availability. They are for all practical purposes infinite resources. A key issue is the impact of increased use of renewable energy sources on reducing CO_2 and other greenhouse gas emissions caused by burning fossil fuels.

This chapter begins by setting the geopolitical background to global energy issues. It is followed by a definition of energy according to the laws of thermodynamics and a review of patterns in global energy consumption including historical and future trends, highlighting current global discrepancies in energy usage. The geopolitical drivers behind global energy policy are identified along with some of the more important results of recent treaties to limit greenhouse gas emissions. The need to develop alternative or renewable energy sources (in particular solar and wind energy) and their role in combating climate change is followed by a discussion on the ethical issues surrounding global energy usage including local versus global needs, and the present inequality in energy production and consumer use. Throughout this review it is important to recognise that the energy debate involves two agendas with conflicting priorities and timescales. The first one is simply pragmatic, acknowledging that by definition, depletables are finite reserves that will one day run out. Replacement technologies for energy generation need to be developed, not as an immediate priority, but over the medium term. The second sees energy policy as the key response to climate change and should be largely driven by the urgencies of that challenge. The key link is to global warming, with the need to move away from dependence on non-renewable sources and towards diversity and security of supply as secondary concerns. Tied in with both are complicating sociological factors including the public perception of risk (especially concerning the use of nuclear energy), and issues relating to scientific uncertainty in, for example, climate change models.

7.2 Global energy issues: the geopolitical background

Energy is regarded widely as a major sustainable development challenge, with close links to the climate change and global poverty agendas. Economic development, in particular the demands placed on societies due to increasing energy consumption, and its relationship to environmental degradation was first addressed in 1972, at the UN Conference on the Human Environment held in Stockholm. After the conference, Governments set up the United Nations Environment Programme (UNEP), in an attempt to provide continued protection for the environment. However, some have argued that little was done in the succeeding years to integrate environmental concerns into national economic planning and decision-making. Campaigns by environmentalist groups and non-governmental organisations (NGO's) during the 1970s and 1980s played a major role in convincing a significant minority of the public in the industrialised nations that the deterioration of the environment was accelerating at an alarming rate, due largely to the combined self-interested policies of Governments and multinational companies, often operating in the developing world. At the same time, scientific studies were identifying other problems such as ozone depletion, global warming and water pollution that could be related to modern patterns of energy consumption, mostly in the industrialised nations. In 1983, the UN set up the World Commission on Environment and Development (WCED). It was here that the concept of sustainable development as that 'which meets the needs of the present without compromising the ability of future generations to meet their own needs' was proposed as an alternative to continuous economic growth. The UN Conference on Environment and Development (UNCED) in 1987 proposed that economic development must run hand in hand with social reform, and that the continued deterioration of the environment would only be prevented through global partnerships between the industrialised and the developing countries. After the 1992 Earth Summit in Rio de Janeiro, 108 governments represented by heads of state or government signed up to three major agreements aimed at changing the traditional approach to development. Amongst them was a wide ranging programme for global action in all areas of sustainable development known as Agenda 21 (section 7.9.1), aimed at preventing global climate change. Among its principles is the statement that individual states have a sovereign right to exploit their own resources but not to cause damage to the environment of other states. Agenda 21 contains detailed proposals for action in social and economic areas and for managing and conserving the earth's natural resources.

7.2.1 Kyoto Protocol

In December 1997, parties to the United Nations Framework Convention on Climate Change (UNFCCC) reached an agreement on a treaty called the Kyoto Protocol, named after the city in Japan hosting the meeting (see Box 2.2). The treaty requires the industrialised nations to reduce their emissions of greenhouse gases according to specific targets and timetables. While these limits vary from country to country, those for the key industrial powers of the United States, Japan and the European Union (EU) were set at 7, 6 and 8% below 1990 emissions respectively. The agreed first budget period will be 2008–2012 and emission targets include all six major greenhouse gases. Significantly, the Protocol allows nations with emission targets to trade greenhouse gas allowances. A market-based component of the Kyoto Protocol relevant especially to developing countries is the clean development mechanism (CDM) which introduces the idea of joint credit implementation, whereby developed countries can use certified emissions reductions from project activities to contribute to their compliance with greenhouse gas reduction targets. Companies will be able to reduce emissions at lower costs than they could at home, while developing countries will receive technologies that allow them to grow more sustainably. Using this mechanism, it is argued that countries can achieve reductions at the lowest cost. The Protocol advances the implementation by all Parties of their commitments under the 1992 Framework Convention on Climate Change. The Protocol also identifies various sectors including energy, where national programmes are needed to combat climate change. It should be noted, however, that the treaty is yet to be ratified.

7.2.2 WSSD Johannesburg, 2002

In August 2002, the second World Summit on Sustainable Development (WSSD) was held in Johannesburg, South Africa. The summit is notable for the fact that for the first time, issues surrounding energy consumption were officially on the agenda. A major outcome appears to be that governments have woken up to the fact that they cannot deliver on their commitments to targets (Agenda 21 and the Kyoto Protocol) alone, and that collaborative working is required between themselves and civil society to effectively progress sustainable development both nationally and internationally. There is also a realisation that full implementation of Kyoto will only slow the present rate of environmental degradation, not reverse it. For this to happen, cuts in greenhouse gas emissions of c. 60% are required, far in excess of the 1% currently being ratified. Russia, China and Mexico announced immediate ratification, with Canada voting towards the end of 2002. It seems that energy proved a controversial issue. Among the headlines relevant to energy are agreement on joint action to improve access to sustainable energy for 2 billion people who lack it, and agreement to phase out energy subsidies which inhibit sustainable development. Agreement was also reached on the need to increase substantially the share of sustainable renewable energy in the global energy mix. Despite this, it is disappointing that no clear targets were set for renewables, or reductions in global energy consumption. Some critics have even declared the final plan of implementation a backward step.

7.2.3 WEHAB

In May 2002, the UN secretary general, Koffi Anan identified five key areas described as critical global challenges of the 21st century. Along with energy, these are water and sanitation, health, agriculture and biodiversity, and are collectively known as WEHAB. According to the WEHAB paper (WSSD, 2002), 'although energy is not in itself a basic human need, it is critical for the fulfilment of all human needs'. These needs include heating, cooling, lighting, cooking, transport, communication and replacing physical labour. Crucially, energy consumption impacts on the rest of WEHAB in ways that are both subtle and overt.

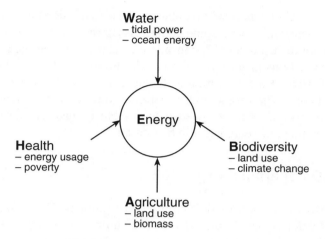

Figure 7.1 Summary of the United Nations' WEHAB (water, energy, health, agriculture and biodiversity) initiative showing some of the links between energy consumption and allied environmental concerns.

It is not hard to see how anthropogenically induced climate change can impact adversely on fragile ecosystems and hence biodiversity. But access to clean energy is also central to health issues in the developing world. For example, it is estimated that biomass (mainly wood and animal dung) provides the main source of energy for cooking and space heating for 2.5 billion people. Inhalation and respiratory disease resulting from smoke inhalation is believed to result in the deaths of 2 million people a year, mostly children under five (ITDG, 2003). Removal of wood from the land also reduces its fertility and increases soil erosion. These problems are further compounded where unsustained population growth places pressure on other natural resources such as water and sanitation. The links between energy and other WEHAB initiatives are summarised in Figure 7.1.

7.3 What is energy?

Energy is defined strictly in physics as the ability of a body to do work. Although it is common to refer to energy **consumption**, this is in fact a misnomer. It is not energy we are consuming (which is neither created nor destroyed), but its ability to do work. Some of the more common forms include kinetic, potential and mechanical energies. All forms of wave motion have energy. The SI unit of energy is the joule (symbol J), and is the work done by a force of 1 Newton moving 1 m in the direction of that force. Thousands of joules are prefixed by k (e.g. kJ or kilojoules), and millions of joules by M (e.g. MJ or megajoules). Examples of energy usage in these units are given in the following sections. Energy, and in particular its transfer and conservation, can be understood formally through the laws of thermodynamics. Imagine a closed system (a region of space of constant mass) with an internal energy, E_i. The change in energy between this and some final state, E_f, defines a measure of the work (W) needed to produce E_f. We can express this as an equation by writing:

$$E_f - E_i = \Delta E = W \tag{7.1}$$

This is the case for an adiabatic system, one which involves no exchange of heat (Sprackling, 1993). The first law of thermodynamics tells us something about the relationship between work and heat (Q). For systems which involve the transfer of heat (such as burning of wood), between the system and its surroundings, equation 7.1 can be rewritten as:

$$E_f - E_i = \Delta E = Q + W \tag{7.2}$$

In the CGS system, the calorie is defined as the amount of heat needed to raise 1 g of water by 1 °C; 1 calorie is equal to 4.186 J. The calorie is still widely used in nutrition, where the energy content of food is expressed as thousands of calories (kcal). For example, the amount of energy needed to sustain an average human is c. 2500 kcal (10 465 kJ) per day for an adult male and 2000 kcal (8372 kJ) per day for an adult female. Another measure of heat sometimes referred to in older UK and some current US texts is the British Thermal Unit (BTU), where 1 BTU = 1.005 kJ = 252 cal. For energy usage on a global scale, much larger units of measurement are required. One such unit is the quad, where 1 quad is equal to 1 quadrillion, or 10^{15}, BTU.

It is also important not to confuse energy with power. Power is defined as the rate at which work is done, and measured in watts, where 1 W is 1 J s^{-1}. Based on average metabolic rates, an average person uses the energy equivalent of a 120 W (120 J s^{-1}) light bulb for basic body maintenance. External power consumption by individuals is most commonly measured in kilowatts (kW). Regional consumption (a town or city) is measured in megawatts (MW) or gigawatts (1 GW = 1 billion watts) while annual global consumption and natural energy fluxes are measured in terawatts (10^{12} W). One million tonnes of oil burned gradually over one year would generate heat at an average rate of 1.33 GW. Finally, a common unit based on the watt that measures electrical

Table 7.1 Conversion factors and energy equivalents

Units of Measurement Energy

1 calorie (cal) = 4.186 Joules (J)
1 kilocalorie (kcal) = 4.187 kJ = 3.968 BTU
1 kilojoule (kJ) = 0.239 kcal = 0.948 BTU
1 British thermal unit (BTU) = 0.252 kcal = 1.055 kJ
1 kilowatt-hour (kWh) = 860 kcal = 3600 kJ = 3412 BTU
1 Quad = 10^{15} BTU = 1.055×10^{18} Joules

energy is the kilowatt hour (kWh), defined as the amount of work done when 1 kW of power is generated in 1 h. Remembering that there are 3600 s in 1 h, 1 kWh equals 3.6×10^6 J. Commonly used units and conversion factors dealing with energy and power are given in Table 7.1.

7.4 Global energy usage

A number of governmental and commercial organisations including the United Nations, the World Bank, International Energy Agency (IEA) and British Petroleum plc regularly publish data on global energy consumption patterns and energy reserves in combination with other socio-economic indicators. Such studies show consistently that the distribution of global energy usage is not divided equally amongst the nations of the world. Factors that determine energy usage relate directly to the state of economic development of a country (measured as a function of gross domestic product (GDP)), its geographical location and population. Some 80% of the world's population live in the developing world, the majority of which have at present little or no direct access to electricity. For example, the global energy *per capita* energy resource averaged over 1 year is (related closely to the manufacture of goods and provision of food) about 0.8 kW, while at the national level the USA tops the league at 10 kW, with Europe at 4 kW and Central Africa at 0.1 kW (100 times less than the US).

7.4.1 Present day energy consumption patterns

At present, some 90% of the industrialised world's energy needs are met using the fossil fuels oil, gas and coal. These natural resources, formed over geological time, are depletable in the sense that they reside in the earth as finite reserves. Estimates of how long they will last, based on current usage, vary from c. 50 to 500 years. Although new discoveries and improved extraction technologies may extend the life span of hydrocarbon reserves, the simple fact is that they will eventually run out. Figure 7.2 shows a map of global primary energy consumption *per capita* for the year 2002 (BP, 2003). The map is shaded according to tonnes of oil equivalent (toe), where one toe equals approximately 40 million BTU. One tonne of oil produces about 4 MW h of electricity in a modern power station. It is clear that the major oil users are the industrialised nations of the northern hemisphere (the exception being Saudi Arabia), and Australia and New Zealand in the south. Data show that the increase in world consumption of primary energy was marginal, growing by 0.3%. Taking hydrocarbons together, coal consumption grew by 1.7% (mainly due to increased demand in China), with oil and gas essentially flat. Consumption of nuclear power expanded by 2.8% globally, despite no real increase in generating capacity (Box 7.1), while hydroelectric power generation fell by 3.7%.

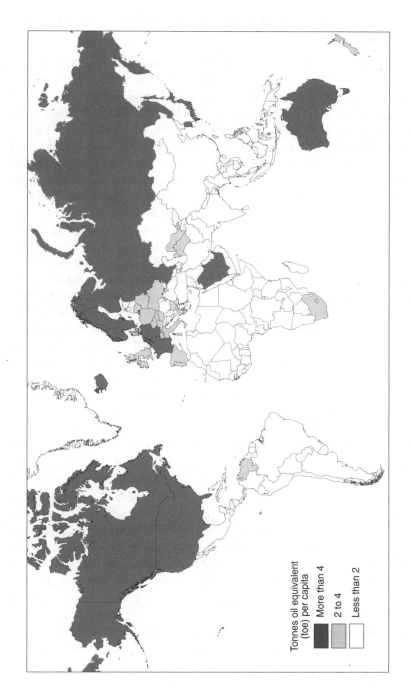

Figure 7.2 World map showing global primary energy consumption in tonnes of oil equivalent (toe). Note the dominance of the northern hemisphere. One toe equals approximately 10 million kilocalories, 42 GJ, 40 million BTU or 12 MW h. (Reproduced with permission from BP, 2003, published by BP plc.)

Tonnes oil equivalent (toe) per capita

More than 4

2 to 4

Less than 2

Box 7.1: Nuclear power – costly and harmful or power for a new age?

Nuclear fission was first discovered in 1939. Fission reactions involve bombarding the nuclei of heavy atomic elements such as uranium with neutrons, causing them to split (or fission) into two or more lighter elements, releasing large amounts of energy in the process. Nuclear reactors are fuelled by uranium, a naturally occurring metal and depletable resource, or plutonium and other artificial by-products of the weapons industry. Using just a small amount of fissionable material, bombardment of the enriched isotope ^{235}U by thermal neutrons can result in a chain reaction that if contained can generate large amounts of energy that can be used to generate electricity. The first public demonstration of nuclear energy was not for peaceful purposes, but after the Second World War commercial nuclear power generation was being hailed by the industrialised countries as a means of providing energy that would be 'too cheap to meter'. Nuclear power is a significant source of carbon-free energy, and as such meets one of the main requirements of the global change lobby in that its continued use does not contribute significantly to atmospheric greenhouse gas emissions. However, a by-product of the fission process is the generation of medium-to long-lived radioactive waste, the high-level form of which is extremely hazardous and long-lived. Although the volumetric amounts of highly radioactive waste are relatively small, after a number of high profile reactor incidents in the 1970s and 1980s, culminating in the 1986 Chernobyl disaster, public opinion, encouraged by anti-nuclear environmentalists, swung against the use of nuclear power on safety grounds. European countries including Belgium and Germany have banned future plant building, while once enthusiastic countries in Asia, including Taiwan and Japan, have scaled back plans for new plant (*The Economist*, 2001). Other countries, including the US and South Africa, are considering new build. So what is the future for nuclear (fission) energy? Commentators are split. Some continue to argue that in the long term, nuclear power must inevitably play a major role in the global energy mix. Others, driven primarily by environmental concerns, argue that no new nuclear power plants should be built until the problem of safe disposal and management of radioactive waste is solved. Even if safe storage can be demonstrated scientifically, the general public may still chose not to accept it. A solution is not a solution unless it is politically acceptable. However, if renewable energies cannot be developed on the scale required to reduce dramatically greenhouse gas emissions, and new technologies involving CO_2 sequestration prove ineffective or overtly expensive, the case for nuclear power, despite its high capital and environmental cost, will be considerably strengthened.

Continued on page 163

Continued from page 162

Global consumption of nuclear (fission) energy has increased from 100 mte in 1976 to 600 mte in 2001. North America and Europe together account for over 70% of the global share followed by Asia Pacific (19.1%) and the FSU (8.5%). Consumption in 2001 rose by 2.8% in all regions except Africa. However, despite the fact that a GW capacity nuclear power station can in principle be built and up and running in 18 months, plant orders are dropping as old power stations reach the end of their lives. A key factor in the success of nuclear power generation has been the generous subsidies granted to energy companies, with critics pointing out that the technology has enjoyed decades of support and subsidies in research, development and operation. Indeed, when the costs of decommissioning old plant and waste disposal are taken into account, nuclear power generation becomes largely uneconomic under present free-market conditions. Advocates claim that despite its bad public image, nuclear power is heading for a renaissance, pointing out that a new generation of reactors will be smaller, safer and more cost-effective (60% efficient by 2030), and will lead the way to a fully fledged hydrogen-energy economy by the end of the 21st century.

Figure 7.3 gives a breakdown of present day (2001) global consumption patterns of primary energy. Primary energy sources are defined as depletable fossil fuels (oil, gas and coal), plus nuclear energy and hydroelectricity. The latter is also a renewable energy resource (section 7.5). When broken down by region, some clear differences in consumption become apparent. Oil remains the single

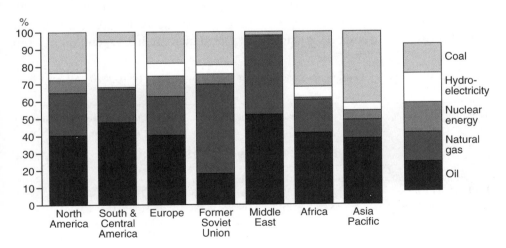

Figure 7.3 Global energy consumption patterns of primary energy by region, 2001, showing oil as the single largest source of energy, but with coal and natural gas marginally ahead in Asia Pacific and the FSU respectively. (Reproduced with permission from BP, 2003, published by BP plc.)

largest source of energy in most of the world, the exceptions being the former Soviet Union (FSU) where natural gas accounts for over 50%, and Asia (Pacific) where coal (40%) is just ahead of oil. Consumption patterns in the industrialised world are similar, although Europe has a slightly greater reliance on nuclear and hydroelectric energy at the expense of coal than North America. The use of nuclear power is variable across the regions, with Europe the largest consumer (c. 12%) and Africa, the Middle East (dominated more or less in equal amounts by oil and gas), and South and Central America the lowest. Hydroelectricity, the only renewable energy resource for which comprehensive data exist, has a variable consumption pattern, accounting for c. 25% in South and Central America, and similar but lower levels of usage (<10%) elsewhere. The projected capacity of other renewable energy sources, including wind, solar and geothermal, is discussed in section 7.5.

7.4.2 Predicted future energy consumption patterns

Data from the International Energy Outlook (IEO) summarising historical and predicted world energy consumption and predicted CO_2 emissions to 2020 are shown in Table 7.2. The data, summarised by region in Figures 7.4a and b, reveal a number of clear trends in energy consumption and emissions that are likely to have serious medium-term environmental and geopolitical impacts. The most obvious is that total world energy consumption is set to increase by over one-third from present day levels (c. 381 quads in 1999) to 612 quads by 2020. Extrapolation of the IEO data shows that the developing countries (Figure 7.4a) are set to overtake the industrialised nations in energy consumption by 2030, with global consumption reaching an estimated 1000 quads by 2050. Coincident with this rise in consumption is a predicted global increase in CO_2 emissions from

Table 7.2 World energy consumption and CO_2 emissions by region, 1990–2020 (Adapted from IEO, 2002)

Region	Energy consumption				CO_2 emissions (million tonnes carbon equivalent)			
	1990	1999	2010	2020	1990	1999	2010	2020
Industrialised countries	182.7	209.7	246.6	277.8	2849	3129	3692	4169
EE/FSU	76.3	50.4	61.8	73.4	1337	810	978	1139
Developing countries	87.2	121.8	184.1	260.3	1641	2158	3241	4542
Asia	51.0	70.9	113.9	162.2	1053	1361	2139	3017
Middle East	13.1	19.3	26.3	34.8	231	330	439	566
Africa	9.3	11.8	15.7	20.3	179	218	287	365
Central and South America	13.7	19.8	28.3	43.1	178	249	377	595
Total World	346.2	381.9	492.6	611.5	5827	6097	7910	9850

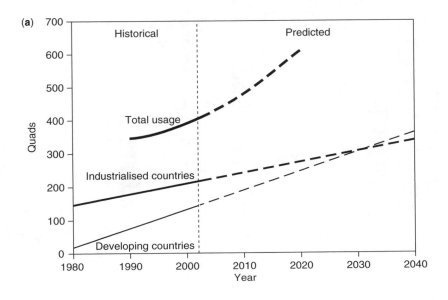

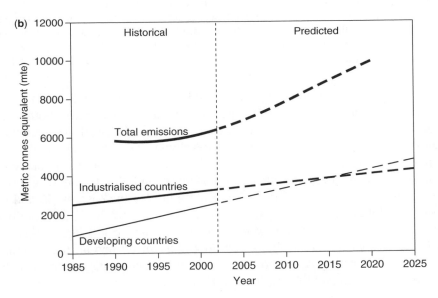

Figure 7.4 (a) Curves showing historical and predicted energy consumption (quads) over the period 1990–2040 for the industrialised and developing countries. Total usage is also shown. (b) Historical and predicted CO_2 emissions expressed as metric tonnes carbon equivalent (MTE) for the period 1990–2025. (Source: IEO, 2002; Table 7.2.)

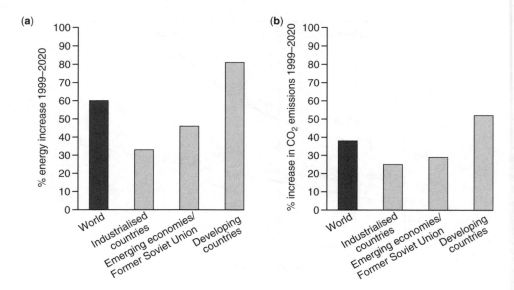

Figure 7.5 (a) Predicted percentage increase in global energy production by region, 1999 to 2020. The biggest percentage increase is in the developing countries, with Asia dominating. The global average increase is estimated at 60% over the period. (b) Percentage increase in CO_2 by region (1999–2020). Biggest predicted increases are seen in the developing world (55%), again with Asia dominating. The world average increase over the period is c. 40%. (Source: IEO, 2002; Table 7.2.)

c. 6000 million tonnes carbon equivalent (1999) to nearly 10 000 million tonnes carbon equivalent by 2020, a 61% increase (Figure 7.4b). While both energy consumption and CO_2 emissions are predicted to rise in the developed nations, the greatest increase will take place in the developing countries, especially Asia, where energy consumption and CO_2 emissions are set to more than double over the next 20 years, followed closely by Central and South America. Energy consumption and CO_2 emissions in the FSU will also increase by 46 and 41% respectively by 2020. The predictions also suggest that by 2015, the developing countries will overtake the industrialised nations as the world's largest producers of CO_2 emissions (Figure 7.4b). When broken down by region, the developing countries, dominated by Asia, show the single biggest percentage increase in both energy consumption and CO_2 emissions (Figure 7.5).

7.4.3 How much fossil fuel is left?

Global consumption of all the three fossil fuels is set to increase over the next two decades, with oil consumption rising by 60%, from 75 to 120 million barrels per day. Much of this demand will be in Asia, with increased consumption used for transport. Issues relating to transport and alternative fuels are discussed in section 7.9.2. Estimates of global reserves of (depletable) fossil fuels vary, but proved reserves of oil (at the end of 2001 exceed 1000 million barrels, of which 65% resides in the Middle East), 165 trillion m³ of natural gas and over 800 billion tonnes of coal (BP, 2003). Rather than looking just at the total reserve, it is often more instructive to consider the ratio of the reserves (the amount of resource yet to be exploited) to present-day production. A plot of global fossil fuel reserves to production (R/P ratio) is shown in Figure 7.6 for the year 2001. Although differences emerge in both the total amount of reserves and their ratios as a function of geography, it is clear that one depletable resource amongst all others, coal, dominates global hydrocarbon reserves (note, however, that oil shale and oil sand deposits are not included in the projections). On current estimates, the ratio of the world's reserves to production for coal is four times that of natural gas and six times that of oil. In the FSU this ratio is higher still, with the coal reserves to production ratio stable until 2500. China and India also have substantial reserves. Unfortunately, this situation is far from desirable. Coal is the most environmentally damaging of the fossil fuels. All aspects of coal power, from mining to transport and to burning to make electricity, are in different ways problematic. When burned, coal emits more CO_2, airborne particulates and sulphur and nitrogen oxides per energy unit than oil or gas combined. Sulphur dioxide has caused severe acid rain problems in parts of the industrialised and developing world, and the large volumes of waste ash often contain high levels of toxic metals.

Finally, this section would not be complete without mention of the more exotic sources of hydrocarbons including coalbed gas, shalebed gas and gas hydrates (marine clathrates). The latter alone exceed the estimated worldwide reserves of conventional gas reserves. Although production is complicated due

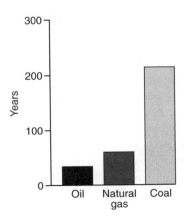

Figure 7.6 Reserves/Production (R/P) ratio by region. The ratio is constructed by dividing remaining reserves by production in a given year. The result is the length of time that those remaining reserves would last if production were to continue at that level. On present day (2001) estimates, the world has approximately 50 years of remaining production of oil and natural gas, and just over 200 years of coal. (Adapted from BP, 2003.)

to the hostile (deep marine) environment in which deposits are found, and also the material properties of gas hydrates, they are attracting considerable interest as primary fuel sources (e.g. Law and Curtis, 2002).

7.4.4 Security of supply (a problem for the industrialised world)

A major problem with the continued dependence of the developed countries on oil relates to the issue of security of supply. As mentioned above, the major proved reserves of oil (c. 686 thousand million barrels, R/P ratio of c. 85 years), along with 36% of global proved reserves of natural gas (as of 2001), are located in the Middle East, a region presently troubled by threat of war and continued tensions between Arab states and Israel. Against a backdrop of ongoing political unrest, some Arab countries have proposed a block on oil exports to the West in protest against British and US foreign policies. As demonstrated in the oil crisis of the mid-1970s, any large reduction in oil supply to the West would have severe consequences for the economies of the industrialised countries.

The price of oil tends to surge to high levels during times of international conflict (Brent crude reached US $40 per barrel during the 1991 Gulf War). The Organisation of Petroleum Exporting Countries (OPEC), the cartel of oil producing nations (which includes six Middle Eastern countries including Iran), can offset high price rises to some extent by increasing production. Currently this will happen if crude oil prices breach their current maximum target of US $28 for more than 20 days. However, increasing oil prices have a destabilising impact on the global economy, especially so where national economies are fragile. The potential for high oil prices to trigger global economic recession is very real, and relates directly to the current high global reliance on hydrocarbons, and oil in particular, as a mainstream provider of energy. Against this is the fact that the cost of providing energy to the world's poor need not be prohibitive. To light up the homes of 1.6 billion people with clean sustainable energy will cost in the region US $9 billion a year for ten years. This compares with the cost between US $250 and US $300 billion a year spent on subsidising fossil fuels and nuclear power.

7.5 Renewable energy

As we have seen, over the next 25 years, the global demand for energy is predicted to grow 2–3% per annum. To achieve this, a new electricity generating plant in the order of 3500 GW will be required. Although these are impressive statistics, there are real problems in deciding how best to achieve this growth in a sustainable fashion while at the same time reducing output of greenhouse gases in accord with international treaties such as Kyoto. This will only be achieved by exploiting alternative sources of energy in place of fossil fuels, which do not result in greenhouse gas emissions. The predicted increase in global demand will only be met by much wider use of renewable energy supplies combined with more efficient use of energy. A renewable energy resource can be defined broadly as 'one that is generated by sustainable energy fluxes operating within the atmosphere, hydrosphere, biosphere and solid earth. The primary source of these fluxes is solar radiation'. Renewable resources generated this way include wind, tidal and solar power, water (including groundwater) and biomass. An important condition is that human use has little or no bearing on future availability. The definition is, however, not universal, and some commentators exclude large-scale hydroelectricity from the list of renewables (Table 7.3), but include landfill gas (methane) and incinerated waste. Other more exotic forms include ocean thermal energy and space-based solar panels. In general, electricity is the best energy vector for mechanical- or solar-generated power. Nuclear energy, initially thought to be a renewable and clean energy source, has now been removed from this category due to the problems with decommissioning nuclear power stations and dealing with nuclear waste. It is dealt with separately in Box 7.1.

However defined, the key renewable sources of wind, solar and tidal power have the potential to provide in principle an infinite amount of energy, free at source, that does not relate directly to the fortuitous location of a sovereign nation above an oil field, or to the degree of industrialisation of a country (although the state of technological development does play a role). For example, each square metre of the habitable surface of the earth receives an average energy flux (a combination of solar, wind and other renewable forms) of some 500 W. A simple calculation shows that if all of this energy could be harnessed 100% efficiently, an area measuring 5×5 m would provide 12.5 kW. Even at c. 1% efficiency, the same area would yield slightly more power, assuming cost-effective extraction methods, than the yearly averaged value for Central Africa of 0.1 kW. Renewable energy sources can for convenience be split into six main categories: solar, wind, geothermal, hydroelectricity, biomass and tidal (Twidell and Weir, 1986). Although useful for accounting purposes, this distinction is somewhat artificial, as most are derived ultimately by energy from the sun. For example, the wind and tidal flows are driven by a combination of solar radiation and gravitational forces, while light from the sun powers photosynthesis in plants. Indirectly,

Table 7.3 Estimated government investment in renewable energy sources (US$ million) (Reproduced with permission from BP, 2003, published by BP plc)

Energy source	2001–2005	2006–2010	2011–2015	2001–2015
Solar water heaters	19.4	18.2	23.2	60.8
PV modules	16.4	8.5	3.2	28.1
Wind power	0.0	9.5	27.0	36.5
Hydroelectricity	10.3	9.1	1.4	20.9
Total	46.2	45.3	54.8	146.3

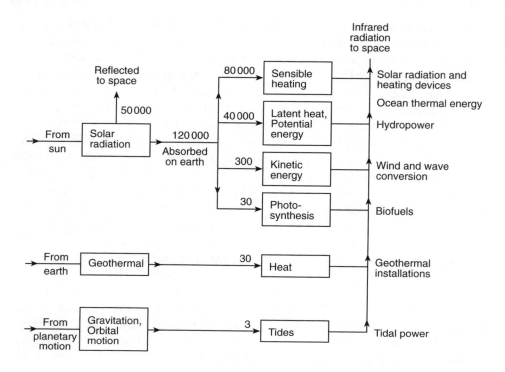

Figure 7.7 Summary of the global solar energy fluxes on earth in terawatts. (Reproduced with permission from Twidell and Weir, 1986; published by E and FN Spon.)

hydroelectricity requires water vapour in the atmosphere, which is derived from solar power (evaporation) at the surface. Only geothermal energy would operate for any length of time if the sun were to stop shining tomorrow (the standdown time of the earth's atmosphere is about two years). Solar radiation dominates the renewable energy budget available on earth. The total global solar flux incident at sea level is 1.2×10^{17} W (1.2×10^5 TW), providing some 30 MW per person on earth at current population levels. The critical natural energy fluxes and global renewable energy systems are summarised in Figure 7.7.

7.5.1 Hydroelectricity and geothermal energy

Some forms of commercial renewable energy including hydroelectric power and geothermal energy are now fairly well established and relatively mature. The IEO estimates that such energy accounted for around 2% of (commercial) primary energy in 1997. Global hydroelectricity consumption has doubled from c. 300 to 600 mte over the period 1976–2000. However, world consumption fell by 3.7% between 2000 and 2001 (Figure 7.8), due mainly to large drops in production in North America (−14%) and Brazil (−11.7%). As of 2001, Europe currently has the highest share of global hydroelectricity consumption at 24%. It remains to be

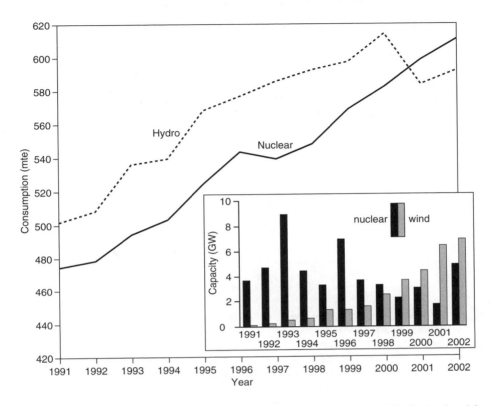

Figure 7.8 Comparison of the global consumption (million tonnes oil equivalent) of hydroelectricity and nuclear power from 1991 to 2001. Usage of both energy sources has increased by c. 20% over the last decade, but with a sharp drop in hydroelectricity consumption during 2001 due to reduced usage in North America and Brazil. Insert shows the changing patterns of consumption of nuclear and wind energies over the same period. (Reproduced with permission from BP, 2003, published by BP plc.)

seen if the 2001 reductions in hydroelectricity use are an anomaly or herald a more widespread downturn in global consumption patterns. It should also be noted that hydroelectricity generation has been criticised by some environmental groups on the grounds that the scale of engineering required to build the plant are damaging to both wildlife and local human populations, the latter sometimes forcibly removed from ancient settlements.

Although the average heat flow at the earth's surface is just 0.06 W m^{-2}, trivial in comparison with other renewable supplies at the surface which total approximately 500 W m^{-2}, geothermal power generation capacity was almost 6 GW in 1990 and expanded by a modest 3% p.a between 1990 and 2000 (Figure 7.9). Geothermal energy is a major contributor to power generation in countries such as the Philippines and Iceland, and also in regions of the United States (e.g. California), where the surface heat flow is high due to geological activity. Geothermal energy is generally of low quality, although

171

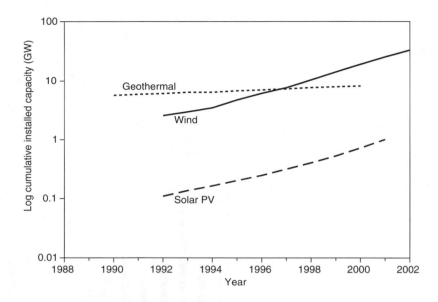

Figure 7.9 Plot showing the cumulative global installed capacity ($\log_{10}$GW) for geothermal, wind and solar renewable energy resources (OECD countries only) for the period 1990–2001. Wind energy (installed wind turbine capacity), shows the biggest increase over the period, from 2.15 GW in 1992 to 24.9 GW in 2001. Solar power (installed photovoltaic power) is also growing at an exponential rate although from a much lower base in comparison with wind (0.1 GW, 1992–0.7 GW, 2000). Growth in geothermal power is linear, showing only a modest increase over the same period. (Source: IEO, 2002.)

temperatures in excess of c. 150 °C can be used to generate electrical power from turbines. While there is potential to develop geothermal power elsewhere, its site-specific nature and reliance on favourable geological conditions is a restriction so that, unlike wind and solar energy, it is not universally available as an energy source.

7.5.2 Wind and solar power

Wind power and solar energy are relatively immature but fast growing. Their rapid growth is due to a combination of technological advances in turbine design and energy conversion systems and supportive policy from governments forced to respond to UN targets relating to climate change, namely reduced greenhouse gas emissions. Increased take-up in the industrialised world above has led to growing commercial markets and falling unit costs. As a result, wind power generation capacity has increased more than tenfold over the last decade (Figure 7.9), and since the beginning of the 21st century, wind power capacity additions have exceeded those of nuclear power, signalling wind's emergence as a mainstream energy source (Figure 7.8). Wind power has been used historically for milling and water pumping, but only since the 1930s have modern machines designed for electricity production in the kilowatt to several megawatt range been available. The global kinetic energy stored in the winds is some 1000 exajoules (10^{21} J), with wind velocity increasing in height above the surface. Available wind power scales with the cube of the wind velocity, so that a doubling of wind speed produces a factor of eight increase in power potential. Power generation from modern wind energy conversion systems requires a minimum wind speed in excess of c. 5 m s^{-1}, with an optimal velocity at 12 m s^{-1}. Currently there is approximately 25 GW of installed wind turbine capacity worldwide, with Europe, the largest consumer, at 18 GW (72% of the 2001 global share), followed by North America at 4.4 GW (18% of global share). Of the developing countries, India has the highest capacity at 1.4 GW with other Asian (including China) and Pacific countries, Southern and Central America and Africa contributing just over 12%. Clearly the future growth potential for wind power in these regions is enormous, while ambitious plans for wind farms in the industrialised world are well underway. In the UK, the windiest nation in Europe with over 60 wind farms producing enough power to run half a million homes a year, there are plans to build one of the largest wind farms in the world in the Thames estuary. Further plans for development include onshore and offshore plants in Scotland and Wales. At the current rate of development, wind energy could account for 8% of the total UK energy supply by 2010.

Wind power notwithstanding, the potential for solar power to meet society's need for energy cannot be underestimated, and usage appears to be growing exponentially (Figure 7.9). Solar energy is used currently either to heat water (solar thermal) or more widely to make electricity (photovoltaics). Photovoltaic (PV) devices, more commonly known as solar cells, work by converting light (photons) directly into electricity. They are silent (unlike wind), produce no emissions, and use no fuel other than sunlight, making it one of the most benign methods of power generation known today. Because photovoltaic devices are self-contained solid-state semiconductors made from silicon and other trace elements with no moving parts, once installed they need no maintenance other than an occasional cleaning (solar cells that contain storage batteries need maintenance similar to that required by a car battery).

Although as a naturally occurring element silicon is technically a depletable resource, it is after oxygen the second most abundant element in the earth's crust, and a major constituent of most naturally occurring planetary materials. As such, it is a resource that is available on cosmological timescales. Furthermore, a typical PV cell will regenerate the energy used in its manufacturing process in 1 to 4 years depending on its application and geographical location. As with wind, installed capacity has increased more than tenfold over the last decade, albeit from a very low base. At present, the PV capacity base is concentrated in a relatively small number of industrialised countries, with Japan, the USA and Germany accounting for about 80% of total installed capacity. For people in the developing world, especially in isolated regions, PV offers perhaps

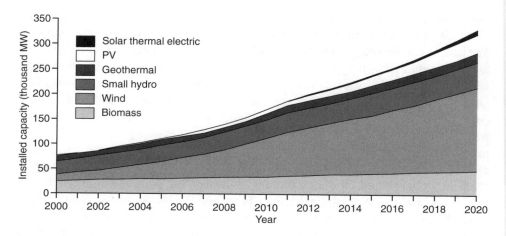

Figure 7.10 Predicted increase in installed capacity (megawatts) of six renewable energy sources (2000–2020). Wind and PV show the highest annual growth rates, with wind energy comprising approximately 50% of installed capacity by 2020. (Reproduced with permission from BP, 2003, published by BP plc.)

the most cost-effective way to supply basic essential needs such as lighting, water pumping, irrigation and refrigeration for vaccines and medications. It is anticipated that within the next 5 to 10 years, PV will become cost competitive with traditional power sources in countries with extensive electrical infrastructure.

It is also important to appreciate that renewable energy is important in ways other than beneficial environmental impact. Taking jobs and the wider economy as an example, we see that most renewable energy investments are spent on labour and materials to build and maintain the facilities as opposed to costly energy imports. Furthermore, in the aftermath of September 11, the issue of energy security is becoming more important, especially in the industrialised countries. As noted earlier, much of the world's oil and gas reserves are in the Middle East, a region currently under considerable political tension. After the oil supply disruptions of the early 1970s, the western world actually increased its dependence on foreign oil supplies. There is a case to be made that for the industrialised nations, this increased dependence is not in their long-term interest. Predicted growth of the main renewable energy sources up to 2020 is summarised in Figure 7.10.

7.5.3 Tidal power and ocean thermal energy

The height of water in the oceans rises and falls in a predictable way, and contains potential energy. Both tidal height and associated tidal flow can be used to generate energy. The global energy dissipation by tidal flows is estimated as 3000 GW, with approximately one-third (c. 1000 GW) available for capture in accessible coastal regions. If seawater can be trapped at high tide by an artificial barrier or dam, it can be made to do work by driving turbines and generating electricity. Approximately 25 major world tidal sites have been identified, with a combined total power of about 120 GW, approximately 10% of the total global hydropower potential. One of the best-known plants of this kind is at La Rance in the Gulf of St Malo, France. Others include the Bay of Fundy, Newfoundland and Kislayaguba, Russia. They have a range of capacities between 200 and 400 MW. Movement of seawater between high and low tides also causes flow and tidal currents, analogous to wind, with tidal power per unit area proportional to the cube of the velocity, but with a density 1000 times that of air. The strongest currents occur near the coast and between islands. One possibility is to combine offshore wind farms with tidal current power generation so that motion of two fluids, air and water, is captured simultaneously by rotors above and below the water line. Schemes piloting underwater tidal farms are currently underway in suitable coastal areas around northwest Europe. Although the behaviour of the tides and tidal streams is well known around the globe due to their importance in navigation, tidal power is very site specific and confined entirely to coastal estuaries. Further problems relate to interference with shipping lanes and damage to fragile marine ecosystems. It thus seems that tidal power is best for local use only in coastal towns and ports, and for diesel-generated electricity on remote island communities.

In contrast to tidal flows, very large energy fluxes (50–70 kW m^{-1} width of oncoming wave) can occur in deepwater waves. Waves, and hence wave power, are created by the wind and are proportional to amplitude and period of wave motion. Waves thus store and transport wind energy over large distances of ocean. Numerous mechanical ways of capturing this energy have been proposed over many years, but have so far proved ineffective in surmounting the many difficulties in harnessing wave power. These include irregular patterns and amplitudes, destructive forces to machinery caused by storms, problems in fixing wave machines to the deep sea floor, and in transporting power to land.

Ocean thermal energy is based on the principle of heat storage and temperature gradients in the oceans. The oceans, covering 7/10th's of the earth's surface, collect most of the incoming solar radiation, and temperature contrasts between the upper layer and deeper waters can be in excess of 20 °C in tropical regions over a depth of 1000 m. An ocean thermal energy system is essentially a floating heat engine, comprised of a closed circuit of pipes circulating a working fluid that takes up heat from the warm water which upon expansion is used to drive a turbine. Although the idea might sound a little outlandish, in fact no new technology is required to generate energy this way, and the thermodynamics is well understood. The Japanese have already built experimental generators on land, and possible sites include the sea around Hawaii and the Gulf Stream of Florida. Disadvantages include cost and scale, with the expense of maintaining the systems and transporting energy onshore, and biofouling by marine organisms that act to reduce heat transfer efficiency over time. As with other sea-based renewable sources, power generation is site-specific, and the full scale of potential has yet to be determined.

7.5.4 Biomass

Approximately half of the world's population relies on non-commercial biomass to provide for its basic energy needs of space heating and cooking. Wood, from trees and bushes, charcoal, crop residues and animal waste together constitute biomass. Bagasse, a by-product of sugar cane milling, is a major energy source in many developing countries (e.g. Peru, Box 7.2). Unlike other primary and renewable sources of energy, biomass is not traded therefore estimates of global

---

Box 7.2: Peru case study: energy and poverty

Peru is the fourth largest country in South America. It has a population of 26 million, over one-third of which live in the capital, Lima. The country is divided into three distinct climatic and geographical regions: a coastal desert, the high Andes mountains and the extensive Amazon basin. Following a decade of severe political crisis and social unrest in the 1980s, the number of Peruvians living in poverty increased from 42 to 55% (some 14 million people), with over half of this portion being the rural population. Fifteen per cent of the national population (3.9 million people) are classified as living in extreme poverty (ITDG, 2002). In an effort to reduce poverty and improve living standards, the Peruvian Government has created a fund (the National Fund for Social Compensation) as a focal point for public sector investment in poverty alleviation. The development and use of renewable energy resources has been championed as a major initiative in helping fight poverty and in encouraging sustainable development in rural areas. Although Peru is well endowed with a climate and geography that make extensive use of renewable energy sources a most attractive option, it is still largely untapped. The barren coastal desert is mainly flat and close to sea level. With an average solar insolation of between 4 and 5 kWh m^{-2} per day, Peru has a high potential for solar energy. Photovoltaic, module-operated telephone systems have already been installed by Telefonica, a Spanish-owned telecommunications company. Socio-cultural factors have been shown to be extremely important in determining the success or failure of projects, with solar cookers proving unpopular as they are unsuited to local cooking methods. Solar drying of food and use of greenhouses is growing, but mainly where foreign technical support is available. The Peruvian coast also has good wind energy potential, with average velocities of 6 to 8 m s^{-1}. There are currently two operational wind plants but there is clearly scope for more. Further inland, the high Andes mountains, with peaks up to 7000 m separated by deep canyons and gorges, provide ideal sites for local hydroelectric power generation. Estimates differ,

Continued on page 177

Continued from page 176

but there are currently between 50 and 250 hydroelectric power plants producing up to 500 kW locally to villages and small towns in the uplands. Geothermal energy is abundant in locations along the Western high Andes, with over 300 hot springs providing water at temperatures of 50 to 90 °C. Biomass, particularly firewood and Bagasse, a by-product of sugar cane milling, is a major energy source. It is a priority of the governments' energy policy to reduce the current dependence of the population on biomass.

biomass energy consumption are incomplete and highly variable. The best estimates are that biomass provides between 7 and 15% of the global energy requirement (Woodward, Place and Arbeit, 2000), with predictions of global usage showing only a slight increase in capacity up to 2020 (Figure 7.10). Although combustion is the most immediate process of energy extraction, fermentation, anaerobic digestion and gasification are also available. Biomass is less site-specific than other renewable energy resources such as wind, or tidal power, as vegetation of some kind can be grown almost anywhere. Furthermore, because it can be stored easily for later use, biomass does not share the same degree of non-dispatchability as, for example, solar energy (section 7.6). Despite this, biomass can only be considered a truly renewable resource if the rate of harvest does not exceed the rate of growth. Factors that act to curtail the use of biomass as an energy source include water availability, soil quality and weather (precipitation) patterns. More acutely, deforestation, in which gathering wood for basic energy services accounts for c. 50% of recent forest and woodland destruction globally, has a number of potentially serious environmental side effects including soil erosion and loss of biodiversity. Biomass energy also differs from other renewables in that, like coal, it is based on chemical reactions (photosynthesis) as opposed to mechanical energy. Like solar power, the primary source of biomass energy is the sun. However, plants are not particularly good at converting light into energy, with only about 1% of incident solar radiation being converted to carbohydrate in leaves, compared to c. 20% in PV cells. It may be possible to improve on this by genetically engineering plants to provide a higher calorific return during combustion, but any such technology will undoubtedly cause concern in some quarters.

7.6 Issues surrounding renewable energy

Developing renewable energies is not without potential environmental consequences and other problems. At a basic level, any component that is artificial will require energies as part of the manufacturing process. Sources of pollution come from emissions involved currently in the production of concrete, steel, glass and other materials required to collect wind, solar, geothermal and hydro-electric energies. Key technological issues for wind and solar energy production are the transient or intermittent nature of the resource (what if the wind stops blowing, or blows too fast?), stress-induced mechanical damage and how best to store the energy. Wind farms in some parts of the industrialised world have also been criticised as noisy and likely to blight the rural landscape. They can also pose a threat to wildlife, and have led to a number of bird kills. A more fundamental technical problem relates to the transport of energy generated at source. For example, energy is only useful if it is available when and where it is wanted. Storage and distribution requirements are different for renewable energy sources compared with most other primary supplies. Indeed, solar and wind energies are classified as non-dispatchable, meaning that it is difficult to make the electricity generated available to people far from the site of generation. In more advanced countries with a national grid set up to dispatch electricity from conventional power stations across the country, there are technical issues that must be met before dovetailing of intermittent electricity generation with the grid system can be done. In developing countries without such systems, it makes sense that energy production via the chosen method is local in the sense that it is produced to meet the immediate needs of those in a geographically small region. Indeed, it is possible to conceive that such a model may become attractive in the developed countries, as current practice of power generation through burning of fossil fuels in a small number of large power stations gives way to increased reliance on renewables. It is also true that a large share of renewable energy is non-commercial in the sense that its production and consumption do not involve a market transaction. Difficulties in collecting data for this sort of energy typically biomass and physical labour mean that it is not generally accounted for in conventional energy statistics, with obvious implications for organisations, such as the United Nations, tasked with target setting and monitoring.

Finally, while renewable energy sources such as wind and solar power are clearly of great benefit to much of the developing world, getting the right mix of energy sources in combination with the most suitable geographical locations continues to prove challenging. For example, in the same country, low-lying regions have little opportunity for harnessing hydroelectric power, but may have good wind potential. Also, while tropical rain forests have clear biomass potential, it would be foolish indeed to turn them into deserts by burning all the wood. Thus, in practical terms, the exploitation of renewable energy must be tailored to its local or regional environment. To this end, pilot studies are currently underway on the solar and wind resources of 13 developing countries (including Brazil, China, Ethiopia and Nepal) under the United Nations Environment Programme (UNEP), to help developing countries locate the best geographical areas to develop their renewable energy capacities.

7.7 Emergent technologies

Before finishing our discussion on renewable energy sources, it is worth considering two emerging technologies that, according to advocates, offer almost limitless potential for providing global energy requirements, with minimal environmental impact. These are the bold claims that might best be treated with caution, especially as they are still under development and their environmental and economic benefits thus untested. Both involve the use of hydrogen, the lightest and cosmically most abundant of the 92 naturally occurring elements. Other more futuristic energy solutions have also been proposed, involving satellite and lunar-based solar power stations that capture the sun's energy in space (Hoffert *et al.*, 2002).

7.7.1 Hydrogen fuel

In 1960, fewer than 4% of the world's population owned a car. This figure is set to rise to 15% by 2020, effectively increasing the number of automobiles from present day levels of 700 million to more than 1.1 billion. While the automobile has transformed the way people live, it is acknowledged within the car industry as an energy inefficient mode of transport. Despite continued advances in engineering design, the internal combustion engine is still only 20 to 25% efficient at converting fossil fuel energy into propulsion. Although better design has helped reduce exhaust emissions by 90% since the 1960s, the cumulative production of CO_2 (along with NO_2, SO_2 and CO) from individual vehicles continues to make a significant contribution to atmospheric loading of greenhouse gasses. Currently, 75% of all automobiles are located in the United States, Europe and Japan. Estimates of future growth suggest that most of this will be in the emerging economies of China, Brazil, India, Korea, Russia, Mexico and Thailand (Burns, McCormick and Borroni-Bird, 2002).

There is a clear imperative to manage this growth as far as possible in an environmentally sustainable way. A deeply attractive solution is the replacement of the internal combustion engine by hydrogen fuel cells. Hydrogen fuel cells use electrochemistry (as opposed to combustion) to convert gas into energy. The process is clean, but not totally benign emitting as exhaust air and water vapour, and has an efficiency of up to 55%. Hydrogen fuel cells thus offer a sustainable way of meeting the market demands for growth in private vehicles and other forms of road transport (including haulage), without compromising the environment by drawing even more on depletable fossil fuels. Their biggest hurdle is not technology, but convincing consumers that hydrogen fuel cell vehicles are not inferior to traditional engines in terms of reliability, performance and personal mobility. Indeed, large scale adoption of hydrogen propulsion may require a paradigm shift in public opinion that can only be brought about carrot-and-stick by a concerted effort on behalf of governments (through increased taxes on fossil fuels), combined with aggressive pricing by the automobile industry. The dilemma is that while extensive consumer uptake of hydrogen cells is contingent upon adequate availability of hydrogen fuel, the infrastructure to create this fuel will not be built unless demand is there in the first place. Furthermore, it takes energy to extract pure hydrogen from other substances such as water. This energy could be provided from renewable sources or nuclear power. Technological issues also surround the storage and distribution of hydrogen fuel. If we get it right, however, consumer adoption of the hydrogen fuel cell may open up the way for the development of a full-scale hydrogen economy.

7.7.2 Nuclear fusion

Nuclear fusion is a subatomic-scale process where two or more atoms of hydrogen fuse together under high temperatures to create helium, while at the same time liberating energy in accordance with Einstein's famous statement $E = mc^2$. The energy liberated during fusion of atomic nuclei is truly enormous – the energy released from the fusion of 1 kg of hydrogen into helium is approximately 10^{14} J, equivalent to burning 20 000 t of coal. Nuclear reactions of this type take place constantly inside the sun (and other stars) and provide the energy flux responsible for the entire earth's climate system and associated energy flows that support all life on the surface of the planet. However, unlike nuclear fission (Box 7.1), nuclear fusion offers the potential to provide clean, unlimited energy using deuterium (the heavy isotope of hydrogen) derived from water as a raw material, with none of the problems associated with the storage of radioactive wastes. But before fusion can compete economically with other energy sources, significant technical barriers must be overcome. For example, peak temperatures over 100 million °C are required to ignite a propagating fusion reaction, while at the same time containing the resulting plasma in a controlled environment. In the sun, gravitational forces provide this confinement. On earth, either magnetic or inertial forces must be used.

Although currently under development, inertial fusion energy (IFE) power plants will probably use steam turbines and generators similar to those used in most coal-fired power plants. Advocates claim that fusion power addresses the primary concerns for nuclear energy sources, while retaining the important environmental benefit of sparse use of natural resources. However, while offering a potential solution for industrialised nations with pre-existing electricity generating infrastructure, it is not clear how such technology, if developed successfully, can be harnessed easily by developing nations with no national grid. It could also be argued that philosophically it is simply an extension of existing practice of highly centralised production. What may work best for the developing world are more numerous, smaller and localised 'energy islands'.

While fusion may provide a long-term replacement of fossil fuels, the technology to achieve it is still several decades away, and it is unlikely that commercial plant will be operational before 2050. More pessimistically, there is a danger that fusion may actually exacerbate poverty in the developing nations if the installation and maintenance of IFEs is prohibitively expensive or requires an extremely specialised work force. We will simply have to wait and see.

7.8 Energy and climate change

The relationship between current energy production, based on burning of fossil fuels, is both complex and contentious (see Chapter 2). Many environmentalists take the view that the measured increase in atmospheric CO_2 is related directly to unsustainable and irresponsible energy policies pursued by the industrialised nations over the last century. It is now an indisputable fact that CO_2 levels are rising, and that almost all parts of the world recorded mean temperatures at the end of the 20th century significantly higher than when it began. Indeed, with global temperatures registering at $0.57\,°C$ above average, 2002 looks set to be one of the warmest years on record, with projected future estimates suggesting temperature rises in the order of $1-2\,°C$ by the end of the 21st century (IPCC, 2001a), far beyond the range of temperature increases experienced over the last 1000 years. It is now possible to distinguish natural variability from human-made change. The realisation that anthropogenic greenhouse gases, as opposed to natural factors such as solar radiation and volcanic eruptions during the last 50 years, have led directly to global warming has been at the cornerstone of recent treaties aimed at reducing emissions. To this end, the Intergovernment Panel on Climate Change (IPCC) has recommended that an atmospheric CO_2 concentration of 550 parts per million by volume (ppmv) should be regarded as an absolute limit and must not be exceeded. This value is approximately twice the pre-industrial level, and 180 ppmv below the present day level of c. 370 ppmv (Figure 7.11). The consequences of global warming are severe and include major variations in precipitation patterns resulting in drought and flooding, abrupt changes in regional average temperatures, and an accelerated rate of relative sea-level rise. If the present day patterns of energy use are responsible for these climate-moderating trajectories, then the case for mitigating action becomes highly compelling.

However, while detailed analysis of climate data over the past few hundred years shows that human activity is the most likely cause of recent global warming (Tett *et al.*, 2002), it is also clear from the geological record that a variety of naturally occurring processes including volcanic activity, hydrothermal outgassing and wetlands emissions can affect levels of atmospheric CO_2, water vapour, sulphur dioxide and methane. Furthermore, abrupt (but natural) changes

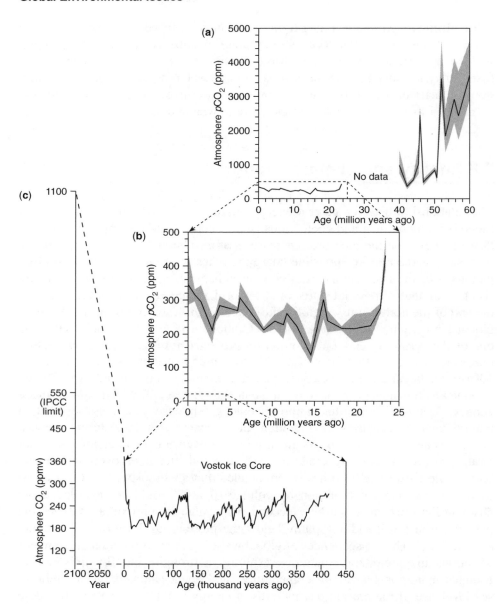

Figure 7.11 (a) Variation in atmospheric CO_2 levels from 60 Ma to the present day and (b) an expansion for the last 25 Ma, after Pearson and Palmer (2000). CO_2 levels for the past 450 ka recovered from the Vostok Ice Core are shown in (c), along with a prediction to the year 2100 based on anticipated perturbations (from Pederson, 2000, reproduced with permission from IGBP, 2000). The IPCC limit of 550 ppmv atmospheric CO_2 is shown for reference. Note the significantly higher values of atmospheric CO_2 in the Late Cretaceous period compared with present day and predicted future values.

in thermohaline circulation in the oceans, particularly the North Atlantic, produced marked changes in northern hemisphere temperatures during the Late Pleistocene. Recent developments in isotope geochemistry now enable geologists to chart these changes over tens of millions of years. The results are instructive. Recent investigations of the chemical composition of planktonic foraminifera from deep sea sediments show that from about 60 to 52 Myr ago (the Late Palaeocene and Early Eocene), atmospheric CO_2 levels were over four times the present day levels, at values in excess of 2000 ppm (Figure 7.11). Only since the Miocene (24 Myr ago) have values stabilised at below 500 ppm. Clearly, there have been times when the earth has been much warmer than today, with global sea levels hundreds of metres higher than now. But more alarmingly, the palaeoarchives also show that past variations in climate have occurred very quickly, often within several decades or less, and that the earth's climate system is sensitive to perturbations well within the range of predicted future anthropogenic forcing. Furthermore, recent studies pointing to periodic release of methane (a particularly potent greenhouse gas), into the atmosphere via submarine melting of methane hydrate (clathrates) buried in sediments along the continental slopes and on land in permafrost, add a new dimension to the threat of global warming (Kennett *et al.*, 2002). The northern wetlands of Siberia, long thought of as a carbon sink, have been shown to be sensitive to relatively small changes in surface temperature, so that a rise of 2 °C would increase methane emissions by an amount equivalent to 80% of the total anthropogenic greenhouse gas emission from the EU. Studies of atmospheric chemistry during the last decade have further identified the distributions of a number of pollutants related directly to energy use and burning of fossil fuels. For example, it appears that biomass burning and organic carbon from fossil fuels lead to net cooling, with radiative forcing by aerosols of comparable magnitude but opposite sign to the greenhouse gases (Bates and Scholes, 2002). The results of these studies are summarised in Figure 7.12, which show the complex relationship that exists between amplification and damping of the climate due to fluxes of different chemical species. Dramatic and sudden climate change will undoubtedly impact on modern biodiversity, which has itself arisen in response to past climatic conditions.

Thus, given the present state of knowledge of the climate system, the continued build up of CO_2 and other greenhouse gases in the atmosphere coincident with the burning of fossil fuels is unwelcome. Irrespective of definitive proof of an exact causal relationship, if, as is likely, the net result is to be potentially dangerous and destructive global climate change, then action must be taken. All the indicators suggest that such action will take the form of incentives, with the major emphasis placed on the governments of the industrialised nations to take the lead. Economic factors are also coming into play, with insurance companies becoming increasingly concerned at the number of claims for losses relating to increased flooding, droughts and storms resulting from climate change. Some possible approaches to the problem are set out below.

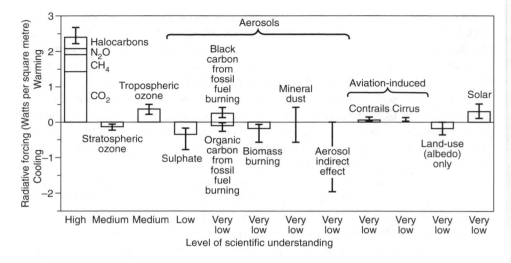

Figure 7.12 Plot showing the estimated radiative forcing (Wm^{-2}) of various greenhouse gases and particulates emitted as by-products of energy consumption and transport for the year 2000, relative to AD 1750. Note that in most cases, the level of scientific understanding of each substance on the global climate system is low. (Reproduced with permission from Bates and Scholes, 2002; published by IGBP.)

7.8.1 Reducing emissions: locking up and taxing

Following the 1997 Kyoto meeting of the parties to the UNFCCC, both the UK and the EU have set clear targets for CO_2 reduction and the increased contribution made by renewable resources in electricity generation. As an example, the UK is now committed to cutting its CO_2 emissions by 20% by 2010, although many commentators feel this is not enough, by a large margin. Concerned NGOs and some politicians inside the governments of the industrialised nations are pressing for much larger cuts averaging 60% of global greenhouse gas emissions under the banner of contraction and convergence (C&C), where the onus is on the developed world to make deeper reductions sooner. While such a reduction in greenhouse gas emissions at the point of emission is acknowledged by most as the best means of reducing emissions, other suggestions involving technological solutions for removing CO_2 from the atmosphere have been proposed. Nature does this most effectively by storing huge quantities in the oceans as limestone rocks. Trees and other vegetation including marine plankton lock up CO_2 as they grow, and release it when they die (or are burned for energy). Although it is sometimes claimed that planting trees will help alleviate the problem, such sentimentality is misplaced as at a global level, forest regrowth would only compensate fractionally towards reducing atmospheric levels. The priority should be on preventing further deforestation, along with other terrestrial carbon sinks such as soils and peat bogs. Recently there has been considerable interest in the idea of disposing of CO_2 in deep geological reservoirs, most typically sedimentary rocks that have previously been drilled for oil. Some preliminary tests on the feasibility of deep carbon sequestration are underway, but it is as yet too early to say if this method of disposal is technologically robust. Even if proven safe and cost-effective, one immediate drawback is that the technique can only be applied to large installations such as power stations, not to the larger share of emissions that in the industrialised world come collectively from homes and vehicles. Other options under consideration, promoted by interest in tradable carbon credits, include fertilisation of the ocean surface with iron, resulting in blooms of phytoplankton that remove CO_2 from the atmosphere, and injecting CO_2 directly into the deep ocean.

It must also be acknowledged that the price that consumers pay for fossil fuels does not reflect the true cost when the potential harm to the environment and climate is taken into account. This is especially true for air travel (section 7.9.2). It is also the case that in many industrial countries, governments heavily subsidise the energy sector. For example, in the US, ExxonMobil (the world's largest oil company) benefits from federal subsidies of c. US $25 billion per year that has the effect of distorting the true economical and environmental costs of energy production from fossil fuels. Similar 'perverse' subsidies also work in the nuclear industry. But it is also true that energy costs in the industrialised world have on average been falling since the 1970s, due in part to shifts in global pricing but also as the result of increased competition between energy suppliers following on from privatisation of the sector during the 1980s. The UK government is to introduce an energy tax called the *climate change levy*. It may be that methods aimed at reducing demand, such as higher fuel taxes, plus improved energy efficiency, especially in the design of new buildings and homes, will offer the best means of reducing emissions over the next few decades.

An example of how legislation is being introduced to combat CO_2 emissions is the EU Emissions Trading Scheme (EUETS). This commits the EU to cutting six greenhouse gas emissions by an average of 8% below the 1990 levels, between 2008 and 2012, thus meeting its commitments under the Kyoto Protocol. However, the EU facilities will be able to trade CO_2 allowances within the 15-nation block to help them meet their emission caps. Failure to meet targets will result in a fine for each tonne of excess CO_2. In summary, these treaties have elevated carbon to the status of a global currency. Greenhouse gas emissions are now a commodity with a market value that can be traded internationally in the same way as, for example, gold. It may be more advantageous for some energy producers to purchase carbon credits rather than invest in energy saving plant. It will be interesting to see how this free market approach to regulation develops over the next few decades.

7.9 Energy efficiency and conservation

7.9.1 Agenda 21

Given that radical scenarios for changing patterns of global energy use are unlikely to be implemented anytime soon, many commentators consider efficiency savings as the best means of reducing consumption and greenhouse gas emissions in the medium term. A major outcome of the 1992 World summit was an ambitious blueprint for sustainable development in the next century, Agenda 21 (section 10.5). This was aimed at promoting energy-efficient technology, alternative and renewable energy sources and sustainable transport systems (UNCED, 1992). Some developing countries are still highly dependent on income from the production, processing and export of fossil fuels, and it may be difficult for these countries to switch to renewables. To this end, Agenda 21 proposals include measures to identify economically viable, environmentally sound energy sources for sustainable development in developing countries, and the use of environmental impact assessments and other national measures for integrating energy, environmental and economic policy decisions sustainably. Development and transfer of technologies for energy efficiency, especially to modernise power systems, and for new and renewable energy systems are also recognised as priorities. Energy generation that involves conversion of different forms of energy by definition results in some loss of ability to do work. That these losses can be substantial is seen in the fact that less than 2% of primary energy produced in a coal-fired plant is converted into light in a regular light bulb via grid transmission. Most ends up lost as low temperature heat. Thus, even small improvements in both downstream and upstream energy efficiency may lead to significant environmental benefits over time.

On the wider scale, Agenda 21 also calls for regional co-ordination of energy plans and studies to determine the feasibility of distributing energy from new and renewable sources and nationally appropriate administrative, social and economic measures to improve energy efficiency. These measures must also involve appropriate national efficiency and emission standards by promoting technologies that minimise adverse environmental impacts. Policies are designed to ensure stricter energy saving standards in new building works, along with subsidies for insulation and installation of solar panels. In the industrialised world, product labelling, in co-operation with the private sector to inform decision-makers and consumers about opportunities for energy efficiency, is also desirable. Electrical goods including refrigerators and air-conditioning units are being designed that are more efficient than the required guidelines, and such good practice should be encouraged, although there is evidence that improvements in technology are commonly offset by an increase in consumer volume. Developing countries are encouraged to promote reforestation for biomass energy and increased use of solar, hydro and wind energy sources.

7.9.2 Transport

Critics say that the use of hydrocarbon fuel in automobile and aviation transport is not only wasteful, but also environmentally damaging. For example, road transportation alone accounts for 25% of all CO_2 emitted in Europe, while at the same time creating airborne particulates and other toxic emissions that are damaging to health. However, efficiency in transport, particularly cars, could result in big savings, with some estimates claiming that a 50% increase in fuel economy could be achieved over the next decade. The case for hydrogen fuel cells has already been made (section 7.7.1). Cheap air travel, especially short-haul flights, also been severely criticised due to their inefficient use of fuel. Environmental problems with increased air transport include vapour trails which can turn to cloud cover. A recent estimate suggests that by 2050, 10% of the sky could be covered by high-level (cirrus) cloud, and that aviation may contribute from 6 to 10% of all global warming in the next 50 years (Royal Commission on Environmental Pollution, 2002). Agenda 21 addresses transport issues by recommending that all countries adopt urban transport plans favouring high-occupancy public transport, non-motorised modes of transport and development patterns which reduce transport demand. It further recommends that in order to limit, reduce or control atmospheric emissions from the transport sector, traffic and transport systems should be better designed and managed. Cost-effective, more efficient, less polluting and safer rural and urban mass transport should be developed and promoted, along with environmentally sound road networks. Technology transfer, and the collection and exchange of relevant information, should be strengthened and national transport and settlement planning properly integrated.

What should we make of all this? In summary, it is easy to see that from an environmental perspective the proposals outlined in Agenda 21 are sound enough. However, in a practical sense they also seem rather glib. How is all this good practice to be implemented, and over what timescale? A successful outcome will be dependent not just upon accepting that change is needed, but in governments having the necessary political will to pursue aggressive and progressive policies that follow through from implementation to delivery.

7.10 Energy and society

An assumption so far unchallenged has been that progress, especially technological advance, is a good thing, and that people in developing countries have a right to increasing standards of living and quality of life that open and

187

continued access to better quality energy can provide. While there are some who would deny this assumption, it is hard for those less visionary to see how such reasoning would not result simply in a return to cave dwelling. This relationship between human development and energy use has been explored recently using UN statistics on longevity, educational attainment and standard of living as measured by a nation's GDP. The resultant indicator, called the Human Development Index (HDI), is shown plotted against annual *per capita* energy use for 60 countries containing some 90% of the global population (Figure 7.13). The data suggest that an annual energy consumption of c. 4000 KWh *per capita* is required for an HDI > 90%, comparable with the richest nations. While by no means a perfect correlation, the inference here is that to reach a global goal of basic human well-being will require a step change in world energy consumption in the developing nations (Benka, 2002). It is clear that governments in the industrialised world have a critical role to play in alerting consumers that their currently wasteful patterns of energy consumption

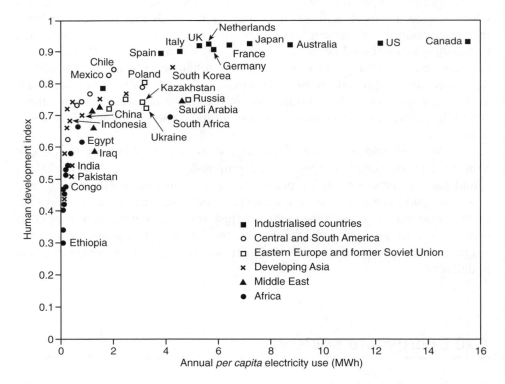

Figure 7.13 Plot of the Human Development Index (HDI) versus annual (*per capita*) electricity use in kilowatt hours (kWh). The data plateau out at approximately 4000 kWh. (Reproduced with permission from Benka, 2002; published by American Institute of Physics.)

are harmful to the environment, while at the same time convincing the energy industry that demand will continue to grow so that investment in new technologies capable of providing a strong renewable energy industry is maintained. We may also ask to what extent does responsibility lie with the individual or state in regulating global energy consumption. While some of the more fundamentalist environmental groups appear bent on making you feel guilty for boiling the kettle, there is good reason to believe that current attempts at promoting public interest in energy conservation in the developed world are largely failing. It is most likely that this is too important an issue to be left to the conscience of individuals. Western governments, with the help of local communities, multi-stakeholder partnerships and commerce as active partners, need to send out a convincing message on energy and climate change aimed at convincing citizens of viable long- and short-term objectives. Ironically, there is almost complete agreement between most governments, multinational oil companies and NGOs that greatly increasing the use of renewable energy is desirable, either now or in the medium term. This changing attitude is reflected in the recent statement by Royal Dutch Shell that it now regards itself as an energy company as opposed to an oil company, and British Petroleum's claim that the famous *BP* brand now stands for Beyond Petroleum. However, some powerful governments including the US appear largely unconvinced by any moves towards tighter regulations. Critics accuse the world's leading contributor of CO_2 emissions of paying lip service to environmental concerns while pursuing 'business as usual' policies with regard to energy consumption. Indeed, until very recently, ExxonMobil has been deeply sceptical of scientific claims relating climate change to energy generation, despite evidence to the contrary by, for example, the IPCC. Some of these large companies also have strong party political links with western governments. Other interest groups with the power to lobby the poorer nations against taking any radical decisions on energy policy include the OPEC countries, which would not benefit from any immediate switch over from oil and gas to renewables. It can also be argued that meeting and managing global energy demand is no longer just about saving the environment. Post-September 11, sustainable development, however delivered, promises a means of tackling the poverty and injustice in which terrorism breeds. To this end, implementation of strategies to provide access to clean energy for the world's poorest people will require a step change in attitudes and increase energy options available for sustainable development (Wackernagel *et al.*, 2002). An expansion of global renewable energy markets, particularly in the industrialised countries, would help to create economies of scale and also stimulate renewable energy markets in developing countries. The present imbalance in global energy consumption between the poor young billions and old rich millions reflects wider divisions that are unjust and some would argue as deeply immoral. It is thus encouraging that this link has now been recognised formally by the United Nations through initiatives such as WEHAB.

7.11 Conclusion

Global energy usage is currently bimodal, with hydrocarbons dominant in the industrialised world and biomass the chief energy source in the developing nations. Given that practically the entire energy infrastructure in the developed countries is set up for fossil fuels, even if stringent renewable energy initiatives were to be introduced immediately, the industrialised nations would still need to rely on hydrocarbons in the short term. Based on present-day production to reserves ratio, these energy sources will run out by the middle of the next century. A by-product of a depletable fuel economy is the release of greenhouse gases into the atmosphere. The recent geological record alerts us to the fact that greenhouse gases can promote rapid (decadal or less) climate change, possibly involving unforeseen feedbacks, and that we should anticipate sudden changes in climate that will be harmful to human societies and biodiversity. Arguably the most serious medium-term environmental consequence of global warming is rapid relative sea-level rise, with those in the developing world being particularly vulnerable (e.g. section 3.2). This problem is being recognised by governments as a pressing issue for both the industrialised and developing nations. The United Nations has taken the lead in rallying for change with treaties including Kyoto and Agenda 21 setting specific targets for the reduction of greenhouse gas emissions. So far, these have not been fully ratified. A range of alternative, clean energy sources could be drawn upon to help plug the gap left by a decreasing global dependence on fossil fuels or dwindling resources. These include wind, solar, tidal, hydroelectric and geothermal energy. Wind and solar energies are driven by terrestrial energy fluxes derived from the sun, and are in principle limitless. The removal of perverse subsidies, amounting to hundreds of millions of dollars, used to support fossil fuel extraction would help make investment in renewables more attractive to the energy sector. Nuclear energy, presently viewed with suspicion and fear by the public nevertheless offers a means of generating large amounts of power without harmful greenhouse gas emissions. Provided that issues relating to safety and waste disposal can be overcome, a medium-term solution to meeting society's growing demand for energy, while at the same time reducing the emission of potentially harmful greenhouse gases, might be an energy mix comprising all forms of traditional renewable energy, along with nuclear power. Nuclear fusion has the potential to provide unlimited clean energy from water, provided technical problems can be overcome. Hydrogen in another guise also offers a new way of providing energy through hydrogen fuel, perhaps making a transition from a fossil fuel to a hydrogen economy by the end of the 21st century. Anti-science environmentalists will have to accept that technological solutions, possibly involving large-scale planetary geoengineering, will play a major part in future energy solutions, and that market forces will be important in changing consumer behaviour. Governments, NGOs, civil society and multi-stakeholder partnerships all have an important role to play in making this happen.

Further reading

Heap, B. and Kent, J. (eds) (2000) *Towards Sustainable Consumption*: *A European Perspective*. London: The Royal Society.
A compact volume containing 17 short essays that bridge the gap between science and society in addressing some of the fundamental challenges faced by Europe as it moves towards a future based on sustainable consumption.

Twidell, J. and Weir, T. (1986) *Renewable Energy Resources*. London: E & FN Spon.
Excellent review of the basic science and technology underpinning renewable energy resources. Packed with facts, figures and calculations that bridge the gap between qualitative surveys of clean and sustainable energy and specialised engineering principles.

Ernst, W.G. (ed.) (2000) *Earth Systems: Processes and Issues*. Cambridge, UK: Cambridge University Press.
Good all round introduction into the relationship between the physical, biological and human sciences and their roles in providing a more integrated understanding of global environmental issues.

Physics Today, **55**, April 2002.
A special issue entitled *The Energy Challenge* that features five review articles on finding long-term solutions to meet the increasing global demand for energy.

Burns, L.D., McCormick, J.B. and Borroni-Bird, C.E. (2002) Untitled *Scientific American*, **10**, pp. 42–49.
An approachable introduction to hydrogen fuel cells as an alternative to the internal combustion engine in automobile transport, dealing also with wider issues of a sustainable, hydrogen-energy economy.

Hoffert, M.I. *et al*. (2002) Advanced technology paths to global climate stability: energy for a greenhouse planet. *Science*, **298**, 981–987.
State-of-the-art review of how advanced technology can help stabilise the global climate while at the same time meeting the increased global requirements for energy.

Part Four

Coping with our Impact

Chapter 8
Managing Urbanisation

Kenneth Lynch

More than half of humankind will live in urban areas by the end of the 20th century, and 60 per cent by 2020... In coming decades, most of the world's poor will be urban, living under conditions that can be worse than those of the rural poor.

World Resources Institute, 1996, p. ix

8.1 Introduction

8.1.1 The growth of cities

As the world entered the 20th century, an estimated 2% of the population lived in urban areas. As we enter the 21st century it is estimated that slightly more than 50% of the population is urban. The world is currently experiencing some of the fastest rates of city growth that history has seen. As we proceed further into the 21st century, all estimates of population change are that the increasing trend towards a more urbanised world will continue. During 2000, the global population changed from less than 50% urban to more than 50% urban. By 2010, it is estimated that the urban population will account for more than 60% of the world's population. Most of this change will take place in the world's poorest countries. It has been argued that 'the planet's future is an urban one and that the largest and fastest growing cities are primarily in developing countries' (Rakodi, 1997, p. 1). There is evidence to suggest that more affluent cities are able to transfer the environmental burden of their lifestyles away from them, while in poorer cities the environmental burden falls mainly on the population. The urban challenge of the near future will be to find ways of the poorest cities meeting their needs, in particular food, shelter, water, while avoiding the worst of the environmental hazards.

The problem of how to manage sustainable urbanisation is therefore an issue of global significance which will become more important as time goes on

Global Environmental Issues. Edited by Frances Harris
© 2004 John Wiley & Sons, Ltd   ISBNs: 0-470-84560-0 (HB); 0-470-84561-9 (PB)

and the world's population becomes more urbanised. This chapter therefore explores the key issues and approaches to understanding the modern concept of sustainable cities and what a sustainable city might be.

The general trends in urban growth show evidence of both historical and geographical variations. Early urbanisation came about as a result of concentration of agricultural surpluses that allowed the concentration of people who could focus their labours full-time on other activities, such as stone masonry, iron working, pottery or garment production. Before the beginning of the 20th century, transport and communications technology was limited and some have argued that this placed a limit on the ultimate size to which urban settlements could manageably grow. The evidence of this is in the relatively small size of the historic centres of many of the world's largest historic cities and the way in which the large cities of the world now incorporate a very large number of previously independent settlements. Indeed at one point, Bangkok was growing so fast that by the time new towns had been completed beyond its outer edge, the outward expansion had already made it yet another suburb (Rigg, 1991). That cities have grown in size is an observable fact. The implications of the growing size of large cities are a subject of intense dispute. Therefore planning and management of the environment and the provision of basic needs to the people living in cities are necessary.

Today, in advanced western economies, the vast majority of the population is based in settlements of some kind. In addition, of those not living in urban areas, a significant proportion live effectively urban lifestyles, even if in a non-urban location, but working, shopping, obtaining services from cities, such as education and health care, and spending leisure time in cities. At the other end of the spectrum in the world's poorest countries, the level of urbanisation is very low, typically 25–30%. For example, Potter and Lloyd-Evans (1998) report that East Africa's urban population grew from 22.9% in 1970 to 41.3% in 2000 and is projected to reach 57.8% by 2025. By contrast, South America already had achieved 60% urban population by 1970, but this grew to 81% by 2000 and is projected to grow to 87.5% by 2025. However, the countries with the lowest levels of currently urban-based population also have the highest rates of urban growth and urbanisation.

8.2 The nature and extent of the 'Urban'

8.2.1 The problem of definition and measurement

Before analysing the key issues of urban management, it is critical to assess the basis for analysis of these issues, the nature of the evidence (Pugh, 2000). Rees (1997) suggests that some see the city as the streets, buildings and other elements of the built environment, while others see the city as a political entity defined by the administrative boundary. Others again see the city as a concentration of cultural, social and educational facilities, while some may see the city as a node of exchange between individuals and corporations, an engine of economic activity. What is often forgotten in all this is that the city is also an ecological entity (Rees, 1997). It is important, therefore, when considering the manifestations of urban development in the built environment, economy or culture, that evidence is also sought for the changes to urban ecology.

There is concern about the reliability of the evidence on which analysis and management of urban environmental problems are based. The problem here is that there is growing concern about the level of accuracy of the raw data, but also of some of the past analyses of these data. This is illustrated by the mid-1970s UN projection of Mexico City's population in 2000 as 31.6 million, which was subsequently revised down to 25.8 million in the 1984–1985 assessment and again to 16.4 million in 1994 (Kjellén and McGranahan, 1997). In part, such projections are based on data that are often seriously compromised by inaccurate census data, by surveys of very large and rapidly growing urban populations and by assumptions about the nature of the population.

Such significant variations can also be accounted for by revisions of definitions of city boundaries. For example, there are wide variations in definition across the world about what constitutes 'urban'. In the main this definition focuses on the technical issue of the size of the settlement (usually measured by population). In this case there is much variation. In a review of definitions of how 147 nations define 'urban', Hardoy, Mitlin and Satterthwaite (2001) found that 21% used population thresholds, 34% used population thresholds combined with other criteria, while 12% specified those settlements that were considered urban. While a country's consistent definition allows for comparison over time, the variation in definition between countries makes cross-country comparisons extremely problematic because it is not possible to compare like with like. In addition, as has happened in China (Knight and Song, 2000) and Thailand (Rigg, 1998, 2001), countries may change their threshold size to reflect changing social and economic circumstances. Hardoy, Mitlin and Satterthwaite (2001) point out that most of India's population live in settlements of between 500 and 5000, which if re-classified from 'rural' to 'urban' would make India's population predominantly urban overnight. This makes temporal and international comparisons extremely difficult. Indeed it could be argued on this basis that too much attention is paid to population size as a defining urban characteristic and not enough attention to economic, political and social characteristics.

Another key concern relates to the issue of what defines the boundary between urban and rural. For example, with respect to southeast Asia, Rigg (1998) cautions against 'pigeon-holing' people into urban or rural categories. He outlines three main difficulties associated with such categorisation:

1. Registration records often do not detect changes in residence.
2. Allocating people to discrete categories such as 'urban' or 'rural' assumes that these categories accurately reflect their realities. Rigg's own empirical

research in Thailand (1998), among other developing countries, has demonstrated the significance of fluid, fragmented and multi-location households to survival strategies. This results in households straddling and moving across the rural–urban interface. Thus, categorisation of them as one or the other makes no sense. In addition, there is the problem of defining the boundaries of urban places which is usually based on some arbitrary definition.

3. Rigg argues that many Asian urban residents do not consider the cities and towns they live in as 'home'. This is because they ultimately intend to return to their rural origins. This, he argues, brings the issue of identities of the individuals into focus and ' "home" and "place" are ambiguous and shifting notions, where multiple identities – both – can be simultaneously embodied' (Rigg, 1998, p. 501).

These factors become important when considering how such issues affect people's vulnerability to environmental hazards and ability to help manage the environment. Lynch (in press), also writing about the developing world, adds a fourth concern to the three above, which refers to the blurred geographical distinction of the divide between what is 'urban' and 'rural'. This is particularly the case when cities physically grow rapidly outwards, advancing their physical extent into the countryside, as well as expanding their influence on urban lifestyles, values and systems of livelihood. Some writers like Iaquinta and Dreschler (2001) take this further arguing that it is not possible to distinguish clearly between urban and rural, but in fact they can identify five types of peri-urban interface with varying levels of rural or urban influence. They argue that identifying the category of interface is key to understanding the environmental and resource management issues. While they propose this framework for the improvement of environmental management of the peri-urban interface, this is also further support for the reservations about international comparisons of cities and generalisations about how to achieve a sustainable city.

8.2.2 The global-environmental dimension

When international agencies such as the World Bank rank the world's economies, those falling somewhere between the most and the least affluent find themselves to be similarly ranked in terms of levels of urbanisation. This correlation has tempted some to suggest that the reason for the poverty of the poorest countries is that the population is still located in rural areas. However, this can be considered in an alternative way: does wealth bring urbanisation or is urbanisation a necessity for the creation of wealth? This is illustrated starkly by Hardoy and Satterthwaite (1990), who drew comparison between Los Angeles and Calcutta approaching 1990, when they were both growing cities of approximately 10 million people. While the populations of both cities were growing quickly – in fact Los Angeles

Table 8.1 The world's largest cities in 1950, 2000 and projected for 2015 (Source: United Nations, 1995)

1950			2000			2015		
Rank	**City**	**Population**	**Rank**	**City**	**Population**	**Rank**	**City**	**Population**
1	New York	12.3	1	Tokyo	27.9	1	Tokyo	28.7
2	London	8.7	2	Bombay	18.1	2	Bombay	27.4
3	Tokyo	6.9	3	Sao Paulo	17.8	3	Lagos	24.4
4	Paris	5.4	4	Shanghai	17.2	4	Shanghai	23.4
5	Moscow	5.4	5	New York	16.6	5	Jakarta	21.2
6	Shanghai	5.3	6	Mexico City	16.4	6	Sao Paulo	20.8
7	Essen	5.3	7	Beijing	14.2	7	Karachi	20.6
8	Buenos Aires	5.0	8	Jakarta	14.1	8	Beijing	19.4
9	Chicago	4.9	9	Lagos	13.5	9	Dhaka	19.0
10	Calcutta	4.4	10	Los Angeles	13.1	10	Mexico City	18.8

was growing faster than Calcutta – the two cities' housing problems were far from comparable. It is therefore necessary to have a more in-depth analysis of the urban management issues than simply assuming it as a question of relative wealth.

One of the ways in which this issue is often illustrated is the data that exist on the rapidly growing cities that are among the largest in the world. As discussed earlier in this chapter, these data are problematic, but one of the constant themes that is clearly illustrated in Table 8.1 is how quickly the historic mega-cities, such as London, New York and Paris have been overtaken by cities of the developing world. However, this is by no means a homogenous trend, and this emphasises one of the key arguments of this chapter. Tokyo maintains its position in the top three, while Los Angeles makes an appearance in the middle column before dropping out of the top ten. One other key trend that is a feature of the global trend towards urbanisation is that the scale of the largest cities in the world has increased over the period, albeit less between the 2000 data and the projection for 2015.

Hardoy, Mitlin and Satterthwaite (2001) argue that it is only in the last ten years that the literature on sustainable development has contained a growing proportion of material that considers cities. It is likely that this is largely due to the fact that many cities are considered almost 'beyond help'. By contrast, Hardoy, Mitlin and Satterthwaite (2001) suggest that there are three main reasons why cities should be considered carefully:

1. A large and rapidly growing proportion of the world's population lives in cities.
2. The distribution of resource use and waste generation is not even. The vast majority of the world's resource use and waste generation is concentrated

in cities. The countries with the highest levels of resource use and waste generation per person are those with the highest levels of urban-based population.

3. Urban policies are very important to the development of policies which promote sustainable development more widely.

The main urban challenge of the future will be in the poorer countries of the world as they contain the cities which are growing fastest, with the least resources, and where sustainability is the greatest challenge. The World Commission on Environment and Development (WCED, 1987 – known as the Brundtland Commission) illustrated this by pointing out that at contemporary trends it was necessary for most developing country cities to increase their services and infrastructure by 65% simply to maintain the existing conditions. However, during this period most cities were to achieve this under conditions of considerable economic and social difficulty, leading the Brundtland Commission to conclude that, 'These projections put the urban challenge firmly in the developing countries' (p. 237). This situation has changed little in the time since the report was published. This chapter therefore focuses on issues relating to cities at an international scale, and the case studies are drawn from Africa, Asia and Latin America where city growth rates are among the fastest in the world, while the governments' and city authorities' abilities to manage them are limited.

8.3 Managing urbanisation sustainably

8.3.1 Current concerns

The current concern with sustainable human settlements is a focus which has developed as a result of an increasing awareness that the problems that face cities are largely global problems. In particular, attention has been drawn to the emergence of cities which have achieved sizes that are unprecedented. The fact that many of these cities are no longer in the world's wealthiest countries is giving cause to question ideas of modernity and development and their link with the role of cities. This change amongst the largest cities is illustrated in Table 8.1.

One positive outcome from the growing global awareness is the growing collaboration between cities across the world which has been increasingly multi-disciplinary.

Before sustainability took a hold, urban sociologists were interested in social structure, class, segregation and various aspects of inequality and poverty. These interests could be applied to housing, the allocation of land, relativities in incomes and access to various urban services. Economists had similar interests, but studied them in terms of theoretical explanation, technical appraisal, measurement and the costs and benefits of policy reform. Architects-planners had regard to macro-spatial form, building technologies, and with relevance to developing countries, some pioneers such as John F C Turner brought self-help housing into relevance for low-income housing policies. Urban geographers...had eclectic interests, often undertaking household questionnaire surveys and adding commentaries on housing, social conditions and urban development.

Pugh, 2000, p. 1

During the 1990s, as the issue of sustainability began to take greater prominence in all the social sciences and urban-related disciplines, there was increasing fusion of ideas as disciplines gradually focused on related concerns. Each discipline lends a useful perspective, but since the concept of sustainable development is largely an outcome of the negotiated relationships between the economic, the political, the social and the environmental disciplines, the need for inter-disciplinarity has become largely accepted.

Satterthwaite (1997) suggests that there is a set of broad categories against which cities can, and possibly should, be monitored. These he identifies as:

- Reducing the health burden on urban populations, by controlling infectious and parasitic diseases and reducing vulnerability to them.
- Reducing chemical and physical hazards at the household, workplace and city scale.
- Achieving high quality urban environment for all those living in the city.
- Minimising the transfer of environmental burdens to environments and communities beyond the city.
- Moving towards sustainable consumption.

The issue of governance – and city governments' role in improving such indicators – has therefore emerged as a key to the ability of cities to achieve sustainability. This is the case, not least because this makes cities more accountable to their residents and more sensitive to the environmental problems affecting them, as well as what they can contribute to their solution.

Although in the West much attention is focused on the growing size of the largest cities, even the smaller cities of the developing world are growing faster than what has been seen in history. For example, '...cities in Africa are growing faster than in any other region. Most of the increase is the result of migration, reflecting people's hopes of escaping rural privation more than actual opportunity in the cities' (United Nations Fund for Population, 1996). There is a growing concern about providing for rapidly growing cities in the developing world, where urban population growth rates are fastest. Much of the research on this issue focuses on the fact that growing urban populations are already outstripping services and infrastructure provision. This chapter will consider issues such as people, food provisioning, environmental hazards, water, sanitation and pollution. Implied within this discussion is the idea that very rapid city growth places enormous pressures on the sustainability of a city and this has implications for the 'growing ecological footprint' (Rees, 1992).

8.3.2 Ecological footprints

As cities grow and develop, their activities tend to become more modern and industrial, which results in an increased demand for raw materials such as food, energy and water. In fast growing cities this has meant that energy production may, to an extent, initially be based on traditional, rural methods and rely on rural supply sources such as wood and charcoal. Where modern power sources have been developed they can impact on rural areas. For example, the construction of dams to create a reservoir and hydro-electric power production plant often forces the relocation of rural populations as well as tapping and redirecting water sources. Rees (1992) developed a method of calculating the sum of waste and resource flows and then used this to calculate an area-based estimate of the land and water appropriated by the urban population to meet its needs. This he called the 'ecological footprint' (section 1.4). Application of this approach has suggested to Rees (1997) that an individual of a high income country uses an area of between three and seven hectares of ecological land. He gives the example of his home city of Vancouver in Canada, which in 1991 had a population of 472 000 occupying an area of 114 km^2. However, applying his method, he was able to calculate that the city required 4.3 ha *per capita* of land in order to support the consumption needs of those living in the city, giving a total requirement of 2 million ha of land. Subtracting the land area of the city from this total land consumption figure indicates that it has a land-based ecological foot print of 1.99 million ha, or 180 times its own land mass. Its marine-based needs are not counted here. One of Rees' collaborators, Wackernagel (1998), applied this technique to Santiago de Chile and found that the ecological footprints of individuals ranged from 0.4 ha per person for those in the lowest 10% income group, compared with 12 ha in the highest 10% income group. His results suggested an overall mean of 2.6 ha per person. This compares with a mean of 2.4 ha per person for the country of Chile. In the case of Chile, the country's ecologically productive land, combined with its long coastline and extensive forest resource, provides considerable additional capacity. However, the level of consumption in Chile is expanding at a faster rate than its population growth rates, as the country experiences economic growth and Chileans take on increasingly high consumption lifestyles that are prevalent in the cities of higher income economies (e.g. Figure 1.1 and related text). As this takes place the city of Santiago's ecological footprint will place an increasing burden on the country's resource base, and many cities experiencing similar growth and development do not have a wider resource base in the country on which to draw. This approach is intended to highlight the impact of cities on resources and ecosystems wherever they are on the earth: the idea that somewhere the ecological load of a city's consumption has to be borne.

An alternative view to this is presented by Main (1995) who questions what he calls the 'environmental demonology' of cities, presenting two main arguments. The first of these arguments is that while cities are certainly the cause of negative environmental impacts on rural people and environments, they are also the cause of positive impacts such as the provision of technologies, materials and increased livelihood opportunities which can benefit both urban and rural populations, and, in turn, environments. Satterthwaite (1997) argues that while it is not possible to quantify all ecological impacts of cities, it is also not possible to quantify the positive impacts that urban areas can have beyond their boundaries. For example, the environmental impacts of a city-based power station may be calculable, but it is difficult to quantify the environmental benefits of areas beyond the city switching from biomass or fossil fuels to electricity for their main energy source. Secondly, Main (1995) argues that such negative conceptualisation of cities often overlooks the fact that, had development to current levels of population and consumption taken place without urbanisation, then the environmental and other impacts would more likely have been far greater for both urban and rural populations. He argues that it is the issue of changes in technology and consumption patterns that are a more important determinant of environmental degradation than either growth or redistribution of the population. Indeed the movement of people to settlements may have facilitated an amelioration of adverse effects on rural environments.

Hardoy, Mitlin and Satterthwaite (2001) take this argument a little further arguing that there are potentially a series of environmental advantages or opportunities to be had from cities:

- Economies of scale and proximity to infrastructure and services.
- Reducing risks from natural disasters, through improved hazard management.
- Water re-use or recycling, through urban water management systems.
- Reduced demand for land, through densification and geographical concentration of populations and activities.
- Reduced heating, through reduced heat loss in terraces and apartment blocks and the opportunities of efficient energy provision through power stations.
- Reduced motor vehicle use. Despite problems of congestion, cities offer opportunities for increasing journeys by foot, cycle and public transport.
- Pollution control and management. Concentrating industrial activities facilitates regulation enforcement and pollution management services.
- Funding environmental management through ease of tax collection where populations and commercial concerns are concentrated.
- Governance: concentrated populations in cities facilitate involvement and mobilisation.
- Potential for reducing greenhouse gas emissions will be greatest in cities where economies of scale facilitate investments in improved technologies.

In most cases these are areas of **potential** that cities represent. For example the concentration of population provides opportunities for 'environmental economies of scale'. In most cases this relates to the declining relative costs of providing infrastructure or services as population density increases. However, increased population can also place a heavy burden on the infrastructure and can have adverse impacts on the city's environment and that of its surrounding areas.

There is therefore much still to be learned about the major environmental problems that face the rapidly growing cities of the developing world. Not least is the extent to which generalisations and comparisons can be made between such a large number of very different cities in different environmental and social contexts. The literature on the urban environmental problems of the developing world is voluminous, covering issues such as population, water, waste, food supply, pollution, energy, disasters, transport and housing. The next section reviews a selection of these issues, highlighting the key problems and some proposed solutions. Waste management and urban pollution are covered in Chapter 9. Other issues are covered in detail in some of the reading material suggested at the end of the chapter.

8.4 Challenges to sustainable urbanisation

8.4.1 People

The headline concern about the fastest growing cities relates to the rapid population growth of the cities of the poorest countries. Ultimately this concern is based on the difficulty of cities even in advanced economies to meet the needs of their citizens. However, the reason for the particular concern as we enter the 21st century is that we are living in a period of time when the world's poorest cities are growing at a rate that is faster than has ever been seen before (Potter and Lloyd-Evans, 1998). Not only this, but most of the city growth is taking place in economies that can least afford to invest in new city infrastructure and with limited employment opportunities for the people living there.

The rapid growth of cities is made up of two main geographic features. The first relates to the demographics of city growth, where the total number of people living in a city is increasing. This is known as urban growth. Urban growth is made up of two main elements: natural growth and migration. Natural growth occurs when the number of births exceeds the number of deaths. Migration is thought to be largely rural to urban; however, increasingly researchers are finding complex forms of migration taking place. For example, Pacione (2001d) summarises migration types as a continuum:

- Circular migration: Where migrants return periodically to their home village in their rural place of origin.
- Long-term migration: This may result from circular migration where migrants eventually settle in the city, particularly where longer term

employment makes frequent return visits expensive and places strain on the family ties.

- Family and return migration: Where the employment may become more secure and where there are opportunities for more household members to obtain employment, migrants may bring families to live with them and settle in the city in order to offset the cost of travelling. This is not necessarily seen as a permanent move and the household may intend to return to their rural origin.
- Permanent urban settlement: In much of Asia and most of Latin America this is more common partly because of pressures on access to rural land and partly due to better urban earning opportunities.

Additional variations have been identified, such as 'step-wise migration' where migrants engage in a series of steps towards the metropolis, moving first to a nearby town followed by moves up the urban hierarchy. Some evidence suggests that much of the urban growth that may be taking place in the developing world is taking place in intermediate cities, rather than in the core cities (Potts, 1997). In addition, some households appear to straddle the urban–rural divide by maintaining close ties with their rural areas and their rural-based relatives. Indeed, some researchers suggest that households actively engage in multi-location strategies, establishing members of the household in both urban and rural locations in order to maximise their livelihood opportunities (Potts, 1997). Thus the household meets its needs by making use of the rural and the urban environments. This complexity in urban population growth makes any simple link between population and environmental degradation problematic.

The second main geographic feature of growing cities relates to the relative balance of numbers of people living in urban and rural areas. Where city growth is very fast, there is likely to be a gradual shift towards a more urban population. This is known as urbanisation. However, urbanisation is also used to refer to the related processes of change that take place in the economy, culture and society, brought about by a growing dominance of urban processes, systems and influences. As a country's population becomes more urban-based, so other aspects of society appear to become urban-dominated. For example, in the mid-1970s Tehran accommodated 13.3% of Iran's population and 28.6% of the country's urban population, but it accounted for 72% of migration between provinces, 44% of migration between urban areas. In terms of the importance of Tehran to the economy, it produced half of the gross national product (GNP) (excluding oil) revenue, it accounted for 40% of national investment, 60% of industrial investment, housed 40% of the large industrial concerns, accounted for 40% of retail employment, 56.8% of hospital beds, 57% of physicians, 64% of newspaper distribution and 68% of vehicle registrations. In addition, income in Tehran was 45% higher than in other large cities and 70% higher than in small towns (Mandanipour, 1998). The issue of environmental sustainability focuses on whether this is a good thing or not. There are three main concerns relating to the process of urbanisation:

1. Health problems often result when a city population increases in density without the infrastructure to safeguard against disease, resulting in over-crowding and poor or even hazardous waste management.
2. Over-urbanisation, where a city grows so big that its size begins to create diseconomies of scale, for example traffic congestion, pollution, housing shortages and infrastructure overreach.
3. Overcrowding is focused in low income areas where the population density is very high and the provision of infrastructure and service is relatively low and the fabric of the buildings is often poor from overuse and poor maintenance. This can leave populations vulnerable to hazards such as rapid disease transmission, a pollution event or land slippage.

From an environmental perspective the concern is that the greater the over-crowding, the greater the environmental impacts on the urban region. One of the ways that this has been framed is Rees' (1992) concept of the ecological footprint (section 8.3.2). However, Main (1995) has argued that the concentration of the population in the city facilitates the provision of services and that the environmental impacts of providing services to such a population in a rural area may have more adverse environmental consequences. In essence, it is not possible to argue that as population increases so environmental impact increases. The links between population and the environment, according to this argument, are far more complex. For example Hardoy, Mitlin and Satterthwaite (2001, pp. 172–173) suggest that there are three main ways in which the inhabitants and the environment are affected by the development of a city:

1. The expansion of the built up area and the transformations this brings, for instance land surfaces are reshaped, valleys and swamps filled, large volumes of clay, sand and gravel, and crushed rock are extracted and moved, water sources tapped and rivers and streams channelled.
2. The demand from city-based enterprises, households and institutions for the products of forests, rangelands, farmlands, watersheds or aquatic ecosystems that are outside its boundaries.
3. The solid, liquid and air-borne wastes generated within the city and transferred to the region around it have environmental impacts, especially on water bodies where liquid wastes are disposed of without adequate treatment, and on land sites solid wastes are dumped without the measures to limit their environmental impacts.

According to McGranahan and Satterthwaite (2000), the issue of scale is important to the consideration of the extent to which population affects the environment. Any measurement of the environmental impact of wealthy cities will score highly if the indicators are based on the extent to which the city meets the needs of their inhabitants. However, they argue that such assessments fail to take account of what they call the 'transfer of environmental burdens' which is illustrated in Table 8.2. That is, these cities meet the needs of their inhabitants by transferring the environmental burdens of this elsewhere. The corollary of this is that in some cities environmental management fails – and therefore fails to meet the environmental needs of their inhabitants – because the environmental burdens of the city are placed to a large extent on that city's own land and population.

The cities of the world are all developing and may progress from low income to intermediate resulting in consequences for the ecological footprint (Table 8.2). As a city in a low income country gains in affluence, it is able to invest in water supply and sanitation, thus reducing the local environmental burden of water demand and the management of wastewater. However, as this affluence increases, this is usually associated with increasing levels of industrialisation which results in increased demand for raw materials and can also be associated with increased production of industrial wastes which, if not managed effectively, can result in increased problems of pollution. Table 8.3 illustrates the scale of impact of three key environmental issues, air, water and waste. Such demands for resources and increased pollution can have local effects, but also increase the scale of impacts. For example industrialisation is associated with an increase in airborne pollution, such as increased carbon emissions from the burning of fossil fuels. This contributes to the global carbon emissions and increases the city's ecological footprint beyond its boundaries. The conclusion

Table 8.2 Stylised indication of the scale of environmental burden of cities according to level of affluence (Reproduced with permission from McGranahan and Satterthwaite, 2002; published by Kogan Page and Earthscan)

Scale of burden	Low income	Intermediate	Affluent
Localised	Severe	Intermediate	Low
City-region	Low	High	Low
Global	Low	Intermediate	High

Table 8.3 Urban environmental burdens at different spatial scales (Reproduced with permission from McGranahan and Satterthwaite, 2002; published by Geographical Association)

	Local	City-Regional	Global
Air	Indoor air pollution	Ambient air pollutions and acid precipitation	Contribution to carbon emissions
Water	Inadequate household access to water	Pollution of local water bodies	Aggregate water consumption
Waste	Unsafe household and neighbourhood waste handling	Unsafe or ecologically destructive disposal of collected wastes	Aggregate waste generation

is that as a city grows in affluence there is a tendency for the city to reduce the environmental burden on its own inhabitants and transfer the burden beyond the city, thus increasing its ecological footprint.

Not only do different cities have different environmental problems, but environmental impacts are not homogenous so that the environmental agendas of cities, and even among city dwellers, differ. McGranahan and Satterthwaite (2000) have helped to bring this into sharper focus by suggesting that there is a 'green' and a 'brown' agenda upon which cities seem to focus. The green agenda focuses on reducing the impact of human activities on the environment, such as reducing water use, resource depletion and reducing pollution (Chapter 9). The brown agenda focuses on problems which are of more immediate concern to city inhabitants such as the way the urban environment affects the health and livelihoods of urban residents.

However, Hardoy, Mitlin and Satterthwaite (2001) express concern that such focal issues are imposed by international aid agencies on the cities of the developing world where the concerns should be increasing access of the low income groups to safe water and minimising the adverse environmental health factors. Part of the cause is that much literature written on environmental problems in cities of Africa, Asia and Latin America is written by authors from Europe or North America. There is, according to Hardoy, Mitlin and Satterthwaite (2001), a growing literature on the environmental problems of single cities in these regions, but it is often by authors who live in the cities concerned and rarely linked with other cities. There appears to be a gap between these and the global overviews written from Europe and North America which generalise about the environmental problems of the cities of the 'developing world'. However, such generalisations are often made on the basis of the analysis of the largest cities. While the problems of the very large cities are of some concern, the reason for the focus on these is perhaps more due to the spectre of the 'mega-city' than that they are home to the largest proportion of the population. In fact the majority of the urban population of Africa, Asia and Latin America live in cities of less than 1 million people – in many countries this is less than 100 000 people (Table 8.4).

Thus in the developing world Hardoy, Mitlin and Satterthwaite (2001) argue that there is a need to focus on the 'brown' agenda. For example Table 8.5 illustrates the results of a survey by Thomas *et al.* (1999) which asked urban residents of Port Elizabeth in South Africa what their environmental priorities were. These are tabulated according to wealth, illustrating the variation in priorities between the poorer households and the wealthier households represented in the survey. Fifty-five percentage of low income respondents reported seeing rats within the previous 24 h, compared with 1.1% of high income households. The average number of people living in each house per room in low, lower-middle and middle income households is 2.5, 2.66 and 2.55 respectively, compared with 1.47 in high income households. In the low and lower-middle income households

Table 8.4 The distribution of the population between settlements of different sizes, 1990 (Reproduced with permission from Hardoy *et al.*, 2001; published by Earthscan)

Region		Proportion of population living in urban areas				
	Rural areas	<1 million	1–2 million	2–5 million	5–10 million	>10 million
Africa	67.9	22.6	4.2	2.7	2.6	–
Asia	68.1	20.2	3.1	3.2	2.4	2.9
Latin America and Caribbean	28.9	43.1	6.0	7.7	4.7	9.4
Rest of the world	27.1	48.4	7.4	10.2	4.3	2.7

Table 8.5 Ranking of main environmental priorities of Port Elizabeth residents by wealth (Reproduced with permission from Thomas *et al.*, 1999; published by Stockholm Environment Institute)

Priority	Wealth				
	Low	Lower-middle	Middle	Upper-middle	High
1 (most common)	Overcrowding	Overcrowding	None	None	None
2	Sanitation	Others (unspecified)	Littering/ dumping	Littering/ dumping	Littering/ dumping
3	Littering/ dumping	Sanitation	Flooding	Need for greening	Other
4	Flooding	Littering/ dumping	Overcrowding	Others (unspecified)	Need for greening

the building materials are dominated by less hardwearing materials such as mud blocks, corrugated iron or zinc roofing (68%) and mud floors (41.5%) (Thomas *et al.*, 1999).

More recently, however, Hardoy, Mitlin and Satterthwaite (2001) have argued that there is a need for a new urban environmental agenda particularly for the cities of the developing world. This focuses on the capacities of the various urban institutions, such as city governments, professional groups, NGOs and community organisations, to identify and address their environmental problems. The argument assumes that institutions for managing the urban environment are already in place in the developing world as they are in the developed. The key is to enable those institutions, by ensuring they have the human capacity, the transparency of organisation, sufficient resources and they are not undermined by the international economy. It is, in fact, the people in the cities and their available assets in the form of economic, social, natural and cultural capital that make cities sustainable. Part of this is the involvement of the institutions as part of the urban context, in the form of communities, community or NGOs and government.

For it is these assets and institutions, and how they are used, that enable the city dwellers to achieve their livelihoods and sustain their cities. This suggests that human resources, people, are a key resource in solving urban environmental problems, a resource that has been overlooked by urban managers until relatively recently. People can therefore be considered to be part of the solution as opposed to the main part of the problem. This interacts with each of the issues that are to be discussed below.

8.4.2 Access to clean water

Water is such a vital part of everyday life and an important resource in a wide range of human activities that it is not surprising that water supply is emerging as a key issue for future urban environmental management. It has been estimated that in Third World cities 1 billion people lack daily access to sufficient amounts of clean water. The issue of water is made more complex by the fact that water resource management includes issues such as water supply, water purification, harvesting, demand management, pollution, flood management, water recycling and sanitation. In addition, Allen (1999) has developed the concept of 'virtual water' to calculate the total water required to meet the needs of a population. For example, the water involved in producing the food for a city is not consumed directly by that city, but by the agricultural activities involved in its production and processing. This 'virtual water' can therefore be calculated and added to the direct water consumption in a way similar to the ecological footprint, giving a clearer sense of the environmental burden being transferred beyond the city perimeter.

Increasingly, there is evidence of a shift in the ethos of water resources management moving from one dominated by finding ways of supplying water, to a situation where other aspects of the water system have become more important. Some see this as a shift from water being 'a public good *par excellence...* to a private sector commodity to be sold to consumers' (Bakker, 2003), while others see this as a way of improving efficiency of water use, in particular through focusing on water conservation through demand management.

Poor quality water supply has been known to be the cause of disease since the early Victorians discovered that contaminated water passed on cholera. It is now well-known that poor quality water supply and the absence of sanitation can lead to the proliferation of easily preventable diseases such as diarrhoea, dysenteries, typhoid, intestinal parasites and food poisoning. Kjellén and McGranahan (1997) argue that the role of water in the transmission of diseases has been overstated and poorly explained. For example, they argue that it is important to distinguish between water-borne, water-washed, water-based and water-related insect-vectored diseases (Table 8.6). It is clear that access to clean water is a key objective in efforts to manage healthy cities. It was estimated in

Table 8.6 Classification and preventive strategies of water-related diseases (Reproduced with permission from Kjellén and McGranahan, 1997; published by Stockholm Environment Institute)

Transmission route	Diseases	Preventive strategy
Water-borne	Infectious skin diseases and louse-borne typhus	Improve quality of drinking water; prevent casual use of unprotected sources
Water-washed (or water-scarce)	Most diarrhoeas and dysenteries	Increase water quantity used, improve accessibility and reliability of domestic water supply, improve hygiene
Water-based	Schistosomiasis, guinea worm, etc.	Reduce need for contact with infected water[a]; control snail population[a]; reduce contamination of surface waters[b]
Water-related insect vector	Malaria, dengue fever, etc.	Improve surface water management; destroy breeding sites of insects; reduce need to visit breeding sites; use mosquito netting

[a] Applies to schistosomiasis only.
[b] The preventative strategies appropriate to the water-based worms depend on the precise lifecycle of each and this is the only general prescription that can be given.

1994 that 300 million city dwellers were without access to safe drinking water. The result is that these urban residents have to collect their drinking water from open sources that are often contaminated by pollution and sewage, or to buy from water vendors which can be expensive and is not a guarantee of water quality. The quality and quantity of urban water supply is a problem that is increasingly international in scale.

Traditional approaches to solving urban water problems have focused on supply in order to ensure that the population has access to clean water. During the 20th century the dominant means by which this was achieved was through state management of water supply systems. This has often involved subsidising the supply of water in order to ensure wide access. There is evidence that water-borne diseases have declined and sanitation improved as a result of this increased and subsidised water supply. This in turn has led to improved health in cities such as São Paulo, Belo Horizonte, Curitiba and Pôrto Alegre (McGranahan and Satterthwaite, 2000). The concern for some has been that increasing supply – especially for very large and very fast growing cities – has resulted in over-exploitation of water resources. In some countries, continuing to increase water supply is placing increasing pressure on the related water catchments. Even for cities where there is an apparent abundance of water, such as Bangkok (Box 8.1),

Box 8.1: Bangkok's water budget

Bangkok was originally built in its location in order to provide a good defensive position from attack through the marshes surrounding the settlement to the north and the sea to the south. As the city has grown to a population of around 10 million today, it has gradually expanded outwards into the marshland, draining it and filling in some of the canals, or *klongs*, that the early settlers constructed for movement around the city. This has resulted in a number of apparently contradictory water-related problems: the city appears to have both too little and too much water.

On the one hand, the increasing urban population demands a greater amount of water for basic residential uses. Increasing urbanisation and modernisation has also resulted in an increase in demand for non-residential uses, such as industrial activities, hydro-electric production and flushing any waste or pollution from the *klongs*. Most researchers report that the *klongs* are so polluted that they are anaerobic, that is the dissolved oxygen has been depleted by organic wastes, making them biologically dead. As a result, the levels of biological oxygen demand in the *klongs* is equivalent to that of normally found in sewage (Rigg, 1991). Tests of other indicators, such as dissolved oxygen, coliform presence and heavy metals, suggest high pollution levels are a constant background feature of life (Bangkok Metropolitan Authority, 2001). Roomratanapun (2001) estimates that up to one million m^3 of waste water is present in the *klongs*, making life for the residents who live and work near them unpleasant and potentially harmful to their health. As the *klongs* flow into the Chao Praya River they are in danger of turning this into a biologically dead river. The link between pollution and water supply is close, as the water supply budget, illustrated in Table 8.7, is very tight. Roomratanapun (2001) estimated Bangkok's water budget as 24.3 billion m^3 supply and 24.1 billion m^3 demand. However, the supply to the dams providing additional supply to the catchment is in decline due to deforestation and increasing demand upstream. This means it is unlikely that the dams supplying Bangkok's water system will be able to continue this level of supply. In addition, all the evidence suggests that as the city's population becomes greater and more modernised the water demand will continue to increase. Hardoy, Mitlin and Satterthwaite (2001) estimated that Bangkok loses approximately 43% of its clean water supply during distribution – mainly through leaking water pipes.

On the other hand, the in-filling of *klongs* has exacerbated the potential for flooding as the canals have also acted as overflow reservoirs during high water periods, such as spate outflow or monsoon sea levels (the city has a mean altitude of around 1 m above sea level and is consequently

Continued on page 213

Continued from page 212

Table 8.7 Water budget for Bangkok, Thailand (Reprinted with permission from Room-ratanapun, 2001; published by Elsevier)

Demand	Billion m³
Industrial and urban water demand estimate	6.0
Required volume to maintain flow and flush out waste and saline water	2.5
Power turbines for electricity production	6.6
Irrigation for two seasons of paddy rice	9.0
Total	24.1
Supply	
Annual rainfall in Chao Praya River	1.4
Dam capacity supplying Bangkok and surrounding agricultural areas	22.9
Total	24.3

vulnerable to raised sea levels). The combination of the construction of buildings, increasing the weight of the city, and the extraction of underground water, reducing the underground volume from a mainly sand and clay soil structure, has resulted in a level of subsidence in the centre of the city that is very high. This has also resulted in the problem of saline water incursion into the city's acquifers. Groundwater abstraction was estimated at 1.1 million m³ per day in 1986, which was thought to be in excess of the 0.8 million m³ per day safe yield. Rates of subsidence were estimated at 0.6 to 5.1 cm/year in 1989 (Phantumvanit and Liengcharernsit, 1989), and more recently estimated at up to 10 cm per annum in some parts of the city (Bangkok Metropolitan Authority, 2001), some of which are now below sea level (Rigg, 1991). The combined effects of increased peak water levels, in-filling of canals and high rates of subsidence have resulted in increased intensity and frequency of flooding.

The problem of water management interacts with other aspects of the environmental management in cities such as Bangkok. For example Jiang, Kirkman and Hua (2001) report that 25% of solid waste is disposed of improperly and that 50–100 t of garbage are disposed of into the city's, *klongs* each day. One aspect of urbanisation is that many urban activities are concentrated in close proximity. Jiang, Kirkman and Hua (2001) also report that three-quarters of Thailand's industrial plants producing hazardous chemicals are located in Bangkok and the neighbouring provinces, including more than 90% of its chemical, dry-cell battery, paint, pharmaceutical and textile manufacturing plants. Phantumvanit and Liengcharernsit (1989) estimated that only 2% of households were connected to the city's sewage system at that time. Therefore when the higher and more frequent floods meet with lower levels of buildings, the resulting flood problems are exacerbated.

Continued on page 214

Continued from page 213

One implication of the difficulty of water supply is the issue of the cost of supply. Hardoy, Mitlin and Satterthwaite (2001) reported that there was a cost differential between the cost charged by private water vendors and the public utility estimated at 5:1. In most cases the public utility supplies the middle and upper income residents of the city, so, in common with many cities in developing economies, the poorest residents pay the most for their clean water supplies (Bakker, 2003).

the problem of water management is acute. As a result some have proposed a shift to a more balanced management of both supply and demand in order to conserve water use, as well as to increase water supply. Increasingly demand management involves:

- Reducing leakage, which Anton (1993) suggested could eliminate up to 20–30% of false consumption in Latin America.
- Introducing water-saving technologies such as smaller toilet tanks and low volume water heads.
- Changing water consumption patterns, through water pricing and voluntary lifestyle changes, which can be influenced by public campaigns.

Consequently, the issues relating to water in cities are the result of a complex set of circumstances that include the environmental context and the social and economic structures. Within the environmental context, the availability of surface water, the availability, nature and level of exploitation of any of the acquifers that may be available to supply a city, the extent to which human activities interact with these water sources and the methods available for the management of waste all interact. Among the human characteristics, water management interacts with the health of the population, providing a defence against diseases, or when managed poorly, a medium through which disease can pass.

8.4.3 Environmental hazards

All cities face risks from environmental hazards some of which are the result of natural processes and some of which are induced by human activities. It is possible to divide the risks into two main categories, disasters, such as earthquakes or flooding, and lower level hazards, such as air or water pollution. Geographers distinguish between these by describing the first as high energy and low frequency and the second as high frequency with low energy. However in the case of some hazards there is a may-be-a-constant background level, for example in the case of water pollution in Bangkok (Box 8.1). In this instance high or sustained levels of background pollution can increase a population's vulnerability to low frequency, high energy hazards such as sudden increases in pollution, flooding or water-borne disease epidemics.

During the 1970s, researchers began to suggest that human vulnerability to hazards rather than the hazards themselves was a more important aspect of understanding the importance of hazards, particularly when considering that a city is a geographical focus for large populations. Further, Varley (1994) reported that human and material losses from disasters had increased during the 20th century. This happened at a time when cities in the developing world grew at unprecedented rates, bringing together large populations of low income urban residents often on hazardous spontaneous settlements making them far more vulnerable to natural and human-induced hazards. It is precisely this issue of vulnerability on which a number of social scientists focused, arguing that although floods, landslides and earthquakes are natural processes, the disasters that can be associated with them are not a result of natural processes, but of human vulnerability. Floods and earthquakes can take place, but only become a disaster if there are people nearby who are vulnerable to their physical effects. Thus it is the interaction between the vulnerability of the people and the physical effects that results in a disaster.

One such hazard is flooding. This is particularly problematic where large-scale urban expansion has taken place without the development of infrastructure to cope with flooding, such as storm drainage systems. The situation where the population of a city has grown very quickly and the development of the infrastructure does not keep pace is common in many fast growing cities in the developing world. Flooding can result in deaths by drowning, but can also leave large populations homeless and contaminate water supplies, leading to epidemics of diseases borne by water or passed on when associated with washing out areas of rat-infestation, waste or sewerage storage. Such problems appear to be greatest for large cities in low income countries located on coastal locations. Cities such as Dhaka, Bangkok (Box 8.1), Jakarta, Shanghai and Alexandria are all reportedly vulnerable to flooding due to the combined effects of subsidence, sea-level rise and inadequate development of storm drainage systems (Pacione, 2001b,d).

Hardoy, Mitlin and Satterthwaite (2001) argue that the research on hazards, such as flooding, suggests that there are three main groups of urban dwellers who are vulnerable to environmental hazards:

- Those who live or work in dangerous locations, where the risk of environmental hazards is higher because of the geography of the area; because health, building or environmental regulations are not enforced; or because the location lacks the infrastructure to cope with a hazard, such as storm drainage.
- Those living in areas which are poorly equipped to respond to disasters, because of either slow or ineffective responses from emergency services.

- Those who are less able to cope with the consequences, such as those who have no access to capital assets or sources of income and are therefore unable to afford medical treatment; or those who have no access to water and shelter; such as pavement dwellers and those who live in open spaces, parks or graveyards, or children and elderly people who are often less able to respond to hazards by moving to safe locations.

They go on to suggest a set of characteristics that influence whether a person is vulnerable to environmental hazards as:

- *Income and assets* – a set of resources which can influence the ability to afford good quality housing which is appropriately located and the ability to provide emergency response and healthcare.
- *Economic or social roles* which can increase or decrease exposure to hazards – for example occupations such as rubbish picking increase possibility of exposure, whereas office work reduces likelihood of exposure.
- *Extent of public, private or community provision of healthcare* – the ability of the community to respond to accident or disease.
- *Individual, household or community coping* – knowing what to do, how to arrange survival strategies.

These sets of assets form a key focus of what Varley (1994) describes as vulnerability analysis, where it is possible to identify which locations and groups of a city's population are most likely to be vulnerable to any kind of natural hazard. Assets provide an indication of vulnerability, regardless of what the environmental or human-induced hazard may be. They give an indication of the population's ability to respond in a disaster situation. This approach, says Varley (1994), is in contrast to an approach which focuses on the physical mechanism of environmental hazards, which while important is limited in its ability to manage a disaster. Hewitt (1997) states that such a focus highlights the fact that the nature of urban growth during the 20th century has magnified the vulnerability of some populations of some cities to hazards. For example, Hewitt suggests that cities provide unique aspects of risk including the nature of the built environment, the danger of dependency on the built environment for survival, the introduction of lethal forces such as modern industry or transport infrastructure, and the difficulties of congestion can exacerbate the human response. However, one should not forget that cities have formed a key part of the responses to disasters, whether they have taken place in cities or in nearby rural areas.

The consequence of this assessment is that although environmental hazards may be naturally triggered, low income and marginalised households are those who are most often vulnerable to such events, many falling into more than one of the vulnerable categories identified by Hardoy, Mitlin and Satterthwaite (2001) above. Indeed McGranahan and Satterthwaite (2000) argue that low income urban residents of economically growing cities can experience what they call a 'double health burden'. They are vulnerable to environmental hazards common in 'dirty' industrialisation, such as respiratory and skin diseases, and vulnerable to communicable diseases common in low income countries. Communicable diseases are most common among infants who are vulnerable to infection. In Calcutta, for example, there is evidence that the incidence of both types of disease is high, suggesting that the population are doubly burdened.

Some researchers suggest that if pollutants are regularly monitored and known to be not harmful, they may in fact increase the yield of urban agriculture by providing chemicals and mineral to the soil (Urban Agriculture Network, 1996). However, this requires that the cultivators have knowledge of pollutants and without this there is a danger that pollutants can not only cause danger to the cultivators, but also pass into the human food chain by contaminating the food cultivated. Figure 8.1 illustrates a storm drain suffering from pollution, litter and subsidence in Kano, Nigeria. However, the population in this area is largely unaware of the dangers to which it is exposed.

The ability of urban dwellers to cope with an environmental hazard is often defined by their income and their ability to mobilise their assets to cope with the disaster. This explains why the same disaster can affect households within a city differently. Poorer households tend to be more vulnerable to environmental hazards, and have less options for coping, than wealthier households. The same is

Figure 8.1 Storm drain suffering from pollution, litter and subsidence, Kano, Nigeria.

true of cities: a disaster can affect two cities very differently. The contrast can be made between the impact of an earthquake – of measurable magnitude – in Los Angeles and in Calcutta. In Los Angeles there are well-equipped and well-resourced emergency services with disaster management plans that are in place and constantly updated and reviewed, the buildings are designed to withstand most earthquakes and the people and the local and national governments are able to mobilise capital assets if required. Calcutta lacks such well-resourced and developed coping mechanisms.

In a survey of residents of Sao Paulo, Brazil (Jacobi, Kjellén and Castro, 1998), air pollution was the most important environmental issue for those residents in central and intermediate areas, while those in the peripheral areas indicated contamination of water supply to be the most important. This suggests that the distribution of environmental hazards varies across the city. Further, when asked about the preferred level of action on such environmental issues, the majority indicated that a governmental response was required for the problems of pollution, indicating the perceived importance of governance to the issue of management of environmental hazards. This suggests that a focus of action on human capabilities and institutions could provide the key to reducing vulnerabilities of urban residents to environmental hazards.

8.4.4 Food

Food is an issue of personal and global concern (see also Chapter 6). Koc *et al.* (1999) estimate that around 35 000 people per year die each day from hunger. Authors such as von Braun *et al.* (1993) are concerned about the particular problems that urban dwellers have in obtaining the food sources they require in the developing world. By contrast, some researchers in the developed world are becoming concerned with the extent of the environmental burden which the consumption patterns practised in cities place throughout the world. This has most clearly been illustrated with the development of the concept of 'food-miles', the number of miles that food has to travel in order to arrive on the dining table.

In cities in the developing world, rapid urban growth and fiscal and foreign exchange constraints, as a result of structural adjustment programmes, have created particular challenges for maintaining urban food security (von Braun *et al.*, 1993). The evidence suggests that where economic deterioration was worst, during the 1980s, the food security of the urban poor was particularly adversely affected. Structural adjustment policies resulted in the removal of food subsidies and redundancies ('retrenchment') among government and parastatal employees, much of the impact of which was focused on the cities. The level of dissatisfaction among urban dwellers was such that the removal of food subsidies resulted in civil unrest, in some cases in the form of what was described as 'food rioting' in the capitals and main cities. Such riots took place in Tunisia (1984), Zambia

(1986), Nigeria (1989), Morocco (1990) and Jordan (1996). There is also some evidence that increased food prices were one of the triggers for the Tiananmen Square demonstrations in Beijing in 1989 (Brown, 1997).

As urban residents become more affluent and as the cities in which they live become generally more affluent there is increased demand for the import of food from greater distances, thus transferring the environmental burden to more distant places (McGranahan and Satterthwaite, 2002). However, this has also meant that as cities have grown, urban markets have become more important as outlets for rural produce and therefore an important resource for the rural population's livelihood strategies which can have beneficial consequences for rural environments (Tiffen, Mortimore and Gichuki, 1994; Mortimore and Tiffen, 1995). However, cities have also begun importing from foreign sources.

One of the ways of combating the increasing ecological footprint (sections 1.4 and 8.3.2) brought about by the need for increasing food supplies is the development of urban-based cultivation (Box 8.2). Some have encouraged the development of urban agriculture in order to reduce food miles, and therefore

---

Box 8.2: Farming food in cities for sustainability?

There is a growing body of research on urban agriculture that is uncovering a wide range of types of cultivation and indicating that urban agriculture is far from homogenous. This is illustrated in Table 8.8, which illustrates one typology of urban agriculture focusing on the types of product and illustrating the distinctions in techniques and locations where such types are typically

Table 8.8 Farming systems common to urban areas (Reproduced with permission from Urban Agriculture, 1996; published by UNDP)

Farming system	Product	Location or technique
Aquaculture	Fish and seafood, vegetables, seaweed and fodder	Ponds, streams, cages, estuaries, sewage, lagoons, wetlands
Horticulture	Vegetables, fruit, compost	Homesites, parks, rights-of-way, roof-tops, containers, hydroponics, wetlands, greenhouses
Livestock	Milk and eggs, meat, manure, hides and fur	Zero-grazing, rights-of-way, hillsides, coops, peri-urban, open spaces
Agro-forestry	Fuel, fruit and nuts, compost, building material	Street trees, homesites, steep slopes, vineyards, greenbelts, wetlands, orchards, forest parks, hedgerows
Other	Houseplants, medicine, beverages, herbs, flowers, insecticides	Ornamental horticulture, roof-tops, containers, sheds, beehives/cages, greenhouses, rights-of-way, urban forests

Continued on page 220

Continued from page 219

found. Such variations in activities and locations have varying implications for environmental management.

Some research on urban cultivation has identified potential environmental problems with encouraging production of food in cities. For example, where the city is an arid zone, the additional demand for water for irrigation, particularly during the driest season, can exacerbate already acute water constraints. Figure 8.2 illustrates a well irrigated urban agriculture plots in Dar es Salaam, a city with water supply challenges. It may be environmentally less destructive to produce the city's food needs in a distant rural area where water is in abundant supply (Alshuwaikhat and Nkwenti, 2002). Alternatively urban cultivation sites may provide useful flood overflow areas. Thus cities can avoid damage to property and infrastructure by keeping sites for cultivation which in turn can benefit from periodic flooding (Lynch, Binns and Olofin, 2001).

Such research outcomes raise two key issues: what is the most appropriate scale at which any test of environmental sustainability should be applied? Given the symbiotic relationship between the city and the

Figure 8.2 Urban cultivation in Dar es Salaam, Tanzania. This photograph was taken in 1994 when this plot was vacant. Since then the road nearby has been upgraded and associated plots have risen in value putting pressure on such use of vacant plots, illustrating the precarious nature of much urban cultivation.

Continued on page 221

Continued from page 220

countryside in terms of environment, economy and society (Lynch, in press) decisions about urban cultivation can impact on rural economies (for example, cities provide rural production areas with markets that are vital for the viability of their economies). While consideration of the ecological footprint is important, the positive as well as the negative effects of urban cultivation beyond the city boundary should be considered.

Additional concerns over the implications of urban agriculture largely highlight issues of environmental or social implications of the development of agriculture within cities. Table 8.9 illustrates briefly some of the advantages and some of the concerns set out in the rapidly expanding literature on this subject. This table is not intended to be a comprehensive summary of all the research, but illustrative of a range of concerns that arise in the research. One of the points that this table illustrates is that in addition to a wide variety of types of cultivation, the environmental and social contexts of the city concerned has some influence on the extent to which urban agriculture has beneficial or adverse impacts on that city and its related rural areas. In addition the evidence of a number of researchers in different cities around the world suggests that there are hazards from locating urban cultivation of food crops close to urban areas, for example because of the danger of contaminated water being used for irrigation.

However, it is important not to make over-generalisations. For example while polluted water can contribute to the productivity of the soil through inputting important minerals, there is considerable risk from heavy metals and pathogens being passed into the human food chain. Bradford (personal communication) reported that peri-urban farmers in Hubli-Dharwad, India, use water from nearby streams which was found to

Table 8.9 Summary table of urban agriculture research (Adapted from Lynch, 2002 with permission of Hodder Arnold)

Advantages	Concerns
Vital or useful supplement to food procurement strategies (Rakodi, 1988)	Conflict over water supply, particularly in arid or semi-arid areas (Mvena et al., 1991)
Various environmental benefits (Lynch, 1995)	Health concerns, particularly from use of contaminated wastes (Lewcock, 1995)
Employment creation for the jobless (Sawio, 1994)	Conflicting urban land issues (Lynch et al., 2001)
Providing a survival strategy for low income urban residents (Lee-Smith and Memon, 1994)	Focus on urban cultivation activities rather than broader urban management issues (Rakodi, 1988)
Making use of urban wastes (Egziabher, 1994)	Urban agriculture can benefit only the wealthier city dwellers in some cases (Smith, 1998)

Continued on page 222

Continued from page 221

contain waste from a nearby hospital, including used syringes. Foeken and Mwangi (1998) reported that, although they observed erosion mitigating efforts among urban cultivators in Nairobi, Kenya, the danger of using polluted water sources was not apparent to the cultivators, even in situations where water courses contained human effluent and during rainy seasons these may flood cultivated land. Lynch, Binns and Olofin (2001) report the problems for urban cultivators when land comes under increasing development pressure from surrounding land uses in the case of Kano in Nigeria, despite the beneficial role some sites play in absorbing seasonal flood water. Smit and Nasr (1992) and the Urban Agriculture Network (1996), however, identify a number of instances where, when carefully managed, polluted water and urban waste can be used to improve productivity of urban cultivation and as part of an urban management strategy tackling issues such as waste and vacant land. However, this requires knowledge of the hazards and strategies for managing them, and there is thus far limited evidence of widespread take-up of such approaches. Urban agriculture can therefore offer opportunities for improving the livelihoods of urban residents, but it is a highly varied activity with a range of benefits and hazards.

reduce the environmental burden involved in the rural production and transfer of food to the city (Urban Agriculture Network, 1996). However, as is suggested above, this may remove an important and lucrative market from the rural population making migration from the countryside more likely as rural livelihoods are undermined. In the words of two key contributors to the *West Africa Long Term Perspective Study*, 'urbanisation can be seen as a major opportunity for agriculture, not only because of its effect on demand and the division of labour, but also because it is mainly city-dwellers who buy land and invest in it' (Cour and Naudet, 1996, p. 21). The links between cities and the related rural areas through food supply networks are complex and it is important to understand their dynamic.

8.5 Strategies for achieving sustainable human settlements

This chapter has focused on evidence of past or present approaches to managing the environment of cities, in particular cities with the fastest growth rates and therefore those with the greatest challenges. This final section will consider the strategies for achieving sustainable human settlements that show promise for the future, providing examples of policies or approaches that have been proposed in the academic literature.

Increasingly, international attention is focused on the growing number of problems that affect our world at a global level. The Earth Summit in Johannesburg in 2002 was just one of a series of international conferences which have attempted to focus the attention of governments, bureaucrats, NGOs and companies on difficulties that can only be addressed at an international level (e.g. Robinson, 2002). One of the key issues that is addressed frequently at such conferences is that of the rapid growth of cities. Whether the conference addresses the issue of cities directly, such as the Conference on Human Settlements, known as HABITAT I, which took place in Stockholm in 1972, and HABITAT II, which took place in Istanbul in 1996, or whether it addresses other issues such as global warming, drinking water and sanitation or desertification, the process of urbanisation is never far from the minds of the participants. For, although cities are not directly responsible for the degradation of land, the destruction of forests or the problem of global warming, there is a generally accepted belief that as people across the world are increasingly located in cities and aspiring to urban lifestyles, this may be accelerating such environmental and economic processes as causes of deforestation or desertification. This chapter has tried to address the current situation with regard to the global development of urbanisation and the evidence of some of the links between this and the environmental problems both inside and outside of cities, but caused by them.

With many ideas and theories about how best to create cities of the future and how to adapt our current cities to more sustainable forms, Pacione (2001c,d) identifies a number of model cities as presented in the literature. These urban utopias are put forward by a range of philosophers, architects, economists and planners, and are summarised in Table 8.10.

One of the reasons for the emergence of such a diverse range of utopian visions of the future city relates to a wide variety of concerns about cities in the past. Each urban utopia has emerged from the particular concerns of the time period and from the cultural context in which it was developed. This is mirrored in a variety of concerns in present day cities. In environmental terms, McGranahan and Satterthwaite's (2002) 'Green Agenda-Brown Agenda' characterisation draws attention to some of this diversity. The Brown Agenda focuses on immediate, local impacts on human health, usually affecting lower income groups. Thus it targets access to and quality of water, air quality, sanitation and waste management. The Green Agenda takes a broader view, considering ecosystem health on a regional or global scale. Rather than addressing immediate, local needs, the Green Agenda is aimed at longer-term problems and benefiting future generations. In very general terms it is possible to contrast the urban environmental health needs of city dwellers in the world's poorest cities with those of the economically wealthy cities. For example, in the developed world there are concerns which focus on the issues of inner-city decline, as the market attracts people and spending increasingly to the edge of cities, there are concerns relating to the problems of traffic congestion and excessive energy consumption. In cities in countries

Table 8.10 Summary of urban Utopia (Reproduced with permission from Pacione, 2001c; published by Geographical Association)

Urban utopia	Description
Green city	Originating with Ebenezer Howard's Garden City concept and Patrick Geddes' theories of urban planning, which were developed later by the landscape architect, Ian McHarg. This relies on abundant space and where the city is large in population, a relatively mobile population.
Dispersed city	The idea of decentralised settlements, illustrated by the work of E.F. Schumacher, author of *Small is Beautiful*. The key concept was that all activities were small-scale and therefore easily managed and with minimal adverse environmental impact.
Compact city	This arose around the time of the ascendancy of the development of high-rise building techniques and is particularly associated with le Corbusier. This emphasises the social and economic benefits of high density urban living, where services can be provided more easily to the population.
Transit metropolis	Focuses on the ways in which mass public transport systems can improve cities not only by removing private transport. This has been applied, at least in part in medium sized, mixed use cities well suited to such planning, such as Singapore and Copenhagen, but may not be a suitable strategy for all cities.
Regional city	Focused on the work of Kevin Lynch who suggested a series of small settlements linked across open recreation space by major roads. A version of this is, to some extent, emerging in the US and parts of the UK, with the development of out-of-town shopping malls and the pressure for more edge of city housing. There are examples of such developments in the megacities of the developing world, particularly Mexico City, Johannesburg and Lagos.
Network city	This is based on recognition of the evolution of networks of 'corridor cities' that have emerged as the functions and activities of cities are merged and distributed. Frequently cited examples include the London-Cambridge corridor, Ranstad in the Netherlands, the Kansai region of Japan. There may be evidence to suggest the emergence of such developments in the Third World's megacities, such as Cairo-Alexandria and South Africa's 'PWV' region, including Johannesburg, in Gauteng Province.
Informational city	Based on the analysis of Castells, this concept is in part based on recognition of the development of cities as post-industrial entities where the focus is increasingly on knowledge and information industries which is facilitated by advances in information and communications technologies. Some have argued that this city will reduce environmental impacts as it focuses on the production and trading of information, which is less environmentally damaging, however, this is likely to underestimate that informational flows may stimulate demand for travel and for goods which may have an environmental burden.
Virtual city	This idea is based on an extension of the informational city, which is currently largely only present in science fiction, whether utopian or dystopian. However, the idea is based on the extension of existing initiatives, such as the use of telematics to attract inward investment (such as city websites), the use of telematics to create virtual communities (such as online gaming or self-help groups). The interesting area of research currently is the link between physical space and cyberspace.

experiencing economic transition there are concerns about increasing pollution as the city attracts increasingly heavy industrial activities in an attempt to emulate the industrialisation of the more economically advanced cities in order to boost employment and economic growth. In cities in the world's poorest countries the concerns focus on rapid population growth rates and failing infrastructure and service provision. These diverse sets of needs and concerns all have unique environmental implications depending on the environmental context of the city. However, it should be noted that this generalisation masks a complexity of Green and Brown Agendas within cities. Cities contain diverse neighbourhoods each with their own development needs. This is particularly true in relation to provision of services such as transport, water and sanitation, housing and medical facilities. Within cities there are also diverse groups with competing interests and concerns. They may be divided by many things such as gender, ethnicity, and wealth, access to services and technology or physical location.

The range of urban utopias (Table 8.10) is an indication of the danger of assuming that the problems and therefore the solutions for all urban management issues are the same. The history of urban management is one that illustrates that imposing theories from above has been shown to be ineffective in managing highly complex structures, forces and systems that have resulted in urban areas. The latest developments in the literature suggest a growing consensus that urban governance is the key to both economic development and achieving an environmentally sustainable settlement. In this aspect the diverse needs of the cities of the developed and the developing world are in agreement. There is evidence of a common approach to such problems as local administrators from all over the world are increasingly seeking out the views and participation of residents in the management and governance of their cities, whether they are based in Kingston-upon-Thames, England or Kingston, Jamaica. However, one of the key issues that needs to be addressed if this approach is to work is the development of the institutional and human capacity of the cities concerned.

Pugh (2002) provides an example where the institutional capacity of local interest groups is imbalanced. He points out the dangers of allowing some interest groups access to the means of governance. He cites the example of the local fisher-people of Folkestone, Barbados, who were given the opportunity to participate in the environmental management of their marine region inspired by the prescriptions set out in *Agenda 21*, one of the main outputs from the Rio Summit in 1992 (UNCED, 1992). In particular Chapter 28 of *Agenda 21*, entitled 'Local Agenda 21', encourages local governments to involve community groups and the public in the decision-making and policy-making processes. A study by the International Centre for Local Environmental Initiatives (ICLEI, 2002) reported that a vast majority of Local Agenda 21 initiatives are based in the developed world. In Folkestone, Barbados, the fisher-people have arguably lost out in the participatory form of environmental management of their coastal zone because their contribution is not seen as credible or as effective as the involvement of the local hoteliers. Such examples illustrate that 'Even though local groups have been established the connections between inequity, empowerment and consensus have not been addressed' (Pugh, 2002, p. 293). This is paralleled, according to Pugh (2002), in other parts of the world where the ICLEI report provides evidence that the business sector make up the largest group included in 83% of the 1487 initiatives reported. By contrast, indigenous groups and ethnic minority groups are least likely to be represented, reportedly having a presence in 22% of local Agenda 21 initiatives. There is then a need for the development of the capacity of all groups to ensure that they are able to contribute to the management of their environment.

This issue is complex, however, and it is important to address the low institutional capacity of local government in developing countries and the low level of empowerment among the poorest urban residents. For some researchers this represents some of the reasons for poverty and disempowerment. It is important for each city to be governed by a transparent and participatory government in order to be able to detect local priorities and to take advantage of the kinds of opportunities cities offer for sustainable development as discussed in section 8.3.2. This is something that the world's so-called global or world cities are still struggling with (Diamond, 2002), so it is no surprise that the poorest cities are finding this challenging.

Hardoy, Mitlin and Satterthwaite (2001) suggest three basic goals for successful cities as:

1. Healthy environments in which the inhabitants can live and work.
2. Provision of infrastructure and services required for healthy living and prosperous economic base for all, including water supply, sanitation, waste collection and disposal, paved roads and footpaths.
3. Balanced and sustainable relationship between the demands of consumers and businesses and the resources, waste sinks and ecosystems on which they draw.

However, achieving such goals relies on an understanding of the complex links between these elements. It also requires a political and administrative system through which the priorities of the urban dwellers can influence the management of the city environment. Rees (1997) suggests that there is much that can be done incrementally in achieving sustainable cities. He suggests a number of basic objectives maximising the potential efficiency that urban environments offer. For example:

- Integrate planning in ways that maximise resource use efficiency.
- Make use of the multi-function potential of green spaces, for example as carbon sinks, for food production or climate modification.
- Maximise livelihood opportunities and self-sufficiency opportunities, for example encourage recycling of waste as compost, or wastewater or rainwater for irrigation.

- Protect ecological integrity of the urban ecology in order to reduce the ecological load imposed on distant ecosystems.
- Aim for zero impact development, and where destruction of ecosystems is necessary compensate by rehabilitation elsewhere.

In an analysis of the environmental management issues in Bangkok, Ross *et al.* (2000) identified three priorities for action in order to address the environmental problems of the city. These can be applied as general principles to cities around the world:

- *Address the nature of decision-making*; encourage all stakeholders to foster a culture of working towards a common interest.
- *Work with the natural ecosystem*, **not against it**; develop methods of developing the built environment to co-operate with natural functions in order to improve urban health. In the case of Bangkok, this involved an understanding of the flood plain systems on which it is constructed and managing future development to maintain and enhance these systems, rather than to overcome them.
- *Understand people's behaviour patterns*; what people do contributes to the nature and extent of environmental problems and the extent to which people are exposed to them.

The challenge is that these initiatives require the collaboration of all urban institutions and individuals to consider the implications of their consumption, development and other urban-based activities. It therefore requires a concerted effort on the part of all concerned to move to more sustainable forms of urbanisation.

Further reading

Satterthwaite, D. (2002) Urbanization and environment in the Third World, in V. Desai, and R. Potter (eds), *The Companion to Development Studies*, London: Arnold, pp. 262–267.
This is one of the six chapters on urbanisation in this Companion that are all concise and accessible introductions to different aspects of this theme.

McGranahan, G. and Satterthwaite, D. (2002) The environmental dimensions of sustainable development for cities. *Geography. Sustainable Development Special*, Robinson, G. (ed.).

Hardoy, J., Mitlin, D. and Satterthwaite, D. (2001) *Environmental Problems in an Urbanizing World*, 2nd edn, London: Earthscan Publications.
This is a comprehensive work which covers a wide range of aspects of the urban environment, particularly in the poorest economies, in more detail.

Pugh, C. (ed.) (2000) *Sustainable Cities in Developing Countries*, London: Earthscan. Provides a comprehensive discussion of managing urbanisation from a range of researchers in the field.

Potter, R.B. and Lloyd-Evans, S. (1998) *The City in the Developing World*, Harlow: Longman.
Provides a good introduction to the geography of cities in the developing world.

Pacione, M. (2001) *Urban Geography: A Global Perspective*, London: Routledge.
This is a comprehensive work covering many key themes in urban geography including many relevant to the discussions in this chapter, with some of the arguments and discussion previewed in a series of articles in the journal *Geography* (Pacione, 2001a,b,c).

Chapter 9
Coping with Pollution: Dealing with Waste

Ros Taylor and Kathy Morrissey

9.1 Why are pollution and waste global environmental issues?

> Pollution Kills. It is as simple as that.
> *John A.S. McClennon, USEPA [cited in Vesilind and Peirce, 1983]*

Pollution also represents wasted opportunities. Polluted environments may become 'no go' areas for human use. Despite increasing legislation the pollution load on our environments continues to rise. This is largely because we have failed to address the root causes of pollution. These are broadly:

- rising human population;
- rising standards of living;
- increased *per capita* resource use;
- increased *per capita* waste generation;
- failure to understand and to respond to the key scientific causes; our relationship to the Global Commons and the basic functioning of ecosystems;
- inadequate models for costing resources.

Though many demographic models predict plateauing of the earth's population by the mid-21st century (e.g. Lutz, 2002), *per capita* consumption of the earth's resources seems set to continue to rise as our living standards world-wide increase. MacKellar *et al.* (1998) give estimates for global population of between 8 and 12 billion by 2050 and between 5.7 and 17.3 billion by 2100. However, this issue is further exacerbated in that as much as 90% of the predicted global population increase may take place in the developing world. As developing

Global Environmental Issues.   Edited by Frances Harris
© 2004 John Wiley & Sons, Ltd   ISBNs: 0-470-84560-0 (HB); 0-470-84561-9 (PB)

societies strive to raise their living standards to those of the developed world, resource consumption and concomitant waste production, environmental degradation and pollution are also likely to rise rapidly. Pollution transfer in the global biogeochemical cycles means that the developed world is not isolated from the polluting effects of increased industrialisation and consumerism associated with raising the living standards in the developing world. Similarly, the developing world is already bearing many of the environmental burdens of industrialisation and consumerism in the developed world. Issues of population growth, living standards, resource supplies and the pathways to development and potential for a sustainable high quality future inextricably link developed and developing countries. They require global co-operation at the highest political levels and in the local actions of all people. Failure to solve these conundra may leave us all with pollution, resource depletion and complex issues for waste management. Avoidance of pollution and resource degradation is a complex global issue which affects us all, together.

Pollution is not new. The earliest pollution legislation in the developed world clearly demonstrates this. Air Pollution legislation was first passed for London in 1273, limiting the use of coal in an attempt to reduce air pollution (Farmer, 1997). Precursors to modern UK legislation predominantly date from the 19th century, reflecting accelerating industrialisation and developing health concerns, for example in the UK, Alkali Acts dating from 1863 and the River Pollution Prevention Act, 1876. However, appreciation of the root causes of pollution, the link with the global biogeochemical cycles and the need for a holistic view and global legislative frameworks has only recently been clearly understood, or at least only recently been acted upon by politicians. The Long-Range Transboundary Air Pollution (LRTBAP) agreement, Montreal Protocol, Rio and Kyoto Protocols, and the formation of the Environment Agency (in the UK) are all examples of this newer integrated approach.

9.2 Pollution defined

An important first consideration is our definition of pollution. It is not a simple concept. Some (authors) include 'natural phenomena' as pollution sources and pollutants. They would, therefore, class dust released from a volcanic eruption as pollution. Others consider that natural phenomena cannot be pollutants; they become so only because of our anthropocentric view of our planetary home. Including these natural phenomena diverts us from the more urgent issue of avoidable pollution, which is the by-product of human activity. In this chapter we will focus on pollution as a human-induced phenomenon; a problem that could be minimised by better understanding of our environment, more appropriate use of planetary resources, or perhaps by an adjustment in our expectations in terms of lifestyle.

A widely used definition of pollution is that proposed by Holdgate (1979, p. 17):

> The introduction by man into the environment of substances or energy liable to cause hazards to human health, harm to living resources and ecological systems, damage to structure or amenity, or interference with legitimate uses of the environment.

The 1996 EU directive on Pollution Prevention and Control uses the following very similar definition cited in Farmer (1997, p. 3):

> Pollution shall mean the direct or indirect introduction as a result of human activity of substances, vibration, heat or noise into the air, water or land which may be harmful to human health or the quality of the environment, result in damage to material property, or impair or interfere with amenities and other legitimate uses of the environment.

Both definitions of pollution stress the link to harm to humans, environment systems or property. This, theoretically at least, is a key distinction from contamination. A contaminant has been defined as the presence of an elevated concentration of a substance in water, air, sediments or organisms but one that does not, apparently, cause any harm; but what of the risks? Our understanding of harm can change, contaminants can bio-accumulate or act synergistically and thus become pollutants and so forth. Last (1987) suggests that this distinction is unwise and unhelpful. Ignoring the risk that contaminants may pose promotes a false sense of security. This viewpoint is adopted in this chapter. Too often we have seen substances which were thought to be harmless contaminants reclassified as pollutants. An example would be our changing understanding of lead in the environment. Identifying natural baselines is important here and difficult to do with certainty. However, this distinction may have considerable practical importance for identifying environments capable of absorbing waste without, apparently, generating pollution, or at least not doing so within several hundred years. An example is the assessment of the potential for waste dumping at sea (Angel and Rice, 1996).

Human additions to the environment may be new substances, for example organochlorine pesticides, which have posed exceptional problems in that their environmental behaviour has proved difficult to predict. Though in recent history, and especially since the publication of *Silent Spring* (Carson, 1962), all such substances were subjected to rigorous toxicological testing, their ecological properties and, in particular, their role in ecosystems was poorly understood. In particular the role of pollution transfer in ecosystems not just at local scales, but at regional, national and international levels was neither understood nor was the scale of potential impacts appreciated. Thus we now see Arctic polar bears, Antarctic penguins and associated fauna with levels of DDT and other organochlorine derivatives in body fat sufficiently high to cause detectable changes in morphology and behaviour, including reproductive success. Linked with the additional hazard of habitat loss due to global warming this may prove very damaging to these biota. Similarly we have experienced thinning of the stratospheric ozone layer from the breakdown of chlorofluorocarbons, used as refrigerants and in aerosols. Scientific analysis of these problems shows a common root; our failure to understand or predict the behaviour of these new substances in biogeochemical cycles. Sadly it also revealed our unwillingness to learn from past mistakes. We replace one disgraced chemical with a new hopeful, for example organochlorines with organophosphates, chlorofluorocarbons (CFCs) with hydrochlorofluorocarbons (HCFCs) or hydrofluorocarbons (HFCs). Rarely do we rethink the problem and adopt a new approach (section 9.5).

Of course not all pollutants are new synthetic substances; some such as the excess nitrates and phosphates that trigger the problems of cultural eutrophication in aquatic ecosystems represent the unusual build up of a naturally occurring substance at a particular point in a biogeochemical cycle, due to human activity. While the immediate and obvious cause of eutrophication in a water body may be discharge of nitrogen fertiliser from neighbouring agricultural land, or phosphates from a sewage treatment plant, the root causes sit further back in the story, in our understanding of biogeochemical cycling (section 9.3.2). Box 9.1 examines eutrophication. The general principle of this story is repeated

Box 9.1: Eutrophication in freshwaters

Lakes and rivers typically show low levels for two key nutrients essential to primary production, namely nitrates and phosphates. These levels reflect catchment characteristics and naturally low levels of nitrate and phosphate in most soils. Lakes can and do naturally increase and decrease in nutrients reflecting catchment changes, such as following flood. As such, the classic story of natural eutrophication as a process of gradual enrichment over time may be oversimplified. The particular issue with culturally accelerated eutrophication (commonly referred to as eutrophication) lies in the rapidity of change and its clear link to human activity (Table 9.1).

Table 9.1 Changing phosphorus loadings in the Norfolk Broads system. The links with agricultural change and intensification and with improved waste disposal for an increased urban population are clear (Adapted from Taylor, 1990; compiled from data in Moss, 1979.)

	μg l^{-1} phosphate
Clear water upland lake	5
Naturally fertile lowland lake	10–30
Norfolk Broads pre-1800	10–20
Norfolk Broads following land enclosure	up to 80
Norfolk Broads following conversion to modern sewage disposal and population increase	max 2000 typically 150–300

Continued on page 233

Continued from page 232

Accelerated eutrophication triggers a number of changes, which are generally undesirable for temperate mid-latitude lakes. Loss of commercial fish species, foul-tasting drinking water, and loss of amenity are among the well-documented changes. In tropical and developing areas, some changes, such as enhanced productivity in a fishery or increased potential for floating garden cultivation may prove temporarily beneficial. Much depends on the exact nature and extent of the burden and the precise ecological details of the receiving system.

In general terms additional primary productivity favours a shift to rapid growing and less palatable phytoplankton species. Excess phytoplankton growth is not all consumed. In shallow lakes it may cast a shadow over bottom growing plants and, by limiting their photosynthesis, remove potential oxygenators from these waters. The excess phytoplankton will sink and decompose and place an oxygen demand on bottom waters. In deeper systems where thermal stratification has set in and hypolimnial waters are locked out of contact with the surface atmosphere, this additional oxygen demand will place pressure on associated fauna and a spiral of death and decomposition may occur. The classic story of eutrophication was based on work on deep mid-latitude lakes, which showed this thermal stratification pattern. The issue was thought less crucial for shallow, unstratified waters and tropical waters. However, the evidence from lakes such as Lough Neagh in Northern Ireland, the Broads in East Anglia, UK, and numerous shallow or unstratified tropical lakes, shows that for these waters eutrophication is also of real concern. In these shallow lakes, the role of sediments as a store of nutrient that can be re-released to waters is very important (Moss, 1998).

The chief causes of eutrophication appear to be nutrient run-off from agricultural land, which is the primary source of nitrates, and waste from treated or untreated sewage, which is the major source of phosphates. As modern problems these both reflect not only rapid human population growth but also our concentrated settlement patterns and therefore concentrated patterns of waste disposal. Also important are our increased expectations in terms of sanitation and our demands for increased agricultural production, which have largely been met by monocultural systems supported by high levels of fertiliser use. In the UK it has long been assumed that while nitrate fertilisers run off land fairly easily, the natural soil deficit in phosphates and different binding mechanisms mean that phosphate runoff from agricultural land is not significant. However, recent work (ENDS, 1996; Hooda *et al.*, 2001) suggests that agricultural soils in the UK are becoming phosphate-saturated and that new fertiliser additions are accompanied by increased run-off of phosphates. If this change is widespread, this would trigger a major rethink of our methods for solving the eutrophication problem.

Continued on page 234

Continued from page 233

Solutions mainly focus on the principle that phosphorus is a point source of pollution and thereby easier to control than nitrate pollution, which comes from diffuse overland flow. Since addition of both is implicit in the eutrophication story, control of one, namely phosphorus, should ensure reduction of eutrophication. This, of course, is only effective where action is pre-emptive or in remedial situations where no phosphate is re-released from nutrient-enriched sediments. The importance of sedimentary release was shown by the restoration experiments at Alderfen and Cockshoot broads in Norfolk. Both the broads were isolated by damming from the nutrient rich river flow. At Alderfen no further action was taken and, despite some initial recovery, renewed phytoplankton blooms were soon re-established. A major factor was phosphate release from sediment within the broad. This was caused by decomposition of natural early Spring phytoplankton growth which triggered mobilisation of phosphate from the bottom sediments. Subsequent extensive summer blooms of blue-green, nitrogen-fixing cyanobacteria occurred. At Cockshoot, where isolation was accompanied by dredging, better long-term recovery was achieved. However, there is some evidence to suggest this was also, in part, due to a decline in predatory fish following disturbance during dredging. This in turn enabled population increase in herbivorous grazers, which helped to control phytoplankton regrowth (Mason, 1996; Moss *et al.*, 1996; Moss, 1998).

Solutions in the natural environment are limited and, as shown, can be complex in practice and may be expensive. In principle, several options that are adopted will depend on careful analysis of the local circumstances. Options include chemical precipitation; removal of predatory fish so that herbivorous grazers of phytoplankton can increase; removal of nitrates and phosphates by biomass harvesting; and removal of nutrient-enriched sediments by dredging. In some cases addition of barley straw has successfully mopped up nutrients, though the precise mechanism for this control is uncertain. In East Anglia, bundles of alder twigs have been used to provide shelter for herbivorous grazers where natural shelter, such as fringing reed beds, has been lost. However, much more effective than remedial action in the natural environment are controls on the release of phosphates and nitrates. These may avoid change in the first place, and in already damaged systems offer the chance of a sustained long-term improvement once remedial work has been undertaken. In the case of nitrate, where 70% of the input is thought to originate from water draining from agricultural land, a change in agricultural practice seems essential. New, European Union (EU), legislation identifying nitrate vulnerable zones (NVZs) was introduced in 1996 and implemented in the UK in December 1998. In the latest 2002 revision, 47% of England falls within a NVZ. The legislation targets areas which contain

Continued on page 235

Continued from page 234

ground or surface waters with nitrate concentrations in excess of 50 mg l^{-1}, or which may soon reach these levels unless preventative action is taken. The immediate aim is to reduce the risk of nitrate rich run-off. To this end a range of restrictions have been placed on farming operations, with a fine of up to £20 000 for non-compliance. These primarily focus on restricting fertiliser use during late summer, autumn and winter times when plant uptake is negligible and run-off in storms may be maximised. Farmers are also required to avoid fertiliser application close to streams and drainage ditches, to leave a 10 m buffer zone near open waters, and to avoid application on steeply sloping lands. They are required to ensure a precise match between their fertiliser use and specific crop needs. The nitrate load in organic manures must also be fully evaluated and use of these is subject to similar restrictions (DEFRA, 2002a). Expert advice for farmers will be needed to ensure effective compliance with these guidelines. However, the potential exists for substantial saving in the farm economy, as well as benefits to wild life and society in general via uncontaminated water supply, with less wasteful use of this key resource.

For phosphate, control via phosphate stripping has seemed the solution and is now fairly routine at sites in sensitive locations in the developed world. Phosphates recovered can be reused as fertiliser thus avoiding the need for new supplies (provided a biologically non-toxic precipitating agent is used). However, the new evidence that agricultural soils have become phosphate saturated and give rise to phosphate-enriched run-off suggests that a more radical approach may be needed here too. Phosphates can be removed from detergents and significant reductions have occurred over the last 15 years. However, as discussed in section 9.2, care is needed while introducing replacement chemicals to ensure that a new hazard is not generated. Since we have a known remedy in this context, it may be sensible to use this approach to avoid water pollution rather than risk inventing a new problem.

worldwide; for example in Lake Washington, USA, the Norfolk and Suffolk Broads in England, Lake Nakuru in Kenya, the Albuferas in Southern Spain. The impacts of waste generated by our intensive agricultural systems and our increasingly urban lifestyles are new and pressing concerns worldwide. The special case of municipal waste forms a focal theme in section 9.4.

Land contamination by mining and mineral prospecting also generates pollution. Again this fits our model of disrupted ecosystems. Mine wastes typically represent materials from depth freshly exposed at the earth's surface. This also represents an excessive addition or release of a naturally occurring substance from the deep inactive reservoir pool into the active cycling part of the biogeochemical cycle. Exposure to oxygen, water and natural acids typically triggers chemical reactions and release of heavy metals, in toxic quantities rather than in the low amounts typical of surface-weathered materials and soils. Contamination of drainage and river systems seems an inevitable result.

9.3 The root causes of pollution

9.3.1 The Global Commons *aka* 'Global Dustbins'

The root cause of pollution sits with the tension between our rising global population, and our rising expectations in terms of living standards, and our failure to pay proper attention to the ideas encapsulated in the 'Tragedy of the Commons' (Hardin, 1968). For centuries we have used the Global Commons, air, water and land, as 'dustbins' for human waste. We have used our Commoners' rights to equal and unfettered access as a waste disposal strategy and we have assumed that the natural absorptive capacity of the Global Commons, especially the atmosphere and oceans, would dilute and biodegrade our wastes until they were of no significance. At the same time we need and expect the Global Commons to provide safe, clean, usable materials. While human populations were small and widely dispersed this behaviour caused few problems. Biodegradation and removal in the natural course of biogeochemical cycling broke down, dispersed and made available for reuse the elements contained within human wastes. As population levels have risen and settlement patterns have changed from small family groups, to village communities to our modern towns and large urban populations, disposal of waste has become a major potential pollution source. The key issue here is the gathering together of resources (food, energy, building materials, etc.) from a surrounding region but with consumption and waste generation focused in one place.

In this context, our 21st-century urban lifestyle poses a major challenge. Cities import resources from across the globe. Their residents consume and dispose of those materials locally (at least in the first instance). As by-products they cause pollution of the local atmosphere and water courses. In particular they generate household and small commercial waste, namely municipal waste. Affluent cities seek to dispose of this waste away from the main centre, that is to export the waste and thereby maintain a good local environmental quality. For example, 72% of London's municipal waste goes to landfill, 90% of which is situated outside Greater London (Figure 9.1). This approach generates additional pressure on the surrounding countryside and in some circumstances, especially in developing countries, may create potential pollution hazards such as groundwater contamination where landfills are not properly sealed. Urban waste disposal may also generate global pollution problems, for example 25% of the UK's methane emissions (a major greenhouse gas) come from landfill (AEA Technology, 2002). The potential for dioxin release in waste incinerators similarly is not just a local

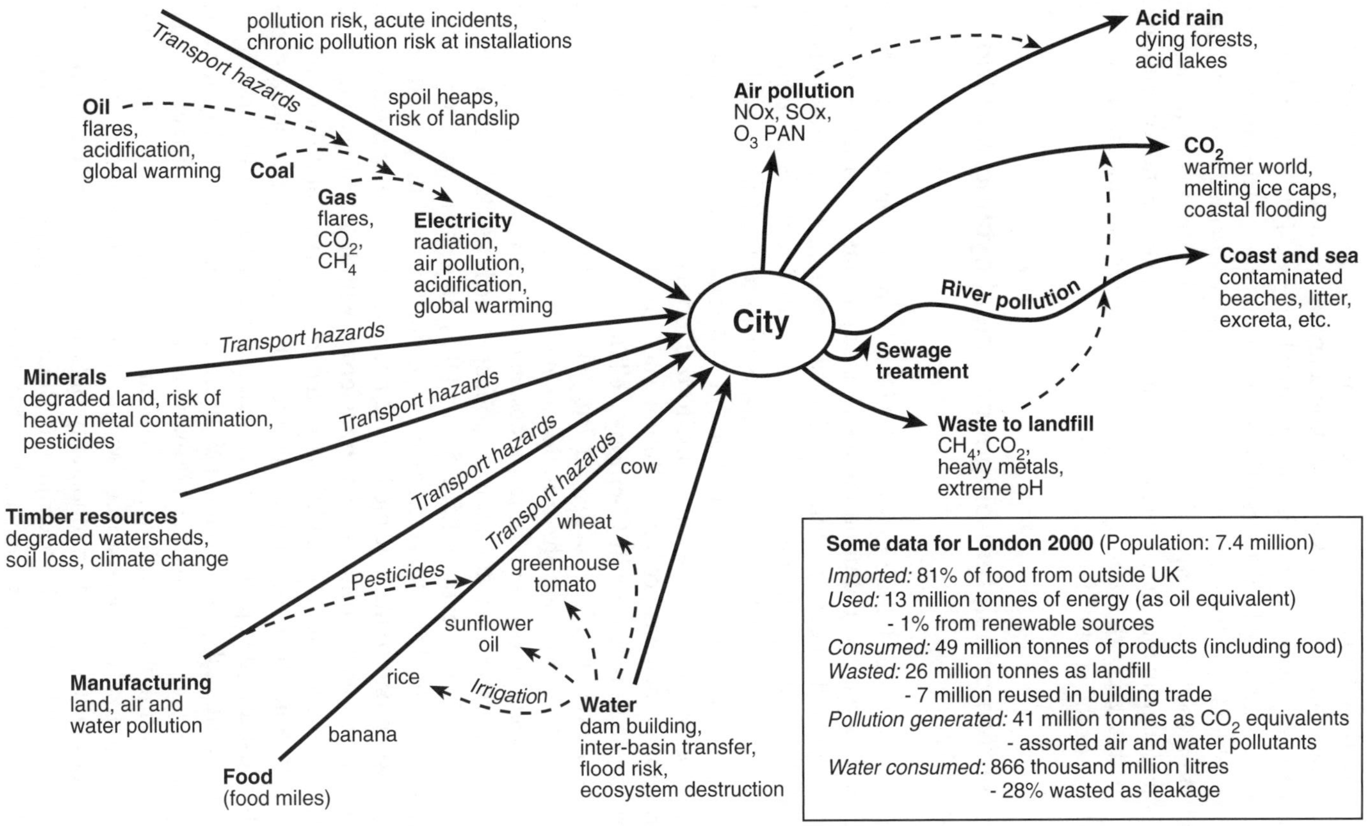
City
pollution risk, acute incidents,
chronic pollution risk at installations
Transport hazards
Oil
flares,
acidification,
global warming
Coal
spoil heaps,
risk of landslip
Gas
flares,
CO_2,
CH_4
Electricity
radiation,
air pollution,
acidification,
global warming
Minerals
degraded land, risk of
heavy metal contamination,
pesticides
Transport hazards
Timber resources
degraded watersheds,
soil loss, climate change
Transport hazards
Manufacturing
land, air and
water pollution
Transport hazards
Pesticides
Food
(food miles)
Transport hazards
cow
wheat
greenhouse
tomato
sunflower
oil
rice
banana
Irrigation
Water
dam building,
inter-basin transfer,
flood risk,
ecosystem destruction
Air pollution
NOx, SOx,
O_3 PAN
Acid rain
dying forests,
acid lakes
CO_2
warmer world,
melting ice caps,
coastal flooding
Coast and sea
contaminated
beaches, litter,
excreta, etc.
River pollution
Sewage
treatment
Waste to landfill
CH_4, CO_2,
heavy metals,
extreme pH
Some data for London 2000 (Population: 7.4 million)
Imported: 81% of food from outside UK
Used: 13 million tonnes of energy (as oil equivalent)
- 1% from renewable sources
Consumed: 49 million tonnes of products (including food)
Wasted: 26 million tonnes as landfill
- 7 million reused in building trade
Pollution generated: 41 million tonnes as CO_2 equivalents
- assorted air and water pollutants
Water consumed: 866 thousand million litres
- 28% wasted as leakage

Figure 9.1 The city as an importer of resources and a generator and exporter of pollution and waste. Data shown for London are adapted from Best Foot Forward (2002). For further discussion of the environmental impacts of cities refer to Chapter 8; Table 4.2; McGranahan and Satterthwaite, 2002.

hazard. Long distance transport and export of these and other waste gases in upper tropospheric wind systems (as exemplified in the acid rain story) (Park, 1987; Bell and Treshow, 2002) may link to transboundary air pollution problems and will almost certainly generate downwind regional effects. Rivers enriched by urban sewage ultimately add a nutrient burden to coastal waters, transferring the pollution problem to a distant environment, but not eliminating it. A key challenge of pollution control is to avoid pollution transfer. Urban wastes present an exceptional challenge in that they represent an unpredictable cocktail of polluting substances. In this context waste disposal is undoubtedly an urgent pollution control priority for the start of the 21st century.

9.3.2 Ecosystem principles: biogeochemical cycling and the flow of energy

A second key issue is our failure to see pollution within the context of the basic functioning of ecosystems. Pollution comes in many forms. We routinely refer to air, fresh water and marine pollution and land contamination, often making little connection between pollution incidents in these different media. Typically, we analyse quite independently, marine oil spills such as the Exxon Valdez off the coast of Alaska in 1989 or the Prestige off NW Spain in 2002; accidental contamination of water supply as at Camelford, Cornwall, UK, in 1988; and major international incidents of air pollution such as occurred following the explosion at the nuclear reactor in Chernobyl, Ukraine, in 1986 or the accident at the Union Carbide chemical factory at Bhopal in India, in 1984. However, a moment's thought shows that these incidents have related causes and consequences. They reflect our demand for resources, and our unwillingness to pay the true environmental costs that attention to concepts such as the Precautionary Principle or Polluter Pays (sections 1.5, 2.5.2 and 9.5) suggests is wise. They also demonstrate one of the key characteristics of pollutants: that they are readily transferred between environmental media and become widely dispersed within the environment. In short, pollutants are transferred within the biogeochemical cycles. They are uncharacteristic additions to the environment and their impact has spread far beyond their original source area. The Chernobyl incident especially clearly illustrates this point. Major contamination of air, marine and freshwater, soils and ultimately food chains has been experienced, not only locally, but across Europe and much of the northern hemisphere. The effects of the accident in 1986 continue to be felt in terms of human health and lost agricultural production and are likely to persist (e.g. Wynne, 1991; Smith *et al.*, 2000). Data from Belarus show 170 cases of childhood thyroid cancer diagnosed in the period 1986–1992 compared with 21 cases in the Belarusian population during the preceding 20 years. On health grounds, 20% of Belarusian arable agricultural land was declared unusable (Ministry of Foreign Affairs, Republic of Belarus, 1994), though this may overestimate the problem due to confusion associated with the breakdown of collective farming linked with changes due to 'perestroika'.

Major effects from Chernobyl radiation have also been seen in northern Sweden and Norway. Here the specialised cationic exchange properties of lichens resulted in accumulation of deposited radionuclides to levels in excess of 100 000 Bq kg^{-1}. Since lichens form the main food supply for reindeer, they in turn showed radioactivity levels far in excess of Swedish food safety guidelines. This rendered the livestock unusable for the Lap populations, not just for food but more importantly as a tradable commodity. Thus the Laplanders' way of life became threatened, a quite unpredictable consequence of distant environmental pollution release (Farmer, 1997). In Britain the passage of the radioactive cloud coincided with intense thunderstorm rainfall over Cumbria and North Wales. To this day, sheep reared in these areas are still subject to special controls with occasional individuals showing levels in excess of the UK food safety thresholds of 1000 Bq kg^{-1}. It is estimated that restrictions on the marketing of sheep reared on hill farms in North Wales and Cumbria may continue for another 15 years. In the acid peaty soils found locally in these areas, caesium does not bind tightly to the soil constituents and so is absorbed by plants and accumulates in the sheep. Though some radiation is removed when sheep are harvested, much goes back to soils during their lifetime, via their faeces and urine and thus the cycle continues.

What this example most graphically tells us is that pollutants are readily transferred between environmental media and that while we may expect the radioactivity of specific particles to decline over time, chemically the materials may be stored in growing biomass, in litter, in soils or in organic sediments. Thus this overall decline may be offset locally by bioaccumulation processes, by natural recycling or by release from temporary soil storage. Radioactivity can also be re-released in chance catastrophic events such as forest fires or more gradually in subsequent biodegradation or acidification. Even today, 16 years after the Chernobyl explosion, in Belarus, Ukraine and parts of Russia unacceptably high levels of radioactivity are present in end of chain foods such as fungi and lake fish and in storage foods such as forest fruits in locations well beyond the 30 km exclusion zone (Figure 9.2).

Chernobyl and its consequences have also highlighted the problem of scientific uncertainty. Much has been learnt about the behaviour of radioactive substances in environmental systems since the Chernobyl accident. It is also clear that much remains to be understood. In 1986 the UK government scientists advised farmers in Cumbria and North Wales that sheep would be back in the food chain 'within a few weeks'. Now it is estimated that restrictions will be necessary at least until 2016, 30 years after the accident at Chernobyl and 100 times longer than the original estimates (Smith *et al.*, 2000). Recent research has also shown that estimates of the effective ecological half life for caesium 137 in terrestrial vegetation and lake waters in Cumbria, and in lake fish in Norway, rose from between 1 and 4 years in the first 5 years post Chernobyl to between 6 and 30 years by 2000. These data are,

Figure 9.2 Roadside selling of forest fruits and fungi in Northwest Belarus. Though just south of the popular health resort of Narach, 500 km from Chernobyl, local radiation hotspots and bioaccumulation processes can render the produce hazardous. Levels in excess of 5000 Bq kg^{-1} have been found in these popular Belarusian products.

perhaps, a major argument in favour of 'renewables' on the basis of the Precautionary Principle when evaluating future energy supply alternatives (Chapter 7).

As noted in section 9.2, human activity has also caused pollution by generating excess of otherwise beneficial, natural substances. A good example is the nitrogen cycle. The active volume of the global N cycle has been increased via industrial fixation of nitrogen from the atmospheric reservoir store in the Haber–Bosch process (Figure 9.3). What is missing here is a clear counterbalancing link. This should cause us concern. Has denitrification speeded up to accommodate this change? Does increased primary production matter? Where is the excess active nitrogen now? Will it cause harm to humans or to the ecosystems on which we depend? In practice the main wastage from the immediately active cycle comes with leaching of fertiliser from excess use on agricultural crops. This waste, as mentioned, may generate problems of eutrophication in waters and thus be reclaimed to the active cycle rather than running to deep-sea sediments and long-term storage or undergoing denitrification. Increased production due to increased nitrates will accelerate the consumption of other nutrients; it will also lead to species change as slower growing species become outcompeted. The effects are not confined to the biogeochemical cycle of a single element.

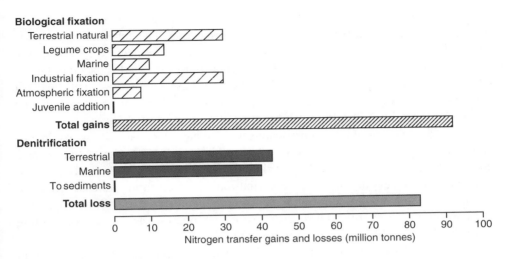

Figure 9.3 A global balance sheet for nitrogen. (Reproduced with permission from The Nitrogen Cycle by C.C. Delwiche. Copyright © 1970 by Scientific American, Inc. All rights reserved.) These estimates show that the rate at which nitrogen is being introduced, in a fixed and biologically accessible form, to the active biosphere from the inert atmospheric reservoir store exceeds denitrification (return to the atmospheric store) by about 9 million tonnes per annum. This implies a build up of nitrogen compounds in the biosphere: in soil, ground water, rivers, lakes and oceans.

The excess primarily reflects industrial fixation and especially fertiliser production. In fertiliser use, not all is taken up by target crops, and nitrogenous products commonly leach into waters draining agricultural land (Box 9.1). There is no obvious counterbalancing mechanism for industrial fixation. One new suggestion is to use GM techniques to introduce denitrification abilities into a wider range of soil and water bacteria. This raises many concerns including the potential for irrevocable ecological change, environmental degradation, loss of important food resources and major economic losses which might arise if these modified organisms spread, or behaved, unpredictably (Fisher and Fisher, 2001). Counterbalancing this view is our uncertainty regarding the long-term consequences of human-induced nitrogen imbalance and the potential to design genetic safeguards against excessive spread of a GM organism.

Related changes in soil acidity, induced by changed nitrogen status, may change the uptake of phosphates and retention of beneficial cations. This may include essential micronutrients, which may be difficult to replace and may, in turn, generate habitat change. Increased use of fossil fuels in combustion has also released additional and varied nitrogen oxides (NO_x) to the atmosphere. NO_x derived from traffic exhausts may be deposited locally where they may enrich soils leading to vegetation changes particularly in ecosystems, such as heaths, which typically occur on nutrient poor soils. NO_x released in large-scale combustion plants can be transported large distances contributing to acid rain and to formation of toxic secondary air pollutants. These changes have major implications for ecosystems and for human health and may involve pollution export from source to recipient country as has been seen between the USA and Canada and between the UK and Norway (Bell and Treshow, 2002; Farmer, 1997).

A similar complex cascade of effects can be seen when the impact of pesticides is examined from an ecological perspective. The schematic summary, shown in Figure 9.4, shows that the effects of pesticide use reach way beyond the relatively simple cases of direct and accidental poisoning and bioaccumulation. When we step back from the detail to examine these major flows and boxes it becomes clear we do not really know what is happening. An analogy with the 'Blind Watchmaker' (Dawkins, 1986) may not be inappropriate here.

That we should be more cautious is clear from developments in our understanding of the global carbon cycle over the past 20–30 years. Models from the 1960s attributed CO_2 build up in the atmosphere almost exclusively to fossil fuel use. Our understanding of the contribution of deforestation, release from soil organic material and drained peatlands is more recent. It is also still incomplete as the current debate over the appropriateness of tree planting as a means of meeting carbon (C) tariff obligations reveals (section 7.8.1).

Less widely discussed, but still relevant, are changes in energy flow in ecosystems. Energy is important as the ultimate driver of the planetary ecosystem and of all systems at smaller scales. Quantitative and qualitative changes, in response to atmospheric changes due to dust pollution or, more profoundly, due to changing global climate regimes have the potential to radically change ecosystems. More locally noise pollution also represents an energy change. Though more commonly examined in terms of human impacts noise pollution can also disrupt wildlife, for example affecting the breeding behaviour of birds whose mating and alarm calls may become indecipherable amid traffic noise.

Harnessing the natural energy of radiation and motion (hydro-electric power, wind, wave and tidal energy) also links profoundly with the patterns of our planetary and regional ecosystems. As far as we are aware no adverse effects on energy budgets are triggered by our temporary harnessing of this energy, and from an ecosystem perspective these renewable energy sources seem less disruptive than fossil fuel and nuclear alternatives. They can also be harnessed locally, close to the point of use. This avoids long distance transport of fuel, which is a major environmental benefit as it reduces the risk of oil spills and chronic pollution associated with tanker transport at sea and pipelines. In short it complies with the Proximity Principle, an important concept in waste management. The recent decision by the EU, following the Prestige incident, to ban single hull oil tankers from European coastal waters seems very belated given that the first major catastrophe in these waters occurred more than 30 years earlier with the wreck of the Torrey Canyon off the coast of Cornwall in 1967. These oil spill incidents emphasise the importance of geographic separation between source areas and consumption points for major resources. They highlight the need for a fundamental rethink of our resource use patterns, a point we will return to later in this chapter. On the other hand, even renewables, such as wind energy, are not without environmental controversy. Issues of noise, bird mortality, aircraft safety and blighted landscapes are frequently made complaints against wind generated electricity.

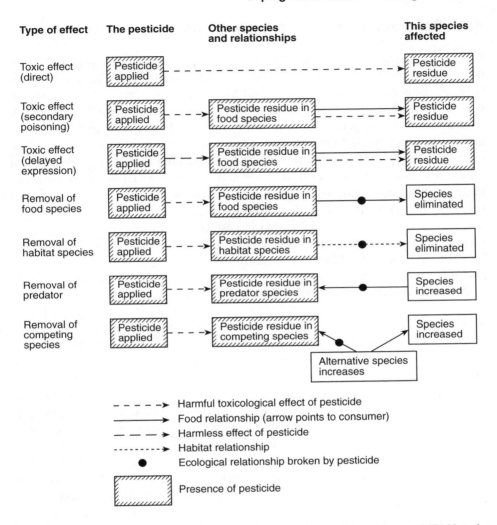

Figure 9.4 A schematic summary of the pesticide problem. (Adapted from Moore, 1967.) Note that direct toxicity of a target (or accidentally poisoned) species is just one possible effect. Pesticides may also eliminate competitor species, changing food chains and leading to rapid population growth of an alternative species and its predators. They may lead to elimination of beneficial habitat species, as in scrub control, or, by affecting a predator species, pesticides may generate population growth of a prey species. In these cases, affected species are not pesticide-contaminated (unshaded boxes). They are affected because key elements of their life strategy have been eliminated. Pesticides may also metabolise within organisms changing from sublethal to lethal forms within the food chain giving a delayed toxic effect. In migratory species, delayed biotic effects may be expressed many hundreds of miles from the original point of application.

9.3.3 Population–resources inequalities

Rising human population has long been considered a major cause of global pollution. Indeed the predictions of Malthus and neo-Malthusians like the Ehrlichs (Ehrlich, Ehrlich and Holden, 1977) and Meadows *et al.* (1972) in the *Limits to Growth* reflect this view. Though these early models are now largely discredited, they importantly triggered our awareness of the finite nature of earth resources and the potential impact that the demands not only of a larger global population but a population all enjoying a western-style level of development might place on the earth's ecosystems. The basic tenet related an exponential rise in human population with an arithmetic rise in available resources, usually evaluated as food supply and, more particularly, wheat. These models are, however, subject to major difficulties in that baseline data are hard to establish with any accuracy. The wide ranging predictions for population growth in the next century, at most 17.5 billion by 2105, at least 5.7 billion and most likely 8.9 billion, mean that as practical planning tools coping with resource consumption, let alone impacts, these models offer little help. Even if we could quantify the potential impact of each individual under current scenarios, we would still have scant basis for predicting the future with any certainty (see section 1.3 for further related discussion).

Amassing reliable data, not just for cereal food consumption but for a typical diverse diet, for other essential commodities such as water and energy consumption and for associated pollution and waste production, presents an almost insurmountable challenge. We will be matching one uncertain data set with another series of dubious reliability. Nevertheless, attempts have been made. Arizpe, Costanza and Lutz (1992) estimated that for a sustainable global economy at present levels of western consumption a population of 2.5 billion would be the maximum possible for everyone to have an average standard of living equivalent to that of Spain in the 1990s. Clearly, we have much more work to do to understand the complexities of human population growth, resource consumption, and sustainable development with equality of living standards. Further, useful reviews and discussion of this complex topic can be found in Adger and O'Riordan (2000) and Graham and Boyle (2002).

A common theoretical approach for estimating the global impact of human population on the environment in terms of resource consumption is shown in the equation (section 1.3):

$$\text{Impact (I)} = \text{Population (P)} \times \text{Affluence (A)} \times \text{Technology (T)}$$

However, in the context of pollution and waste, technology is both a generator of impacts and a source of solutions. Again this testifies to the complexity of the population to resource use relationship. Since over 80% of the increase in predicted global population will come from just seven rapidly developing countries namely, China, India, Nigeria, Indonesia, Bangladesh, Pakistan and Brazil

(Adger and O'Riordan, 2000), improving our understanding of this relationship is of major significance. If these people are to enjoy a standard of living equivalent to our current western lifestyle then it would seem to follow that there will be increased consumption of resources, with associated pollution impacts and increased challenges for waste control. The way forward may be through acknowledging that affluence, and high consumption of natural resources, does not necessarily bring improved quality of life.

More important perhaps is our human response to this potential threat. Do we ignore potential problems (Arenostruthius), do we see this as Gaia controlling a troublesome pest, namely us, are we Cornucopians, optimistic for a technological fix, or are we Armaggedonists certain of a degraded planetary future but without the comfort of a Gaian purpose? Holdgate (1994) suggests these different scenarios exist because the data for resource consumption, pollution and our models for human population growth are so uncertain. They lack a theoretical framework such as an ecological perspective offers. However, simple transfer of population ecology models to human populations encounters many difficulties. Our perception and use of resources constantly change as we develop. Thus the models never stabilise. Attempts to define carrying capacity from these models have failed. Holdgate suggests that the way forward lies in changing our focus, in an alliance of ecology and economics: in short, in the proper costing of resources.

9.3.4 Costing the earth?

It has often been claimed that markets are good at setting prices but incapable of recognising costs (Tickell, 1996). In this short subsection it is not possible to engage in a full debate about the cost of pollution and more importantly how 'improper costing' of resource use contributes to pollution. We can only highlight some key points for thought. Fuller analyses of this important, developing subject can be found in Adger (2000).

Prices are indicators. We need them to tell us the truth about costs. But market prices rarely tell us the true environmental cost of goods. Even where these are acknowledged, they may be difficult to measure in equivalent terms. Tickell (1996) uses the analogy of the price of coal pointing out that what it does not include are the costs to individuals and societies which arise from burning coal. We may assume it does include the costs of mining and the costs of fuel and infrastructure for transport and distribution. Recent legislative controls mean we may also include the costs of flue gas cleaning. Missing, however, are the health costs, which stretch to the on-costs to disrupted families and society caused by illness and premature death. Also excluded are the wider pollution costs, in this instance most notably costs of acidification, which are not just damaged habitats and lost potential but also include damage to agricultural crops and commercial forestry, rivers, lakes and fisheries and may also extend to accelerated erosion of building materials. Other effects will include contributions to global warming from CO_2 emissions. Analysing the full picture from resource extraction to manufacture, use and disposal including the side effects presents a Herculean task. Wider social costs as well as the immediate technological and distributional costs need evaluation. Conventional economics too often treats environmental costs as externalities. Even where costs are internalised through imposition of legislative controls and taxes, these usually focus on end of pipe solutions and costs of remediation. The more hidden costs of resource extraction and its environmental consequences, perhaps thousands of miles distant from the point of use, the issues of transport and the problems of packaging and its disposal are less often considered or included in the price. The aggregates levy introduced in the UK in 2002 is one example of an attempt to address these externalities.

Environmental economics is only beginning to devise and refine models to account for environmental costs. One alternative approach is ecological footprinting which estimates how much land or water area is required to produce all the resources consumed and to absorb all the waste generated by an individual, product, institution or country (sections 1.4, 8.3.2). Schmidt-Bleek (1994) developed the concept of 'ecological rucksacks' as a practical means of evaluating on-costs of minerals. von Weizsäcker et al. (1998), using this approach, estimated that a 10 g gold ring carries a rucksack of 3 tonnes. All goods and services have ecological rucksacks, in other words costs and impacts associated with extraction of raw materials, its processing and transport. For example, a catalytic converter weighs less than 9 kg but carries a rucksack of more than 2.5 t mainly due to the platinum that is used. This burden or rucksack could be reduced by using recycled platinum as opposed to virgin material (von Weizsäcker et al., 1998).

True costs are politically difficult to use if consumers are unwilling to pay. Can learning true costs focus us more clearly on distinguishing between needs and luxuries, and are we willing to forgo established luxuries for the sake of other's needs? This underpins a major part of the north–south debate, over inequality in resource use, food supply, water, energy and so forth. It is also relevant in the context of pollution and waste. Are we willing to accept distant pollution and pesticide manufacture and its use in food production, which we would reject in our own back yard? Can we exchange our pollution or waste burdens when expedient to encourage our own development? Does development have to mean more resource exploitation? Must it be measured in Gross Domestic Product (GDP) or can we revise our assumptions and place a greater premium on quality of life?

9.4 The special case of waste

A visitor from outer space, examining modern industrial economies, might well decide that their main purpose was to turn raw materials into waste.

UNEP, 1999

Waste is a locally arising problem with both local and global consequences. Society's need to dispose of its waste products creates a major source of pollution. Air pollution from transport originates from the waste products of a vehicle's combustion process; discharges of effluent contain substances that can no longer be used in an industrial process; even agricultural run-off contains wasted nutrients from fertiliser application. As cities pull in resources from around the globe to service their infrastructure (Figure 9.1), they produce significant amounts of waste which often cannot be dealt with within the city and which has to be transported beyond the city perimeters. Management requires safe collection and disposal, but sustainable management requires an integrated system. This includes large increases in efficiency, investment, economically feasible recycling with a stable market for end products and shouldering of responsibility by governments, industry and the individual.

9.4.1 Waste production

Waste is produced by all activities in a product's lifecycle: mining raw materials, transportation between all stages, refining, manufacturing, packing, selling, using and final disposal. In 1990, over 1 billion tonnes of municipal solid waste (MSW) was produced globally, approximating to 2/3 kg per person per day (Beede and Bloom, 1995, cited in van Beukering, 1999). Definitions of MSW vary across the world, making detailed data comparisons difficult, but it is primarily household waste, and assorted other wastes typically collected by local authorities. Although MSW is often a relatively small fraction of total waste produced, with mining, construction and demolition contributing the major share, all waste production is ultimately driven by the consumer's lifestyle and requirements of goods, services and energy (Table 9.2).

There is a significant difference in both quantities and constituents of waste produced by prosperous and developing countries. Overall, the USA produces approximately 10 billion tonnes of solid waste per year, 33% of the world's solid waste from only 4.6% of world population (Miller, 1998). In 2000 the USA produced 2 kg *per capita* per day of municipal waste alone (EPA, 2002). By contrast, waste production in India is estimated to be at 0.33 kg *per capita* per day of

Table 9.2 Waste production in England, 1998–1999 (Data source: DEFRA, 2000)

Source	Quantity (million tonnes)
Construction, demolition, mining, quarrying, agricultural wastes and sewerage sludges	294
Industry and commerce	78
Municipal waste	28
Total	400

household, commercial and industrial waste (van Beukering *et al.*, 1999). Many studies have demonstrated that waste production increases with affluence and as developing countries, especially urban areas, undergo rapid population growth and industrialisation, there is the potential for all these countries to reach American levels of waste production.

Accurate information on waste production in developing countries is difficult to obtain given the lack of records and often informal nature of waste management and disposal. However, as a country becomes more prosperous, not only does the quantity of waste increase, but as more luxury items are desired and produced the content of the waste reflects this with more packaging and electronic goods. Whereas the primary content of western economies is paper and card, in developing countries household waste consists of mainly kitchen waste, due to the fact that most food is prepared with fresh ingredients – western societies rely much more on convenience and processed foods and have more luxury waste. Figure 9.5 shows the percentages of organic waste and paper and card for three locations of differing prosperity. The content of municipal solid waste in all areas can vary widely and can often be unpredictable and hazardous in nature. Many hazardous materials are disposed of everyday in general household waste including aerosols, paints, oils, bleaches, batteries and electronic goods. Europe and the USA have been driven to implement strict industrial hazardous waste controls following expensive mistakes in the past such as Love Canal in the USA (e.g. Miller, 1998). Industrialising countries could benefit from the lessons learned as the USA is still paying the price today, spending more per year on dealing with past pollution than on current hazardous waste management (International Solid Waste Association (ISWA), 2002).

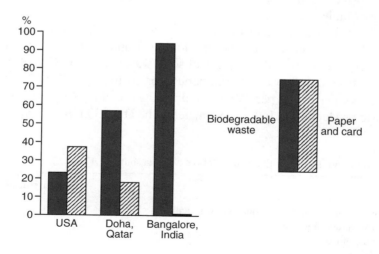

Figure 9.5 Organic and paper/card content of household waste in three locations (Data sources: EPA, 2002; van Beukering, 1999; PROWA, 2000, as cited in al Kawari, 2002.)

9.4.2 Waste management options

Solid waste management regulation and implementation vary around the world (Figure 9.6). European Union countries generally have the most advanced waste management practices (ISWA, 2002), with a range of fiscal instruments and legislative controls not only for how waste management facilities are managed but also on quantities permitted. There is a concerted effort in Europe to move away from landfill as the primary disposal route and to move to a more integrated approach that involves incineration, recycling, and composting as well as implementing reduction measures. The UK lags behind in recycling and composting with a recycling rate of 12% for municipal waste whereas countries like the Netherlands and Austria recycle up to 50% of their municipal wastes. In developing countries in Africa, Asia and Latin America, waste management can be non-existent, and the main disposal route for waste is uncontrolled dumping around the perimeter of the cities or open burning causing many health and environmental problems. Although governments have realised that this problem will only be exacerbated with rapid industrialisation and population growth, lack of policy, finances, legislation and enforcement mean that solid waste management and, more worryingly, hazardous waste management are left wanting. In low-income economies, the capacity to manage urban services is being outgrown by population growth (as was the case in the 19th century in the USA and western Europe). Whilst Western economies are fine-tuning their waste management, the majority of countries in the world are struggling with the basics of sufficient collection services and control of disposal sites (ISWA, 2002).

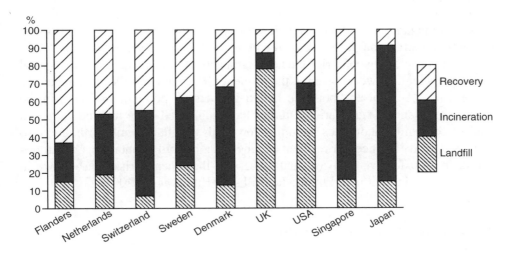

Figure 9.6 Waste management practices in some developed countries (Data source: Environmental Protection Agency, 2002; Strategy Unit, 2002; Bai and Sutanto, 2001.)

9.4.3 Landfill

Until now, the most commonly employed solid waste disposal solution was land-fill. However, this is falling increasingly out of favour as the main waste disposal option, particularly in Europe with issues of space and increasing environmental regulation, as well as concerns over the sustainability of taking valuable resources out of the economic system. However, there will always be a need to landfill some waste, and carefully managed landfills can minimise pollution and even produce a source of energy generation. The main environmental impacts arise from leachate and landfill gas emissions, both of which can be managed in state-of-the-art-engineered landfills, although there are also aesthetic impacts, impacts of transportation and health concerns for those living in close proximity to landfill sites.

The global warming potential of landfill gas is significant. Twenty-five per cent of the UK's 2.4 million tonnes of methane emissions originate from landfill sites (Salway *et al.*, 2002). As methane is 21 times more damaging than CO_2 as a greenhouse gas, simply flaring methane, producing CO_2 has a reduced impact on greenhouse gas emissions. Many modern landfills also use methane as an energy source.

In developing countries, landfills are often merely uncontrolled dumps presenting considerable health and environmental problems. However, given the inadequacies in basic infrastructure of water supply, wastewater treatment and solid waste collection, upgrading dumps to sanitary landfills (i.e. a liner and leachate treatment, compaction of material and soil covers) has not been considered a priority, despite awareness of the problems (Johannessen and Boyer, 1999). Unmanaged landfill gas presents an explosive risk as well as being toxic to workers, and methane quantities are accentuated by the high organic content of the waste. Where control is practised at all, it is mainly passive venting of landfill gas but no treatment nor energy recovery is normally undertaken. Another characteristic of the landfills and dumps in developing countries is the presence of waste pickers who sift through the waste for items of economic value (section 9.4.6).

In developing countries the health impacts of uncontrolled waste disposal are very evident. However, even in the affluent UK, with controls and safeguards, concern has been raised about the health impacts on populations living in the vicinity of a landfill site. Various studies have suggested there may be increased risks of congenital anomalies including low birth weight and still birth, and also increased risk of cancer in residents living near landfills containing hazardous waste. In an extensive study, a small excess of these health risks was shown in populations living within 2 km of a landfill site (Elliott *et al.*, 2001).

9.4.4 Moves to reduce landfill

Major economic drivers for reducing reliance on landfill in Europe and some other countries such as Singapore are the restriction on space and concern over the increasing volumes of waste generated. These countries have instigated a wide range of incentives to encourage alternative management options including banning MSW from landfill altogether.

Singapore has under gone rapid industrialisation and economic development over the past 50 years and once the second of Singapore's two landfill sites was full in 1999, they were left with no other suitable site on the mainland. Initiatives on recycling and reduction, and a focus on incineration as the preferred disposal method have shifted the emphasis dramatically. In 1976, 98.9% of waste was landfilled but by 2000 the figure was just 12.8% and only non-incinerable waste and incinerator ash are now landfilled at a purpose built offshore site (Bai and Sutanto, 2001).

Countries and regions in Europe which have banned MSW from landfill include Denmark and Flanders. Other methods employed include a landfill tax charged on waste entering a landfill site. In the UK the rate of £13 (€20.8) per tonne is low compared to other countries, for example Denmark, with a tax of €50 per tonne (Strategy Unit, 2002). This tax level, along with direct charging, is currently under review by the UK government. Direct charging has been very successful in some countries in reducing waste for disposal, providing it is accompanied by an effective recycling infrastructure. It is widely practised in Austria and Luxembourg and in some states in Australia. It has been difficult to judge how effective the landfill taxes have been as they are often implemented alongside a package of other measures.

The EU Landfill Directive, implemented in 2001, addresses both the need to reduce pollution from landfill and the need to address the ever-increasing flow of waste to landfill. The directive prohibits the co-disposal of hazardous and non-hazardous wastes as well as banning certain materials from landfill altogether including tyres, liquid wastes and clinical wastes. The directive also set targets for reducing biodegradable waste in landfills, which will have a major impact on waste management across the EU, particularly the UK where biological treatment is low. Biodegradable waste reductions required by the EU Landfill Directive are as follows: by 2010 – 75% of 1995 levels; by 2013 – 50% of 1995 levels; by 2020 – 35% of 1995 levels.

9.4.5 Incineration

Incineration is increasingly being employed, particularly in Europe and other developed economies with the vast majority producing energy from waste. Some high-income countries in Asia such as Singapore (section 9.4.4) and Japan rely heavily on incineration with 13 incinerators in Tokyo alone (International Environmental Technology Centre, IETC, 1996). Although incineration has the advantage of dramatically reducing the volumes of waste (Table 9.3), the practice has attracted much opposition from environmental pressure groups, particularly

Table 9.3 Incinerator residues as a percentage of waste input at three UK incinerators (Data source: Environment Agency, 2002)

Location	APC residues (%)	Bottom ash (%)
Lewisham	3.38	24.65
Edmonton	3.16	31.47
Sheffield	3.21	38.45

with reference to emissions of acidic gases containing dioxins and heavy metals such as mercury and lead. However, strict legislation including the updated European Directive on Incineration means that modern incinerators in Europe have significantly reduced air emissions and many old plants have closed (Environment Agency, 2002). However, as flue technology has improved, the main waste problem now for incinerators is dealing with contaminated air pollution control (APC) residues (referred to as fly ash in the past). Incineration in low-income economies is rarely a practical or economic option as the high moisture content of organic waste and low qualities of combustible materials (e.g. paper and plastics) mean that additional fuel is needed instead of producing excess energy (IETC, 1996).

Another significant drawback of incineration is that when it is also used to generate energy it potentially creates demand for waste, detracting from recycling and waste minimisation. Waste incineration must be employed as part of an integrated waste management strategy and must be restricted to dealing with residual MSW to avoid cramping out recycling or encouraging non-optimal resource use (McLanaghan, 2002). In Surrey, UK, one side effect of the announcement of the proposals in 2001 for new waste incinerators was a significant increase in public participation in local authority recycling and waste reuse initiatives. Responses such as this call into question the whole issue of the acceptability of incineration as a waste disposal strategy. If this initial response can be maintained and reproduced elsewhere, we are rising up the hierarchy of preferred waste management solutions (Box 9.2).

Box 9.2: Waste hierarchy

The waste management options discussed in this chapter form a hierarchy of waste solutions, which along with the Proximity Principle (waste dealt with as near as possible to where it is generated) has been formally acknowledged by the EU and the UK government. It forms a central theme of the UK National Waste Strategy (DEFRA, 2000) with landfill generally accepted as the worst environmental option and reduction at the top.

Continued on page 253

Continued from page 252

If we consider the hierarchy on an international level (Figure 9.7), we can add dumping and open burning to the bottom of the tree, as they are a reality in developing countries (van Beukering *et al.*, 1999). As we have seen above, except for reduction, each level of the hierarchy has environmental impacts, but in theory and usually in practice the impacts reduce as we proceed up the hierarchy. However, in addition to environmental and health impacts of the various options the costs and benefits of each option must be evaluated in the context of local situations and geographical environments. For example, in an isolated environment incineration with energy recovery may be a better environmental option than transporting waste over large distances for reprocessing. An integrated waste management strategy based on this hierarchy, which also includes education, continuing research and investment is the way forward. Reliance on a single strategy is inflexible and may have undesirable effects. For example, we need to ensure that our reliance on incineration is not so great that it becomes a waste sink, with the requirement for a constant supply of waste to fuel energy demands detracting from recycling and the principal aim of reduction.

Although some parties strive for zero waste economies, in reality this seems a long way off, particularly as we work to improve the quality of life for developing countries. What **is** needed is a marked change from end-of-pipe fixes and minor changes to business as usual to a major re-evaluation of resource use, industrial practices and particularly lifestyle expectations.

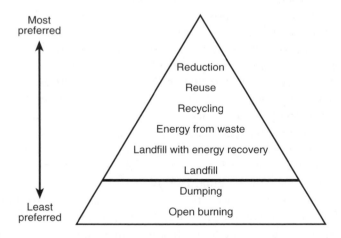

Figure 9.7 International waste hierarchy. (Adapted from DEFRA, 2000; van Beukering, 1999.)

In the case of well-managed incineration, solid wastes consist of bottom ash, recovered iron and steel, and air pollution control residues. Air emissions consist primarily of CO_2, approximately 1000 kg per tonne of waste input (National Society for Clean Air and Environmental Protection, NSCA, 2002a). Bottom ash is the hot residue produced during combustion and from which iron and steel are separated out by electro-magnets for recovery. It contains low levels of dioxins and is not classed as hazardous waste in the UK. It can, therefore, be disposed of to landfill or used as landfill cover but increasingly it is being used for various construction applications including as a road base or construction blocks. Air pollution control (APC) residues, which derive from the gas cleaning process, contain lime used to neutralise the gas acidity, carbon, dioxins and heavy metals. Although concentrations of dioxins and heavy metals are higher than those found in bottom ash, it is the extreme alkalinity that necessitates special disposal.

9.4.6 Recovery

It is often stated that reuse and recycling is not a new concept and has been practised formally and informally for many years. However, Western, throwaway societies seem to have lost sight of these ideals and practices, whereas reusing and recycling materials are often an essential part of everyday life in developing countries. In low-income countries, resource scarcity and poor pay drives thriving repair and second-hand markets operating on different levels (IETC, 1996). Householders reuse many items including glass, plastics and organic material which is fed to animals or sold to poultry or pig farmers. Very little is wasted. In urban areas, waste pickers (or scavengers) recover items from the streets and waste dumps for use or sale and make a significant contribution to reuse and recycling. The workers are exposed to considerable health risks from exposure to toxic and infectious materials, human and slaughterhouse wastes and they also hinder compaction activities. Their role is often not formally recognised and they receive meagre incomes and are considered of very low status by the general population. It has been estimated that in developing countries, up to 2% of the population survive by recovery of waste. Some municipalities have formalised this practice by employing residents from nearby slums. In some of the poorer countries, waste pickers have been helped by NGOs to form waste co-operatives to formalise their work, and provide a more structured opportunity to sell waste materials to industry. These schemes remove the need for the middlemen who often exploit waste pickers (Medina, 1997).

In industrialised countries, recycling is increasingly driven by a growing awareness of environmental and sustainability issues. However, formal recycling and reprocessing is not without environmental impacts; transport produces emissions with local and global impacts, location of bottle bins has caused objections due to noise and transport and there are inevitably impacts from energy and resource use in the reprocessing of materials. High levels of material recycling have been achieved by European countries such as the Netherlands, which recycles 91% of the glass in the waste stream (NSCA, 2002b).

Lifecycle analysis (LCA), evaluating product impacts from 'cradle to grave', and other methods have been used to examine the economics, energy use and transportation impacts of material recycling. Some materials have a very clear economic value such as scrap metal but other reprocessing operations need government support at least in the first instance. Glass is a mature recycling industry around the world with many EU countries recycling large quantities. Although crushed glass cullet is more expensive than soda ash (virgin material) as a raw material, energy savings are significant. Approximately 1.5 times more energy is required to produce a tonne of virgin glass than a tonne of recycled glass (NSCA, 2002b). Studies have also shown that the energy used to recycle plastic bottles is 8 times less than the energy required to manufacture the same virgin polymer, and that emissions of CO_2 and nitrogen oxide are reduced by all methods of plastic recycling (RECOUP, 2000).

Given the biodegradable waste targets laid out by the European landfill directive, composting and anaerobic digestion have been identified as having a significant role to play in diverting organic waste from landfill. Composting involves aerobic degradation of organic material, with significant bulk reduction, releasing CO_2, whereas anaerobic digestion involves enclosing waste without oxygen, producing a gas, which can be used as an energy source, and a digestate, which can be used as fertiliser or compost (McLanaghan, 2002). Centralised composting is increasingly practised in many countries. Home composting is practised around the world, and encouraged in Europe, Australia, Japan and New Zealand and by some NGOs in developing countries (IETC, 1996). However, it is difficult to assess the contribution of home composting to reducing waste to landfill.

9.4.7 Reduction and resource efficiency

Despite advances in waste management techniques and sustainability thinking, the amount of waste we produce is still increasing. In the UK, MSW is increasing by about 3%/year (DEFRA, 2002b). In developing countries, waste production is increasing dramatically especially in industrialising urban areas; for example, waste generation in China more than doubled between 1986 and 1995 (ISWA, 2002).

Numerous strategies have been employed to encourage waste reduction. Product taxes (i.e. taxing resource use as opposed to waste) on certain goods have had some success in reducing waste or shifting consumption patterns. Examples include a tax on carrier bags in Ireland, Sweden and Finland, taxes on light bulbs and tyres in Denmark and taxes on beverage containers in Belgium and Norway

(Strategy Unit, 2002). The key to the effectiveness of product taxes is accessibility and affordability of alternatives, otherwise the consumer is penalised with little shift in product use. European producer responsibility legislation for packaging, electronics and cars is aimed at encouraging reduction as well as recycling.

Obviously the best solution to our waste problem is not to produce the waste in the first place. This places the emphasis much more on the waste generator to seek inputs to their homes or industrial processes that produce less waste. Considerable cost savings have been made in industry through waste minimisation initiatives (Phillips *et al.*, 2001). However, there is no escaping the personal responsibility to reduce the waste created by our lifestyles, be it consciously looking for products with less packaging, buying recycled products (Box 9.3) or making a weekly trip to the bottle bank, but as part of another journey.

Box 9.3: Closing the loop – developing markets for recycled products

The key to make recycling an economic, practical and sustainable waste management option is a developed market for products manufactured from the reprocessed material (secondary material). This is fundamental to the whole waste management issue and has often been overlooked in favour of improving collection and reprocessing infrastructure, which though essential are unsustainable without the demand for products.

There are many products which are, or can potentially be made from, secondary materials, but there are barriers that hinder progress, such as the public perception of recycled products (Table 9.4).

Closing the loop means that our waste re-enters the economy as a useful product having undergone some degree of reprocessing. This reduces our reliance on virgin materials and thus reduces the environmental impacts

Table 9.4 Examples of end products of secondary materials (reprocessed wastes) (Data source: WRAP, 2001)

Material	Products
Wood	Particle board production, woodchips used on gardens, golf courses, children's playgrounds, animal bedding, mulches and composting
Plastic	Fibre, stationery, garden containers, traffic cones
Paper/card	Paper, cardboard, insulation, animal bedding, MDF
Glass	Road aggregates, shot blasting, tile-making, water filtration
Biodegradable waste	Compost, energy

Continued on page 257

Continued from page 256

of obtaining and producing raw materials. At the same time it removes waste from landfill and incineration and reduces energy use (section 9.4.6). Figure 9.8 provides a simplified life cycle of material in the closed loop, demonstrating that all the stages are interrelated and that underdevelopment of one stage can jeopardise the whole cycle.

Manufacturers and developers must be confident that secondary materials are reliable in supply and quality and that there is a market for products made using these materials. Often manufacturers and developers have restrictions governing what material they should use. These standards and specifications, whether statutory or voluntary, are important, particularly in a global market where the buyer may never meet the seller (WRAP, 2001). However, they often discriminate unnecessarily against secondary material or do not exist at all, for example for biodegradable waste and compost in the UK. The consumer must in turn be confident of the quality and increasingly the environmental impact of the product. There is a perception that products made from reprocessed waste materials are inferior in quality to those made from virgin materials. Energy use and environmental impact of reprocessing must not be greater than those of the virgin product (WRAP, 2001; Hogg *et al.*, 2002).

The UK has recognised that much work is required to improve the situation. In 2001, the UK government established the Waste and Resources Action Programme (WRAP) with £40 million over three years to address these issues and barriers. They have ambitious targets to increase recycling and markets and are looking at procurement, standards and financial mechanisms to stimulate investment in recycled products and reprocessing capacity.

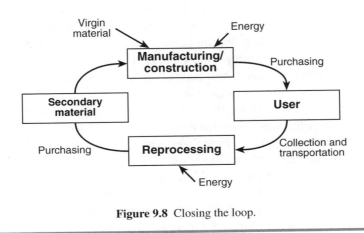

Figure 9.8 Closing the loop.

9.5 Conclusion: Issues for the management of pollution and waste

Must development mean pollution? What are we to do with our waste? Do we pollute this river or that sea? How do we decide between competing claims in different environmental media? In conclusion we explore some of the issues relevant to the management of pollution and waste. While avoidance is the ultimate solution, in practical terms we have to deal with the waste and pollution we are producing now; with the pollution and waste that we have inherited; and with the increase in pollution and waste that will be likely consequences of a rising global population, accompanied by rising living standards and concomitant resource consumption. If all earth's people are to share an equally good quality of life, with the opportunities enjoyed by the developed world today, then practical ways for managing and minimising pollution and waste are priorities. We should avoid disposal, which merely contains the problem until some future date. The case of radioactive waste is particularly worrying. The UK government is still deciding on the long-term disposal options for 10 000 t of radioactive waste which are currently in temporary safe storage. Even if no more nuclear plants are built in the UK, this figure could rise by as much as 500 000 t during the next 100 years as material currently in use is finished with, and in decommissioning and cleaning up existing plants (DEFRA, 2001). Until 1997, the UK was still dumping low and medium level waste in the North-East Atlantic Ocean after its European neighbours had agreed to cease to do so under the London Dumping Convention (1972). This waste was dumped in concreted steel containers, which are expected to last up to 300 years but some of the contents such as plutonium-239 have a half-life over 20 000 years (Holliday, 1995).

That we are already aware of these urgent issues is shown in developing pollution control and waste minimisation legislation and in green taxation, such as the landfill tax. These legislative and fiscal instruments attempt to internalise costs so that the benefiting manufacturer pays the cost. This is the principle underpinning Polluter Pays. These pollution prevention costs are preferable to the costs of damaged health, lost resources and increased frequency of climatic hazards. This is not just about making manufacturers more environmentally responsible; it is also about influencing consumer choice.

Consumers, it sometimes seems, are more environmentally concerned than businessmen and politicians. They have considerably influenced both commodity provision and national policy, through ethical investment, support for organic farmers, local production and initiatives such as 'Fair Trade'. The stark contrast in consumer response to genetically modified (GM) products in foods between Europe and the USA is reflected in their respective governments' policies. If we look at countries such as India, we can see this debate is further complicated. Some see GM products as a panacea for feeding a burgeoning population but others argue that this puts Indian farmers at the mercy of global biotechnology companies (section 6.4.2). Can we only afford to be environmentally caring once we have developed?

No one would suggest that poverty, hunger and lack of clean water should be ignored as urgent priorities, yet it is important to avoid repetition of mistakes and be certain that solutions proposed are sustainable. Inter-basin transfers of water, as proposed in Spain (WWF, 2002), may solve an immediate water shortage problem but on the other hand they may maintain high water consuming intensive agricultural production, with its wide array of detrimental environmental impacts. They may also work to the serious disadvantage of the area from which water is abstracted as has been dramatically seen for the Aral Sea (Moss, 1998). The issues are complex and require full environmental and economic analysis at donor and recipient ends. Rethinking our approach to resource use may give a more prudent and productive perspective.

Technological innovation may be part of the solution to this dilemma offering an affordable way forward for all. An example of this is the development of 'Greenfreeze' technology in response to the refrigeration dilemma (Greenpeace, 1997). As noted in section 9.2, replacing CFCs with HCFCs or HFCs was an almost knee-jerk response to the revelation of the ozone depletion problem generated by CFCs. Though less ozone-damaging, these new gases still have a potential long-term impact and are also significant greenhouse gases. Gains from their use will almost certainly be offset by the increased demand for refrigeration from the developing world. In a 'back to the future' technological development the NGO, Greenpeace, worked with research chemists and the East German Company DKK, Scharfenstein, to develop new hydrocarbon refrigerants based on propane and isobutane. These substances do not deplete ozone and have a much lower global warming impact. Similar refrigerants were used in the 1930s, but at that time large quantities were needed and gave less internal cold space for refrigerator size than subsequent CFC technology enabled. Furthermore, the larger amounts of coolant required in the original technology gave potential fire and explosive hazards in the case of accidental leakage. Resolution of these issues with 1990s technological innovation has seen the rapid introduction and use of 'Greenfreeze' technology by major manufacturers, including developments to replace use of CFCs and related chemicals in the associated insulating foam. This technology has been taken up by the developing world including China, the fastest growing world market for refrigerators.

This is a very positive example of innovation giving a market lead and reducing a major pollution threat. Public sensitivity to the issues of ozone depletion and global warming meant that manufacturers were keen to employ the new technology since they otherwise stood to lose their share of market sales. Governments have since legislated to ban the use of CFCs and their derivatives in refrigeration. Response in Europe was much quicker than in the USA where companies which had invested heavily in the production of HCFCs and HFCs were slow to accept these changes.

A curious side effect of this development was seen in 2002 with the introduction throughout the EU of a new Directive on Ozone Depleting Substances (ODS). This required the safe disposal of CFC- and HCFC-containing refrigerators. Disposal or reuse of coolant is not a technical problem but the CFCs in the insulating foam presented a greater challenge. Since this is only necessary for refrigerators using this older technology, this places a limit on the utility and profitability of any facility developed to process this waste. In the UK no facilities were available when this legislation became law and local authorities were required to store domestic refrigerators. As an interim solution these were exported to Germany where the relevant technology was already in place. A substantial market for second hand refrigerators in the developing world was also blocked. Hence, this legislation aimed at resolving a problem with ozone and global warming, created a demand for additional virgin resources for refrigerator manufacture. The need to think through legislative solutions before their enactment is very important. Similar problems for the UK may lie ahead with legislative proposals over disposal of 'End of Life Vehicles' and safe disposal or reuse and recycling of tyres. There is also concern that too tight legislation may encourage illegal dumping and fly tipping.

Environmental NGOs have an important role leading and educating public and manufacturer opinion. Although the public is aware that these issues are contentious, people need and seek expert advice. The role of the media is important as an accessible source of debate and advice, as the growing popularity of interactive environmental discussion programmes testifies. Significant here is the search for deeper understanding, particularly the linking of science and technology with environment and with economic and social expectations.

Two emotive issues are hazardous waste disposal and the use of the sea for waste disposal. Angel and Rice (1996) argue that since oceans cover 71% of the earth's surface it is illogical to exclude them from our waste disposal options. They argue that it is only in the shallow coastal waters of the continental shelf (top 200 m) that biomagnification, and contamination of food chains and physical resources are real hazards. This is just 5% of the earth's surface leaving 300 million km^2 with waters typically 3000–6000 m deep. If we also avoid the continental slopes, the area up to 3000 m deep, this leaves 53% of the earth's surface available for use for waste disposal.

It is suggested that the evidence of wrecks, discarded shoes and cans and similar debris, all in good condition, testifies to the inert nature of these areas. Thus it seems that sealed canisters of waste would remain sealed indefinitely. It has also been observed that where thermal vents in the deep ocean floor emit heavy metals, mixing action readily disperses the metals, and micro-organisms present in these areas apparently thrive (Angel and Rice, 1996). This implies that even if a leak occurred no significant pollution problems would result. Countering these points is the realisation that we have only recently discovered the variety of life at great depth in the oceans linked with hydrothermal vents suggesting that, once again, we might be about to interfere before we fully understand.

Nevertheless, we have to do something now. Perhaps we should not rule out deep-sea disposal on purely emotive grounds. Part of Angel and Rice's (1996) argument relates to the fuel energy required to support future world development, and feed a growing population with universally improved living standards. Nuclear energy is proposed as the solution to this and, since nuclear fusion is not yet available as a practicable technology, this will mean greater reliance on existing nuclear fission technologies and the associated and currently unresolved waste disposal problems. This scenario of a nuclear energy future may prove unfounded, and would bring pollution risks and waste disposal problems that alternative investment in renewables would avoid. However, the point being made, the potential for use of the deep ocean in waste disposal, has equal relevance to the safe disposal of incinerator fly ash or to concentrated sewage effluent, and merits further evaluation.

Understanding the nature of disposal at sea was also relevant to the case of the Brent Spar oil storage platform. In this case Shell proposed to dispose of a redundant platform by sinking it in waters 2200 m deep to the west of Scotland near Feni Ridge (Greenpeace, 1998). Shell and the UK government agreed this was the Best Practicable Environmental Option (BPEO). The proposal triggered a wide-ranging debate on policy of ocean waste dumping. Greenpeace aggressively opposed the proposal on the grounds that too little was known about the likely consequences of the particular proposal or about waste disposal in deep sea in general. They also claimed that disposal at sea for the Brent Spar would set an unwelcome precedent given the high number of oil platforms in the North Sea and globally. It might also undermine existing bans on dumping of radioactive and industrial wastes at sea.

However, the adverse publicity generated positive alternative solutions. Brent Spar was towed to Norway, where it has since been used to build a quayside. Reuse of the steel may be considered a sensible environmental solution. The platform was made of 6700 t of high-grade steel for which there is demand as scrap metal. Recycling was another obvious alternative. It would take four times as much energy to create this steel from virgin materials, as to dismantle and recycle. Against this is the hazard associated with towing to land, the energy demand and the potential oil leaks and pollution. The Brent Spar controversy prompted new (1998) EU legislation banning the dumping of oil and gas installations in the North Sea and North Atlantic.

Evaluating risks and finding the best solution is self-evidently a complex issue. There will be no single best solution to waste management since the environmentally friendly, least damaging option will vary according to the geographical location and the socio-economic context in which the waste is generated and/or disposal proposed. Learning to evaluate risk and give a fully rounded environmental impact assessment of pollution and waste is a major challenge for the 21st-century environmental scientist. Multi-disciplinary skills are needed and must be brought to bear if we are to safeguard our future. Equally, we must not discard public feeling nor ignore the fact that the environment has an intrinsic value and that to some people, certain actions are just wrong, even if the science says otherwise.

Further reading

O'Riordan, T. (ed.) (2002) *Environmental Science for Environmental Management*, 2nd edn. Harlow: Pearson Education.
Good wide-ranging discussions of themes raised in this chapter can be found in this text, including sustainability, ethics and the scientific management of the environment.

Johnson, R.J., Taylor, P.J. and Watts, M.J. (eds) (2002) *Geographies of Global Change*, 2nd edn. Oxford: Blackwell.
This text presents geographical perspectives on economic, political, social, cultural and environmental change.

Farmer, A. (1997) *Managing Environmental Pollution*, London: Routledge.
A useful background text discussing the more technical aspects of pollution and its management as well as related social and political concerns. This text covers the broad spectrum of pollution.

von Weizsäcker, E.U., Lovins, A. and Lovins, H. (1998) *Factor Four. Doubling Wealth, Halving Resource Use*, London: Earthscan.
This text gives a thought-provoking insight into how we can make a step change in reducing waste generation.

ENDS *Reports* published by Environmental Data Services Limited.
These are a useful source of comment and analysis on new environmental legislation and its relevance to waste management and pollution control.

Some useful web sites are listed below.
For up-to-date comment and data on waste and its management see:
WRAP www.wrap.org.uk
International solid waste association www.iswa.org

For useful topical comment and views, which sometimes challenge established orthodoxy, see the websites of leading environmental NGOs such as:
Greenpeace www.greenpeace.org
Friends of the Earth www.foe.org
WWF www.panda.org

Governmental Agency web sites are useful sources of national data. In the UK, the DEFRA and Environment Agency web sites are particularly useful.
www.defra.gov.uk
www.environment-agency.gov.uk

Part Five

Conclusion

Chapter 10
Sustainable Development

Frances Harris

10.1 Introduction

Global environmental issues are complex problems. Although perceived as environmental problems, political, social and economic issues are in some cases the root causes of environmental problems, and in all cases barriers which must be overcome to ensure more sustainable use of the natural resources of our planet. In this book we have considered human–environment interactions in a broad sense (Chapter 1) followed by discussions of eight global environmental issues. Each of these chapters has described the scope of the global environmental issue from a scientific point of view before going on to discuss the factors which make the environmental issue difficult to resolve. These factors include:

- Uncertainty about the environmental science underpinning the issues.
- The different ways in which the environmental problem affects different parts of the globe.
- The range of actors who are stakeholders in the issues.
- The different ways in which the problems are perceived by varying social and cultural groups.
- The varied economic and political driving forces affecting global environmental issues.
- Costs associated with changed technologies/practices or remediation of the problem.

The aim of this book is to help people understand environmental issues and enable them to make an assessment for themselves of the conflicting information

Global Environmental Issues.   Edited by Frances Harris
© 2004 John Wiley & Sons, Ltd   ISBNs: 0-470-84560-0 (HB); 0-470-84561-9 (PB)

which is available. The challenge for environmental scientists is twofold: Firstly, to assess the validity of the information available about a global environmental issue, which includes questioning the values of the source of the information to identify potential bias. Secondly, to integrate all the information concerning a global environmental issue, be it scientific, political, economic or social, to come to a balanced view about how to achieve a more sustainable way to manage resources. Policies for sustainable development must be based on an assessment of available evidence. The challenge for environmental scientists is to have the tools for making these assessments. Without this, environmental policy may be misguided, resulting either in irreparable damage to our environment, or changes to our way of life being made unnecessarily.

10.2 Distinguishing between natural and human-induced environmental change

As we consider sustainable development, it is important to distinguish between natural environmental change and that which is aggravated by humans. Several of the chapters in this book have pointed out that environmental change is not a new phenomenon, but an ongoing and dynamic process. As mentioned by Hulme (Chapter 2, p. 21), climate exhibits 'interannual, multidecadal and millennial variations'. This has had an effect on sea level and on the biosphere. As climate changes, the range of flora and fauna is affected, so land cover varies. As sea level varies, coastal environments, including coral reefs, change. Environmental change is not always a bad thing: it can create new opportunities and new habitats. As Nunn (Chapter 3, p. 49) says, 'early Holocene sea level rise both destroyed ecosystems and created opportunities for new ones to develop'. However, it is generally agreed that environmental change is now occurring more quickly than previously, and also that anthropogenic disturbances are now a significant force of global environmental change, and it is this anthropogenic effect on the environment which sustainable development seeks to curb.

Such rapid rates of environmental change are more challenging to ecosystems, and to humans. The resilience of an ecosystem is defined as its buffering capacity or 'the ability of a system to absorb perturbations' (Berkes and Folke, 1998, p. 6). Extremely rapid rates of change leave little time for ecosystems, or the people living within them, to adapt to change. There is concern that we may be stressing environmental systems beyond their resilience. When the pace of gradual change is accelerated to the point where change occurs as shock events, ecosystems and human livelihood systems may be destroyed. It is important to recognise that the environment is dynamic and changing. The challenge is to work alongside environmental change to ensure sustainable development.

10.3 Interpretations of sustainable development

During the 1980s and 1990s sustainable development was the buzzword to be used wherever needed; by environmentalists, politicians and economists; in the media, in speeches and in international agreements. However, it does not always sound as if they are talking about the same thing. This is because each group has refined its own definition of sustainable development, with the result that what deep green environmentalists mean by sustainable development is not the same as what economists mean. Box 10.1 shows some definitions of sustainable development, beginning with the often-quoted definition from the World Commission on Environment and Development (WCED), but including others from international reports and treaties, philosophers and economists. This small sample of interpretations of sustainable development shows that although originally used by environmentalists, 'sustainable development' can have economic, social, political or environmental dimensions, depending on the values and goals of the individual. Its broadening use (and some would say misuse) has meant that it is increasingly hard to define, as it has come to mean all things to all people (Redclift, 1991). Indeed, its amorphous definition may be one of the reasons why it is so widely used, particularly in political circles.

In seeking to understand sustainable development, the first question to ask is what is to be sustained? Is the goal of sustainable development to sustain the environment, people's livelihoods, or economic growth? Originally, the idea of sustainability was used in an environmental sense, and referred to saving the environment, that is maintaining habitats, ecological systems and natural resources (IUCN, 1980). Such a 'green' interpretation sees the environment as being of greater importance than mankind, so that our needs and demands on the environment must be secondary to the need to preserve the environment. Human activities must be environmentally sensitive with respect to natural resource use and pollution. To achieve this will require significant changes in our exploitation of the environment, and therefore our lifestyles, livelihoods and global economic growth.

A less ecocentric interpretation of sustainable development sees environmental conservation as important, but tempered by human needs: thus humans are of equal importance to the environment, and both natural resources and livelihoods should be conserved. For many people, environmental issues are secondary to personal comfort. Hence, they seek to conserve their lifestyle. Thus attempts to move towards sustainability can only be considered if this does not jeopardise personal comfort and lifestyle. This interpretation of sustainable development relates to what humans can justifiably take from the environment. However, our level of resource use is debatable in itself. Meeting basic needs (water, food, shelter) places much less stress on the environment than the more comfortable lifestyles (particularly of the west) where consumption, the impact

---

Box 10.1: Some definitions of sustainable development

Development which meets the needs of the present without compromising the ability of future generations to meet their own needs.

WCED, 1987, p. 43

Improving the quality of life while living within the carrying capacity of supporting ecosystems.

UNEP et al., 1991

Sustainable economic development involves maximising the net benefits of economic development, subject to maintaining the services and quality of natural resources.

Barbier, 1989

A social state of process by which underdevelopment (poverty, disease, illiteracy, high infant mortality, low life expectancy, low productivity, poor medical and health facilities) is reduced or averted, leading to the attainment of health, literacy, low infant mortality, higher life expectancy, productivity, medical and educational facilities, sufficient prosperity to allow the evils of underdevelopment to be held at bay.

Attfield, 1999, p. 98

The satisfaction of basic needs, including the needs for individual autonomy and making meaningful contributions to society.

Attfield, 1999, p. 98

A comprehensive economic, social, cultural and political process, which aims at the constant improvement of the well-being of the entire population and of all its inhabitants on the basis of their active, free and meaningful participation in development and in the fair distribution of the benefits resulting therefrom.

UN, 1986

Sustainable development is a process in which the exploitation of resources, the direction of investments, the orientation of technological development and institutional change are all in harmony, and enhance both current and future potential to meet human needs and aspirations.

WCED, 1987, p. 46

---

on the environment (land cover change, urbanisation, food production, energy use) and pollution are much greater.

These interpretations of sustainable development raise further questions. Where humans have already had some impact on the environment, should the environment be restored to its previous condition (although the natural dynamism of environmental change means that deciding what is the original state of the environment is very difficult), or conserved as it is, preventing further change? (Attfield, 1999, section 5.10). This also raises the fundamental question of whether natural resources can be used sustainably. The United Nations Environmental Programme (UNEP) definition of sustainable development (Box 10.1) tempers any improvement in quality of life with the concern that the carrying capacity of the environment is not exceeded. Is there a renewable level of off-take which can be harvested from the environment without reducing production levels? If there is, how do we decide what this level of off-take is? As already stated, environmental systems are dynamic, with fluctuations from year to year as well as longer-term cyclical changes in productivity. Therefore, the yield from ecosystems (whether it be timber, grassland, wild foods and agricultural crops, or fish) will also vary. Models of carrying capacity of land have been replaced with ideas of more responsive levels of off-take from natural systems, depending on fluctuations in yield (Benkhe and Scoones, 1992). In view of this, setting a level of resource use which is deemed to be sustainable may be inappropriate. Instead, demands on resources should be more responsive to fluctuations in supply. In addition to considering levels of resource use, we must consider the capacity of the environment to absorb the effects of our use of the environment such as increased CO_2 emissions, run-off of nitrogen and phosphorous fertilisers (Box 9.1) and pollution. Technology has a role to play here, for example adoption of renewable energy sources can reduce CO_2 emissions.

In practice we rely on the environment to do more than meeting basic needs, or even a comfortable living style. In most countries natural resources are also their key to economic growth, and this is especially true of developing countries. Economic growth may well be at a cost to the environment, either because it is natural resource-based (e.g. selling hardwood from tropical rainforest), or damaging to the environment (e.g. relies on fossil fuel energy leading to climate change, atmospheric pollution...). Several of the definitions of sustainable development in Box 10.1 refer to sustainable development as economic development. The quote from Barbier (1989) refers to maintaining income from economic development, but makes clear that this is dependent on conserving the services and quality of natural resources.

It is due to this that there has been an increased willingness to 'cost' the environment. There are many limitations to costing the environment (see sections 9.3.3 and 9.3.4). The environment has different values to different stakeholders, and its value changes over time, and increases with scarcity. As Nunn points (section 3.5), the use of the 'Common methodology for assessing vulnerability to sea level rise' was not successful as it was hard to assign financial value to coastline used for subsistence purposes by local communities. However, in some cases it is seen as the only way to provide a counter-argument to those who under-value the environment.

So far, sustainable development has been considered as an environmental problem. Although levels of natural resource use and pollution are key to sustainable development, as each chapter (2–9) has shown, global environmental issues mask underlying social and political issues which need to be addressed in order to resolve the environmental problem. Thus Attfield's definitions in Box 10.1 describe achieving sustainable development as a social process, through changing deprivation to better social and health conditions, and a political process, as they refer to the importance of autonomy and the ability to contribute to society as components of the satisfaction of basic human needs.

The United Nations Conference on the Human Environment in Stockholm in 1972 linked environment and development. At the World Summit on Sustainable Development held at Johannesburg in 2002, the link between poverty and environment was reinforced. Environmental management for long-term sustainability is not a priority for communities and households which have immediate needs for food, shelter or money. In these cases, environmental misuse, which may be acknowledged to be wrong for the long term, is the only means of short-term survival. Political and economic issues such as alleviation of poverty, and international debt relief, are also issues which need to be addressed as part of sustainable development. International debt puts pressure on countries to misuse natural resources to meet debt payments. For example, countries may need to sell tropical timber, so decreasing forest cover and biodiversity. Campaigns for ethical trade (section 6.2) also see changing economic systems as key to sustainable development.

10.4 Challenges to sustainable development

10.4.1 Stakeholders

The range of definitions and interpretations of sustainable development are put forward by many different people, with varying interests and concerns. Agenda 21 (section 10.5) identifies nine stakeholder groups with regard to sustainable development. These are women, children and youth, indigenous people, non-governmental organisations (NGO), local authorities, workers and trade unions, business and industry, the technological community, and farmers. However others could be added, such as governments, who are required to implement Agenda 21 and make decisions concerning priorities among the many issues encompassed in Agenda 21. Another stakeholder group is future generations, who will be dependent on the decisions made by this generation and the next. Can such a broad range of stakeholders, with differing priorities, agree upon and work towards a common goal?

10.4.2 Uncertainty

The complexity of global environmental systems means that, despite the large amounts of research dedicated to understanding these environmental problems, there is still uncertainty concerning each global environmental issue, and the manner in which global environmental issues interact to contribute to the global ecosystem. The data on which decisions are made are also uncertain. In Chapter 4, Boyd and Foody explain how changes in the threshold level of land cover can affect whether an area is classified as forest or not (section 4.3). A decision on whether a forest should be defined by 8 or 10% cover can have ramifications for many people. Debates about the accuracy of data are made to undermine discussion of actions needed to be taken (Box 3.1 provides an example concerning sea-level gauges).

In addition to uncertainty about scientific facts, there is also uncertainty regarding what is to come. We cannot be exactly certain of what the future will bring, in relation to population growth, the full effects of global warming, diseases, or new technologies which might replace some of the environmental resources and services on which we currently rely. The precautionary principle guides decision-makers towards taking defensive action concerning the environment: to avoid any actions until it is certain that they will not be bad for the environment. Some argue that we should go on as we are, until we can be certain of the environmental effects of our actions, while others say we should change our activities now, because initial evidence suggests that what we are doing is detrimental to the environment. However, scientists and governments face urgent problems, and the need to make decisions based on the best information available now. 'All that is certain is our uncertainty' (Harrison, 1993, p. 59), yet governments are faced with making hard decisions based on uncertain science (Newby, 1991).

10.4.3 Consumption and lifestyles

Figure 1.1, of ecological footprints of nations, and the discussion in section 1.4, showed that there is considerable disparity between nations regarding their impact on the environment. This was related to levels of development, consumption and pollution in different countries. The definitions of sustainable development, ranging from meeting basic human needs, through to sustaining economic growth, reflect the levels of consumption and lifestyles which different stakeholders hope will be upheld in any programme to achieve sustainable development. Developing countries aspire to the more comfortable living standards of developed countries. However, as pointed out in section 9.3.3, if the global population adopts the lifestyle currently seen in the developed countries, the environmental effects would be enormous. Thus any rise in living standards in developing countries should be counterbalanced by lowering the ecological footprint of developed countries. However, there may well be considerable resistance among those who do live comfortable lifestyles if they were asked to make concessions to accommodate the general rise in living standards in developing countries. Both

President George Bush Senior and President George W. Bush of the US have stated that they will not countenance environmental regulation which would mean a lowering of the standard of living of US citizens (Tickell, 2001).

10.4.4 Arguments over cause and responsibility

This north–south divide is exacerbated by the different burdens that nations are seen to place on the global environment. The issue of sea-level rise illustrates this problem. Sea-level rise is widely seen as being the result of global warming, a problem attributed to the wealthy developed nations of the west. The victims are inhabitants of low-lying coastal areas and islands (e.g. South Pacific see section 3.4.1). Wealthy nations can afford to build coastal defences, whereas poorer countries cannot (Box 3.2 on Bangladesh). Nunn describes an interesting example of how controversy over the exact magnitude of the problem is used to avoid taking further measures to resolve it, or assist vulnerable countries in protecting themselves against sea-level rise (Box 3.1).

The issue of biodiversity conservation also illustrates the north–south divide. The Convention on Biological Diversity (CBD) outlines a programme of activities to develop national strategies. Yet this initial activity, as well as the implementation of measures which might follow (e.g. creation of national parks, termination of logging and conservation of forests), places tremendous financial burdens on many of the countries which are wealthy in biodiversity, but not economically.

10.5 Achieving sustainability

Since the UN Conference on the Human Environment in 1972, there has been a growth in international action to resolve global environmental issues. Mitigation of these issues requires action at the local level, but this must also be reinforced by collaboration between nations to work towards an agreed goal. The conference in 1972 was the first such international meeting. Not only did it discuss and debate the meaning of sustainable development (Box 10.1), but it also made more explicit the link between environment and development. Twenty years on, the UN conference on Environment and Development (UNCED) held at Rio de Janeiro raised awareness of environmental and development issues as we approached the turn of the millennium, and produced Agenda 21. Agenda 21 considers sustainable development from social and economic perspectives, tackling, among other issues, poverty, consumption patterns, population growth, health and settlements, and also from a natural resource management perspective. (Section 7.9.1 provides a discussion of energy issues in relation to Agenda 21.) Agenda 21 illustrates the all-encompassing nature of the phrase 'Sustainable Development', and the enormity of the task ahead if all these problems are to be addressed. Ten years on from the development of Agenda 21, the World Summit on Sustainable Development (WSSD) in Johannesburg was criticised as an expensive talking shop by some. However, the interaction between local level and global level environmental decision-makers, resulting in better mutual understanding, may prove to be valuable. The 2002 summit highlighted the north–south divide: representatives of developed countries were keen to work on global environmental problems such as climate change and the Kyoto agreement, whereas developing nations were more interested in tackling immediate problems of poverty and the provision of clean drinking water to poor and rural communities. While the north argues over the long-term prognosis of our environment, the south is concerned about the immediate problems resulting from social inequities caused by current patterns of global development.

A criticism of large international meetings is that it is difficult to turn meetings into agreements, and later it is difficult to implement agreements on the ground, at the local level. Certainly, the stakeholders at international meetings are varied, and each delegate comes with their own agenda. Negotiating trade-offs and compromises amongst such a large and diverse group, and under the glare of international media, is extremely difficult. The Kyoto Protocol (Box 2.2; section 7.2.1) illustrates the many problems associated with agreeing to, and implementing, international treaties for environmental protection. Although it aims to control global emissions of CO_2 levels, 80% of the world's population live in countries which are not signatories to the agreement (Nunn, personal communication). Furthermore, the US, the largest contributor to CO_2 emissions, refuses to sign. Agreeing to the Kyoto Protocol would require changes in the US standard of living and lifestyles, which is politically unacceptable. Furthermore, the beneficial results of implementing any environmental policies are not likely to be seen within the lifetime of the existing government, so that environmental issues have a low priority in the short-term horizon of politicians. Even among countries that have agreed to the Kyoto Protocol, there is debate over which nations should make the most cuts to emissions. Developed countries are widely believed to have caused the problem, and less developed countries feel that they should make the most effort to solve it. The counter-argument is that developing countries, with larger populations, will become larger contributors to environmental problems if their populations are successful in attaining the standard of living to which they aspire. Therefore they should make efforts now to prevent their environmental impact increasing. However, such efforts will cost money, and developing countries feel that this burden will limit their economic development.

Ultimately, the international conferences and treaties are aimed at changing environmental management practices at the local level. One difficulty is turning global environmental goals into implemental local strategies for action. A subsequent difficulty is assessing their success. Agenda 21 is now reaching local government structures in the UK. National decisions regarding energy sources (renewable, fossil fuels . . .) for power influence national CO_2 emissions. In the UK, customers at the household level also have some choice, as 'green electricity' providers compete with conventional sources of energy. Personal choices regarding transport (whether to travel on foot, bicycle, car or public transport) influence national outputs of CO_2 and other pollutants. These choices are influenced by wealth, culture, lifestyle decisions (to live in cities or engage in rural–urban commuting), national strategies for public transport provision and road networks. National desires to control waste are related to the environmental effects of pollution and costs of finding places to dispose of waste (section 9.4.3). Landfill taxes in the UK are driving local and district councils to encourage householders to reuse and recycle. In this case, market forces are driving change, although 'blue box' recycling schemes also encourage recycling through social pressures.

Immediate and visible environmental problems encourage quicker decision-making and action. For example, the Montreal Protocol concerning ozone depletion has been implemented relatively quickly, as it is based on a visible hole in the ozone layer, rather than a prediction of environmental problems. Even so, it will take time for planned reduction in CFCs to be achieved as stocks of appliances containing CFC coolants are gradually replaced (section 9.5).

10.6 Conclusion

The previous chapters (2–9) show that in addition to the science underpinning the environmental issues there are many other factors which come into play when decisions are made about how to manage the environment. Different stakeholders desire different outcomes from environmental management. Figure 1.2 shows how environmental knowledge passes through many minds before it reaches decision-makers. At each stage there is the potential for reinterpretation and misrepresentation. Thus environmental science is influenced by power struggles between stakeholders as it is turned into policy.

Whether naturally occurring, artificial, or a combination of the two, globally, society must learn to cope with environmental change. Adaptation to environmental change is an ongoing process. Hulme (section 2.2.4) describes 'reflexivity in social behaviour' and this co-evolution of environment and society is the basis of human–environment interactions. Coping strategies are varied. The simplest coping strategy, to move to a more favourable environment, is becoming less possible as the planet becomes more crowded. Furthermore, as environmental change increases, the sheer numbers of environmental refugees are increasing, and it is harder and harder for existing communities to absorb these newcomers. For example, as Nunn (Box 3.2) states, a 1 m rise in sea level could displace 11% of Bangladesh's population from their homes. A 1 in 10 displacement might, in theory, be manageable, but in practice this is more than 13 million people. They need not only to be rehoused, but also to find new livelihoods, if they are not to become a burden on society.

An alternative to relocation is to develop coping strategies to deal with environmental change *in situ*. For coastal communities, this can mean raised housing (on stilts) and dykes or coastal sea defences. In developing nations, ultimately, inundation requires moving. In forested areas coping can mean changing from slash-and-burn farming techniques to more permanent, residential farming systems such as short-fallow farming systems. Of course, predicting change is difficult due to the complexity of the global ecosystem and interactions between different aspects of environmental change. People's ability to adapt also depends on the resources available to them: for example financial and technical resources to build coastal defence structures, or knowledge and social capital to exchange seeds and farming techniques to adapt food production to changing climatic conditions. For governments, supporting adaptation to environmental change competes with other development issues such as food security, and provision of health and education services.

Until recently, the environment has been seen as separate from government decision- and policy-making. As environmental issues have become more prominent and threatening to economies (i.e. natural resource shortages, public health problems associated with environmental issues such as a hole in the ozone layer, pollution) governments have begun to take notice. Since Rio, the precautionary principle has become more prominent, but is still not adopted conscientiously. There is still a 'react and regulate' (Rees, 1997) approach subsequent to environmental damage, rather than a 'predict and avoid' approach. Even when science is fairly certain about environmental damage, decisions to mitigate or avoid environmental problems are tempered by negotiation, in which governments engage in what Robinson (2002) has called 'destructive conflict' and 'political intransigence'. Arguments about the exact nature of science can stall negotiations when the general pattern of change is clearly evident. Nunn asks whether it is worth arguing whether sea-level rise will be 35 or 55 cm when it is clear that either figure will displace many from their homes and livelihoods, and therefore urgent action is necessary (section 3.4.1). Many believe that technological solutions will find an 'alternative' to the environment. However, technological alternatives cannot replace some things, such as lost biodiversity. Furthermore, the cost of replacing global climate control systems would be enormous, even if the technology were there.

At the beginning of this book three reasons were proposed to explain why global environmental issues are so hard to resolve: the complexity of environmental problems; the many stakeholders involved; and the need for change in our own patterns of consumption and pollution of the earth. It is hoped that the perspectives on global environmental issues presented by the authors in this book will have broadened readers' understanding not only of the environmental issues, but also of the many factors which affect our understanding of the global environmental issues. We all need the analytical tools to interpret the science, its social implications, and the politics of addressing and resolving these environmental issues.

Further reading

Attfield, R. (1999) *The Ethics of the Global Environment*. Edinburgh Studies in World Ethics, Edinburgh: Edinburgh University Press.
Explores the ethical issues relating to environmental problems, starting with philosophical approaches to the environment. It then applies these concepts and theories to topics such as global resources, sustainable development, population and poverty and biodiversity conservation. It also discusses questions of global justice.

Harrison, P. (1993) *The Third Revolution. Population, Environment and a Sustainable World*, London: Penguin.
A very readable book considering the implications of population growth, rising consumption and damaging technologies for the environment. It debates the effect of man's environmental impact on sustainable development.

Redclift, M. (1987) *Sustainable Development: Exploring the Contradictions*, London: Methuen.
A short, readable text examining the contrasting interpretations of sustainable development.

World Commission on Environment and Development (WCED) (1987) *Our Common Future*, Oxford: Oxford University Press.
A seminal text defining sustainable development.

List of Acronyms and Abbreviations

AEA	Atomic Energy Authority
AEP	Agri-environment Policy
APC	air pollution control
AVHRR	Advanced Very High Resolution Radiometer
BPEO	best practicable environmental option
Bq	bequerel
C&C	contraction and convergence
CAMPFIRE	Communal Areas Management Programme for Indigenous Resources
CAP	Common Agricultural Policy
CBD	Convention on Biological Diversity
CDM	clean development mechanism
CFC	chlorofluorocarbon
CGIAR	Consultative Group on International Agricultural Research
CITES	Convention on International Trade in Endangered Species
CO_2	carbon dioxide
CSS	Countryside Stewardship Scheme
DDT	dichlorodiphenyltrichloroethane
DEFRA	Department of Environment, Food and Rural Affairs
EIT	Economies in transition
EMAS	Eco-Management Audit Scheme
ENDS	Environmental Data Services
ENSO	El Niño/Southern Oscillation
ESA	environmentally sensitive area
ETM	Enhanced Thematic Mapper
EU	European Union
EUETS	European Union Emissions Trading Scheme
FAO	Food and Agriculture Organisation
FRA	Forest Resources Assessment

Global Environmental Issues.   Edited by Frances Harris
© 2004 John Wiley & Sons, Ltd   ISBNs: 0-470-84560-0 (HB); 0-470-84561-9 (PB)

List of Acronyms and Abbreviations

FSU	former Soviet Union
GATT	General Agreement on Tariffs and Trade
GDP	Gross Domestic Product
GIS	Geographical Information Systems
GM	genetically modified
GMO	genetically modified organism
ha	hectare
HCFC	hydrochlorofluorocarbon
HDI	Human Development Index
HEP	hydro-electric power
HFC	hydrofluorocarbon
H_2S	Hydrogen Sulphide
HYV	high yielding variety
ICLEI	International Centre for Local Environmental Initiatives
ICM	integrated crop management
IEA	International Energy Agency
IETC	International Environmental Technology Centre
IFE	interial fusion energy
IFF	Intergovernmental Forum on Forests
IGBP	International Geosphere Biosphere Programme
IEO	International Energy Outlook
IPCC	Intergovernmental Panel on Climate Change
IPF	Intergovernmental Panel on Forests
ISWA	International Solid Waste Association
ITDG	Intermediate Technology Development Group
ITTA	International Tropical Timber Agreement
ITTO	International Tropical Timber Organisation
IUCN	International Union for Conservation of Nature and Natural Resources
JI	joint implementation
ka	thousand years
LCA	life-cycle analysis
LEAF	Linking Environment and Farming
LFA	less favoured area
LRTBAP	long-range transboundary air pollution
Ma	million years
MODIS	Moderate Resolution Imaging Spectroradiometer
MSW	municipal solid waste
MTE	metric tonnes equivalent
Myr	million years
NAO	North Atlantic Oscillation
NGO	non-governmental organisation
NH_3	ammonia

NOAA	National Oceanic Atmospheric Administration
N_2O	nitrous oxide
NO_2	nitrogen dioxide
NO_x	nitrogen oxides
NSCA	National Society for Clean Air and Environmental Protection
NTF	National Tidal Facility
NVZ	nitrate vulnerable zone
ODS	ozone depleting substance
OECD	Organisation for Economic Co-operation and Development
OPEC	Organisation of Petroleum Exporting Countries
R/P	reserves/production
PAL	Pathfinder AVHRR Land Programme
ppmv	parts per million by volume
PV	photovoltaic
RECOUP	Recycling of Used Plastics
RIL	Reduced Impact Logging
RSPCA	Royal Society for the Prevention of Cruelty to Animals
SO_2	sulphur dioxide
SPRU	Science and Technology Policy Research
TM	Thematic Mapper
TNC	trans-national corporation
UK	United Kingdom
UN	United Nations
UNCCD	United Nations Convention to Combat Desertification
UNCED	United Nations Conference on Environment and Development
UNCTAD	United Nations Commission on Trade and Development
UNDP	United Nations Development Programme
UNEP	United Nations Environment Programme
UNFCCC	United Nations Framework Convention on Climate Change
UNFF	United Nations Forum on Forests
US	United States
USDA	United States Department of Agriculture
USEPA	United States Environmental Protection Agency
WCED	World Commission on Environment and Development
WCFSD	Word Commission on Forests and Sustainable Development
WEHAB	Water, Energy, Health, Agriculture and Biodiversity
WRAP	Waste and Resources Action Programme
WRI	World Resources Institute
WSSD	World Summit on Sustainable Development
WTO	World Trade Organisation
WWF	World Wide Fund for Nature (formerly World Wildlife Fund)

References

Adams, W. and Hulme, D. (2001) Conservation and Community. Changing narratives, policies and practices in African conservation, in D. Hulme and M. Murphree (eds), *African Wildlife and Livelihoods. The Promise and Performance of Community Conservation*, Oxford: James Currey, pp. 9–23.

Adger, W.N. (1999) Social vulnerability to climate change and extremes in coastal Vietnam. *World Development*, **27**, 249–269.

Adger, W.N. (2000) Environmental and ecological economics, in T. O'Riordan (ed.), *Environmental Science for Environmental Management*, 2nd edn, Harlow: Pearson Education, pp. 93–118.

Adger, W.N. (2001) Social capital and climate change. *Tyndall Centre Working Paper*, **8**, Norwich, UK: Tyndall Centre, UEA.

Adger, W.N. and O'Riordan, T. (2000) Population, adaptation and resilience, in T. O'Riordan (ed.), *Environmental Science for Environmental Management*, 2nd edn, Harlow: Pearson Education, pp. 149–170.

Adger, W.N., Huq, S., Brown, K., Conway, D. and Hulme, M. (2002) Adaptation to climate change: setting the agenda for development policy and research. *Tyndall Centre Working Paper*, **16**, Norwich, UK: Tyndall Centre, UEA.

AEA Technology (2002) UK Greenhouse Gas Inventory, 1990 to 2000. National Atmospheric Emissions Inventory, UK www.naei.org.uk

Agnew, C.T. (1995) Desertification, drought and development in the Sahel, in A. Binns (ed.), *People and Environment in Africa*, Chichester, John Wiley, pp. 279–293.

Agnew, C.T. (2002) Drought, desertification and desiccation. *Geography*, **87**, 256–267.

al Kawari, M. (2002) *A Review of Waste Management in Qatar*. Unpublished Coursework Report for BSc Hons Environmental Science Honours, Environmental Science, UK: Kingston University.

Allen, J.A. (1999) Water stress and global mitigation: water, food and trade. *Arid Lands Newsletter*, **45**, Spring/Summer 'Water in Cities'. http://ag.arizona.edu/OALS/ALN/aln45/allan.html

Allen, P.A. (1997) *Earth Surface Processes*, Oxford: Blackwell.

Allen, M.R. and Ingram, W.J. (2002) Constraints on future changes in climate and the hydrologic cycle. *Nature*, **419**, 224–232.

Allen, P., van Dusen, D., Lundy, L. and Gliessman, S. (1991) Integrating social, environmental and economic issues in sustainable agriculture. *American Journal of Alternative Agriculture*, **6**, 34–39.

Global Environmental Issues.   Edited by Frances Harris
© 2004 John Wiley & Sons, Ltd   ISBNs: 0-470-84560-0 (HB); 0-470-84561-9 (PB)

Alley, R.B., Mayewski, P.A., Sowers, T., Stuiver, M., Taylor, K.C. and Clark, P.U. (1997) Holocene climate instability: a prominent widespread event 8200 yr. ago. *Geology*, **25**, 483–486.

Alshuwaikhat, H.M. and Nkwenti, D.I. (2002) Developing sustainable cities in arid regions. *Cities*, **19**, 85–94.

Altieri, M.A. (1987) *Agroecology: The Scientific Basis of Alternative Agriculture*, Boulder, Colorado: Westview Press.

Anderson, J.R., Hardy, E.E., Roach, J.T. and Witmer, R.E. (1976) A land use and land cover classification scheme for use with remote sensor data. *Geological Survey Professional Paper* **964**, Washington DC: US Government Printing Office.

Angel, M.V. and Rice, A.L. (1996) The ecology of the deep ocean and its relevance to global waste management. *Journal of Applied Ecology*, **33**, 915–926.

Anton, D. (1993) *Thirsty Cities – Urban Environments and Water Supply in Latin America*, Ottawa: International Development Research Centre.

Apuuli, B., Wright, J., Elias, C. and Burton, I. (2000) Reconciling national and global priorities in adaptation to climate change: with an illustration from Uganda. *Environmental Monitoring and Assessment*, **61**, 145–159.

Arizipe, L.R., Costanza, R. and Lutz, W. (1992) Population and natural resource uses, in J. Dooge, G.T. Goodman, T. O'Riordan, M. de la Riviere and M. Lefevre (eds), *An Agenda for Science for Environment and Development into the 21st Century*, Cambridge: Cambridge University Press, pp. 61–78.

Arnell, N.W. (1999) The effect of climate change on hydrological regimes in Europe: a continental perspective. *Global Environmental Change*, **9**, 5–23.

Attfield, R. (1999) *The Ethics of the Global Environment. Edinburgh Studies in World Ethics*, Edinburgh: Edinburgh University Press.

Bai, R. and Sutanto, M. (2001) The practice and challenges of solid waste management in Singapore. *Waste Management*, **22**, 557–567.

Bakker, K. (2003) Archipelagos and networks: urbanization and water privatization in the south. *Geographical Journal*, **169**, 328–341.

Bangkok Metropolitan Authority (2001) *Bangkok State of the Environment 2001*. Bangkok Metropolitan Authority, Bangkok. United Nations Environment Programme website http://www.eapap.unep.org/reports/soe/bangkoksoe.cfm/

Banks, J. and Marsden, T.K. (2000) Integrating agri-environmental policy, farming systems and rural development, Tir Cymen in Wales. *Sociologia Ruralis*, **40**, 466–480.

Barbier, E. (1989) *Economics, Natural Resources, Scarcity and Development*, London: Earthscan.

Barnett, J. (2001) Adapting to climate change in Pacific Island countries: the problem of uncertainty. *World Development*, **29**, 977–993.

Barnett, J. and Dessai, S. (2002) Articles 4.8 and 4.9 of the UNFCCC: adverse effects and the impacts of response measures. *Climate Policy*, **2**, 231–240.

Barr, C., Benefield, C., Bunce, B., Rinsdale, H. and Whitaker, M. (1986) *Landscape Changes in Britain*, Grange-over-Sands, UK: Institute of Terrestial Ecology.

Bates, T. and Scholes, M. (2002) Working towards a new atmospheric project within IGBP. *IGBP Global Change Newsletter*, **50**, 11–14.

Bell, D. and Valentine, G. (1997) *Consuming Geographies: We are What We Eat*, London: Routledge.

References

Bell, J.N.B. and Treshow, M. (2002) *Air Pollution and Plant Life*, 2nd edn, London: Wiley.

Belward, A.S., Estes, J.E. and Kilne, K.D. (1999) The IGBP-DIS global 1-km land-cover data set DISC over: a project overview. *Photogrammetric Engineering and Remote Sensing*, **65**, 1013–1020.

Benka, S.G. (2002) The energy challenge. *Physics Today*, **55**, 38–39.

Benkhe, R.H. and Scoones, I. (1992) *Rethinking range ecology: implications for range-land management in Africa*. Issues paper no. 33, International Institute for Environment and Development. London: IIED.

Bennett, A.J. (2000) Environmental consequences of increasing production: some current perspectives. *Agriculture, Ecosystems and Environment*, **82**, 89–95.

Bennett, B. (2001) What is a forest? On the vagueness of certain geographic concepts. *Topoi*, **20**, 189–201.

Berkes, F. and Folke, C. (eds) (1998) *Linking Social and Ecological Systems. Management Practices and Social Mechanisms for Building Resilience*, Cambridge: Cambridge University Press.

Berkhout, F. and Hertin, J. (2000) Socio-economic scenarios for climate impact assessment. *Global Environmental Change*, **10**, 165–168.

Berkhout, F., Hertin, J. and Jordan, A. (2002) Socio-economic futures in climate change impact assessment: using scenarios as 'learning machines'. *Global Environmental Change*, **12**, 83–95.

Berkoff, J. (2001) World Bank water strategy: some suggestions related to agriculture and irrigation. *Discussion Paper 1*, SOAS Water Issues Group.

Best Foot Forward (2002) *City Limits: A Resource Flow and Ecological Footprint Analysis of Greater London*, Oxford: Best Foot Forward.

Billing, P. (1996) *Towards Sustainable Agriculture – The Perspective of the Common Agricultural Policy in the European Union*. Unpublished paper presented to the Workshop on Landscape and Nature Conservation, Stuttgart, 26.9.96.

Bilsborrow, R.E. (1987) Population pressures and agricultural development in developing countries: a conceptual framework and recent evidence. *World Development*, **15**, 138–203.

Bishop, K. and Phillips, A. (1993) Seven steps to the market – the development of the market-led approach to countryside conservation and recreation. *Journal of Rural Studies*, **9**, 315–338.

Blaikie, P.M. (1985) *The Political Economy of Soil Erosion in Developing Countries*, London: Longman.

Blakemore, W.F. (1989) Bovine Spongiform Encephalopathy and scrapie: potential human hazards. *Outlook on Agriculture*, **18**, 165–168.

Blench, R. (1998) Biodiversity conservation and its opponents. *Natural Resources Perspectives* No. 32, London: ODI.

Blum, E. (1993) Making biodiversity conservation profitable: a case study of the Merck/INBio agreement. *Environment*, **35**, 16–45.

Boardman, J., Foster, I.D.L. and Dearing, J.A. (eds) (1990) *Soil Erosion on Agricultural Land*, Chichester and New York: Wiley.

Bolin, B. and Kheshgi, H.S. (2001) On strategies for reducing greenhouse gas emissions. *Proceedings of the National Academy Sciences*, **98**, 4850–4854.

Bonan, G.B. (1997) The effects of land use on the climate of the United States. *Climatic Change*, **37**, 449–486.

Bond, I. (2001) CAMPFIRE and the incentives for institutional change, in D. Hulme and M. Murphree (eds), *African Wildlife and Livelihoods. The Promise and Performance of Community Conservation*, Oxford: James Currey, pp. 227–243.

Bongaarts, J. (1994) Population policy options in the developing world. *Science*, **263**, 771–776.

Boserüp, E. (1965) *The Conditions of Agricultural Growth. The Economics of Agrarian Change under Population Pressure*, London: Allen and Unwin.

Boserüp, E. (1981) *Population and Technology*, Oxford: Blackwell.

Bowler, I.R. (2002a) Sustainable farming systems, in I.R. Bowler, C.R. Bryant and C. Cocklin (eds), *The Sustainability of Rural Systems: Geographical Interpretations*, Dordrecht: Kluwer, pp. 169–188.

Bowler, I.R. (2002b) Developing sustainable agriculture. *Geography*, **87**, 205–212.

Boyce, M.S. and Haney, A. (eds) (1997) *Ecosystem Management*, New Haven, CT: Yale University Press.

BP (2003) *BP Statistical Review of World Energy*, BP plc, 40pp.

British Biogen website (www.britishbiogen.co.uk/bioenergy).

Broadus, J., Milliman, J., Edwards, S., Aubrey, D. and Gable, F. (1986) Rising sea level and damming of rivers: possible effects in Egypt and Bangladesh, in J.G. Titus (ed.), *Effects of Changes in Stratospheric Ozone and Global Climate, Volume 4, Sea Level Rise*, Washington DC: UNEP and US Environmental Protection Agency, pp. 165–189.

Broadway, M. (1997) Alberta bound: Canada's beef industry. *Geography*, **82**, 377–380.

Broadway, M.J. (2000) Planning for change in small towns or trying to avoid the slaughterhouse blues. *Journal of Rural Studies*, **16**, 37–46.

Broecker, W.S. (1987) Unpleasant surprises in the greenhouse? *Nature*, **328**, 123–126.

Brönnimann, S. (2002) Picturing climate change. *Climate Research*, **22**, 87–95.

Brown, L. (1997) *Agriculture – the Missing Link*. Worldwatch Paper 136. The Worldwatch Institute, Washington.

Brown, M.B. (1993) *Fair Trade. Reform and Realities in the International Trading System*. London: Zed Books.

Brown, S., Sathaya, J., Cannel, M. and Kauppi, P. (1996) Management of forests for mitigation of greenhouse gas emissions, in R.T. Watson, M.C. Zinyowera and R.H. Moss (eds), *Climate Change 1995, Impacts, Adaptations and Mitigation of Climate Change*, Report of Working Group II, Assessment Report, IPCC, Cambridge: Cambridge University Press, pp. 773–797.

Browne, A.W., Harris, P.J.C., Hofny-Collins, A.H., Pasiecznik, N. and Wallace, R.R. (2000) Organic production and ethical trade: definition, practice and links. *Food Policy*, **25**, 69–90.

Brundtland Commission Report (WCED) (1987) *Our Common Future*, Oxford: Oxford University Press.

Bryant, R.L. (1998) Power, knowledge and political ecology in the third world: a review. *Progress in Physical Geography*, **22**, 9–94.

Bryson, W. (1993) Britain's hedgerows. *National Geographic*, **182**, 94–117.

Burger, J. (2000) Landscapes, tourism and conservation. *Science of the Total Environment*, **249**, 39–49.

References

Burns, L.D., McCormick, J.B. and Borroni-Bird, C.E. (2002) Untitled. *Scientific American*, **10**, 42–49.

Cabanes, C., Cazenave, A. and le Provost, C. (2001) Sea level rise during past 40 years determined from satellite and in situ observations. *Science*, **294**, 840–842.

Campbell, H. and Coombes, B. (1999) Green protectionism and the exporting of organic fresh fruit and vegetables from New Zealand: crisis experiments in the breakdown of Fordist trade and agricultural policies. *Rural Sociology*, **64**, 302–319.

Campbell, J.B. (2002) *Introduction to Remote Sensing*, 3rd edn, London: Taylor & Francis.

Canter, L.W. (1986) *Environmental Impacts of Agricultural Production Activities*, Chelsea, Michigan: Lewis Publishers Inc.

Carson, R. (1962) *Silent Spring*, Boston: Houghton Mifflin.

Chapin III, F.S., Zavaleta, E.S., Eviner, V.T., Naylor, R.L., Vitousek, P.M., Reynolds, H.L., Hooper, D.U., Lavorel, S., Sala, O.E., Hobbie, S.E., Mack, M.C. and Diaz, S. (2000) Consequences of changing biodiversity. *Nature*, **405**, 234–242.

Chase, T.N., Pielke, R.A., Kittel, T.G.F., Nemani, R.R. and Running, S.W. (2000) Simulated impacts of historical and cover changes on global climate. *Climate Dynamics*, **16**, 93–105.

Cihlar, J., Beaubien, J. and Latifovic, R. (2001) *VGT Land Cover of Canada*, Ottawa: Natural Resources Canada.

Cincotta, R.P., Wisnewski, J. and Engelman, R. (2000) Human population in the biodiversity hotspots. *Nature*, **404**, 990–992.

Clunies-Ross, T. (1990) Organic food: swimming against the tide, in T.K. Marsden and J. Little (eds), *Political, Social and Economic Perspectives on the International Food System*, Aldershot: Avebury, pp. 200–214.

Committee on Global Change (1990) *Research Strategies for the US Global Research Program*, Washington DC: National Academy Press.

Convention on Biological Diversity (CBD) (1992) *The Convention on Biological Diversity*. Secretariat for the Convention on Biological Diversity, United Nations Environment Programme (http://www.biodiv.org/convention.articles/).

Conway, G.R. (1997) *The Doubly Green Revolution: Food for all in the 21st Century*, London and Cornell University Press, New York: Penguin Books.

Conway, G.R. and Pretty, J.N. (1991) *Unwelcome Harvest. Agriculture and Pollution*, London: Earthscan Publications Ltd.

Costa, M.C. (1996) Tropical forestry practices for carbon sequestration: a review and case study from southeast Asia. *Ambio*, **25**, 279–283.

Cour, J.-M. and Naudet, D. (1996) West Africa in 2020. *The OECD Observer*, **200**, June/July, 20–26.

Cronin, T.M. (1983) Rapid sea level and climate change: evidence from continental and island margins. *Quaternary Science Reviews*, **1**, 177–214.

Cullet, P. (2001) Property rights regimes over biological resources. *Environment and Planning C: Government and Policy*, **19**, 651–664.

Davidson, D.A. and Harrison, S.J. (1995) The nature, causes and implications of water erosion on arable land in Scotland. *Soil Use and Management*, **11**, 63–68.

Dawkins, R. (1986) *The Blind Watchmaker*, Harlow: Longman.

DeFries, R.S. and Los, S.O. (1999) Implications of land-cover misclassification for parameter estimates in global land-surface models: an example from the simple biosphere model (SiB2). *Photogrammetric Engineering and Remote Sensing*, **65**, 1083–1088.

DeFries, R.S. and Townshend, J.R.G. (1994) Global land cover: comparison of ground-based data sets to classifications with AVHRR data, in G.M. Foody and P.J. Curran (eds), *Environmental Remote Sensing from Regional to Global Scales*, Chichester: Wiley, pp. 84–110.

DeFries, R.S., Hansen, M., Townshend, J.R.G. and Sohlberg, R. (1998) Global land cover classifications at 8 km spatial resolution: the use of training data derived from Landsat imagery in decision tree classifiers. *International Journal of Remote Sensing*, **19**, 3141–3168 (http://www.tandf.co.uk/journals).

Delwiche, C.C. (1970) The Nitrogen Cycle, in *The Biosphere*. Collected papers from *Scientific American*.

Demeritt, D. (2001) The construction of global warming and the politics of science. *Annals Association American Geographers*, **91**, 307–337.

Department of Environment, Food and Rural Affairs (2000) *Waste Strategy 2000*, London: DEFRA Publications.

Department of Environment, Food and Rural Affairs (2001) *Managing Radioactive Waste Safely: Proposals for Developing a Policy for Managing Solid Radioactive Waste in the UK*, London: DEFRA Publications.

Department of Environment, Food and Rural Affairs (2002a) *Nitrate Vulnerable Zones in England*, London: DEFRA Publications.

Department of Environment, Food and Rural Affairs (2002b) *Municipal Waste Management Survey 2000/01 Results of the Survey*, London: DEFRA Publications.

Department of the Environment and Ministry of Agriculture, Fisheries and Food (DoE/MAFF) (1995) *Rural England: A Nation Committed to a Living Countryside*, Command Paper **3016**, London: HMSO.

Devoy, R.J.N. (ed.) (1987) *Sea Surface Studies: A Global View*, London: Croom Helm.

Diamond, D. (2002) Managing the Metropolis in the global village. *Geography*, **87**, 305–315.

Dixon, R.K., Brown, S., Houghton, R.A., Solomon, A.M., Trexler, M.C. and Wisniewski, J. (1994) Carbon pools and flux of global forest ecosystems. *Science*, **263**, 185–190.

Doering, O. (1992) Federal policies as incentives or disincentives to ecologically sustainable agricultural systems. *Journal of Sustainable Agriculture*, **2**, 21–36.

Douglas, B.C. (2001) Sea level change in the era of the recording tide gauge, in B.C. Douglas, M.S. Kearney and S.P. Leatherman (eds), *Sea-Level Rise: History and Consequences*, San Diego: Academic Press, pp. 37–64.

Douglas, B.C., Kearney, M.S. and Leatherman, S.P. (eds) (2001) *Sea-Level Rise: History and Consequences*, San Diego: Academic Press.

Douglas, I. (1994) Human Settlements, in W.B. Meyer and B.L. Turner II (eds), *Changes in Land Use and Land Cover: A Global Perspective*, Cambridge: Cambridge University Press, pp. 149–170.

Dower, R., Ditz, D., Faeth, P., Johnson, N., Kozloft, K. and Mackenzie, J.J. (eds) (1997) *Frontiers of Sustainability: Environmentally Sound Agriculture, Forestry, Transportation and Power Production*, Washington DC: World Resources Institute.

Duckham, A.N. and Masefield, G.B. (1970) *Farming Systems of the World*, London: Chatto and Windus.

References

Dunlap, R.E., Beus, C.E., Howell, R.E. and Waud, J. (1992) What is sustainable agriculture? An empirical examination of faculty and farmer definitions. *Journal of Sustainable Agriculture*, **3**, 5–40.

Dwivedi, O.P. (1986) Political science and the environment. *International Social Science Journal*, **38**, 377–390.

Dyson, T. (1996) *Population and Food: Global Trends and Future Prospects*, London: Routledge.

Easterling, D.R., Meehl, G.A., Parmesan, C., Changnon, S., Karl, T.R. and Mearns, L.O. (2000) Climate extremes: observations, modeling and impacts. *Science*, **289**, 2068–2074.

Eaton, D., van Tongeren, F., Louwaars, L., Viser, B. and van der Meer, I. (2002) Economic and policy aspects of 'terminator' technology. *Biotechnology and Development Monitor*, **49**, 19–22.

Economist, The, 19 May 2001. Nuclear power – a renaissance that may not come, 28–31.

Edwards, R.L., Beck, J.W., Burr, G.S., Donahue, D.J., Chappell, J.M.A., Bloom, A.L., Druffel, E.R.M. and Taylor, F.W. (1993) A large drop in atmospheric $^{14}C/^{12}C$ and reduced melting in the Younger Dryas, documented with ^{230}Th ages of corals. *Science*, **260**, 962–968.

Egdell, J. (2000) Consultation on the countryside premium scheme. Creating a 'market' for information. *Journal of Rural Studies*, **16**, 357–366.

Egziabher, A.G. (1994) Urban farming, co-operatives and the urban poor in Addis Ababa, in A.G. Egziabher, D. Lee-Smith, D.G. Maxwell, P.A. Memon, L.J.A. Mougeot and C.J. Sawio (eds) *Cities Feeding People: An Examination of Urban Agriculture in East Africa*, Ottawa: International Development Research Centre, pp. 67–84.

Ehrlich, P. (1968) *The Population Bomb*, New York: Ballantine Books.

Ehrlich, P.R. and Ehrlich, A.H. (1990) *The Population Explosion*, London: Hutchinson.

Ehrlich, P.R., Ehrlich, A.H. and Holden, J.P. (1977) *Ecoscience*, 3rd edn, San Francisco: W.H. Freeman & Co.

Elliott, P., Morris, S., Briggs, D., Dehoogh, C., Hurt, C., Jenson, T.K., Maitland, I., Lewin, A., Richardson, S., Wakefield, J. and Järup, L. (2001) *Birth Outcomes and Selected Cancers in Populations Living Near Landfill Sites*. Report to the Department of Health Small Area Health Statistics Unit (SAHSU), London: Imperial College.

El-Raey, M., Nasr, S., Frihy, O., Desouki, S. and Dewidar, Kh. (1995) Potential impacts of accelerated sea-level rise on Alexandria Governorate, Egypt. *Journal of Coastal Research, Special Issue*, **14**, 190–204.

Environment Agency (2002) *Solid Residues from Municipal Waste Incinerators in England and Wales; A Report on an Investigation by the Environment Agency*, UK: Bristol Environment Agency.

Environmental Data Services (1996) Phosphate pollution shock for arable farmers. *ENDS Report*, **263**, 10.

Environmental Protection Agency (2002) *Municipal Solid Waste in the United States: 2000 Facts and Figures*, USA: Office of Solid Waste and Emergency Response.

Estes, J.E. and Mooneyhan, D.W. (1994) Of maps and myths. *Photogrammetric Engineering and Remote Sensing*, **60**, 517–524.

Evans, N.J. and Morris, C. (1997) Towards a geography of agri-environmental policies in England and Wales. *Geoforum*, **28**, 189–204.

Evans, N.J., Morris, C. and Winter, M. (2002) Conceptualizing agriculture: a critique of post-productivism as the new orthodoxy. *Progress in Human Geography*, **26**, 313–332.

Evans, R. (1996) *Soil Erosion and its Impacts in England and Wales*, London: Friends of the Earth.

Fairbanks, R.G. (1989) A 17,000-year glacio-eustatic sea level record: influence of glacial melting rates on the Younger Dryas event and deep-ocean circulation. *Nature*, **342**, 637–642.

Fairbridge, R.W. and Krebs, O.A. (1962) Sea level and the southern oscillation. *Geophysical Journal*, **6**, 532–545.

Fairhead, J. and Leach, M. (1996) Enriching the landscape: Social history and the management of transition ecology in the forest-savanna mosaic of the Republic of Guinea. *Africa*, **66**, 14–36.

Farmer, A. (1997) *Managing Environmental Pollution*, London: Routledge.

Feyt, H. (2001) Protecting intellectual property rights of living species: history and current debates around plant varieties. *OCL – Oleagineux Corps Gras Lipids*, **8**, 514–523.

Fisher, D.E. and Fisher, M.J. (2001) The nitrogen bomb. *Discover*, **22** (4), 8pp.

Flierl, G.R. and Robinson, A.R. (1972) Deadly surges in the Bay of Bengal, dynamics and storm tide table. *Nature*, **239**, 213–215.

Foeken, D. and Mwangi, A.M. (1998) *Farming in the City of Nairobi*. Working Paper No. 30, African Studies Centre, University of Leiden. Available at http://asc.leidenuniv.nl/pdf/wp30.pdf [Last accessed: 4/12/03].

Food and Agricultural Organisation (2001) *State of the World's Forests*, FAO statistics, www.fao.org/ (accessed 15 June 2002).

Foody, G.M. (1999) The continuum of classification fuzziness in thematic mapping. *Photogrammetric Engineering and Remote Sensing*, **65**, 443–451.

Foody, G.M. (2002) Status of land cover classification accuracy assessment. *Remote Sensing of Environment*, **80**, 185–201.

Foody, G.M. and Atkinson, P.M. (2002) Current status of uncertainty issues in remote sensing and GIS, in G.M. Foody and P.M. Atkinson (eds), *Uncertainty in Remote Sensing and GIS*, Chichester: Wiley, pp. 287–302.

Foster, C. and Lampkin, N. (2000) *Organic and In-conversion Land Area, Holdings, Livestock and Crop Production in Europe*. Report to the European Commission, FAIR3-CT96–1794, University College, Aberystwyth.

Francis, C.A. and Younghusband, G. (1990) Sustainable agriculture: an overview, in C.A. Francis, C.B. Flora and L.D. King (eds), *Sustainable Agriculture in Temperate Zones*, New York: John Wiley and Sons, pp. 1–23.

French, G.T., Awosika, L.F. and Ibe, C.E. (1995) Sea-level rise and Nigeria: potential impacts and consequences. *Journal of Coastal Research, Special Issue*, **14**, 224–242.

French, H. (2002) Reshaping global governance, in L. Starke (ed.), *State of the World 2002: A Worldwatch Institute Report on Progress Towards a Sustainable Society*, London: Earthscan Publications Ltd, pp. 174–198.

Friedl, M.A., McIver, D.K., Hodges, J.C.F., Zhang, X.Y., Muchoney, D., Strahler, A.H., Woodcock, C.E., Gopal, S., Schneider, A., Cooper, A., Baccini, A., Gao, F. and Schaaf, C. (2002) Global and cover mapping from MODIS: algorithms and early results. *Remote Sensing of Environment*, **83**, 287–302.

References

Fuller, R.M., Groom, G.B. and Jones, A.R. (1994) The land cover map of Great Britain: an automated classification of Landsat Thematic Mapper data. *Photogrammetric Engineering and Remote Sensing*, **60**, 553–562.

Furukawa, K. and Baba, S. (2000) Effects of sea-level rise on Asian mangrove forests, in N. Mimura and H. Yokoki (eds), *Proceedings of APN/SURVAS/LOICZ Joint Conference on Coastal Impacts of Climate Change and Adaptation in the Asia-Pacific region*, Kobe: Asia-Pacific Network for Global Change Research, pp. 219–224.

Futureforests website (www.futureforests.com/sundaytimes).

Gabriel, Y. and Lang, T. (1995) *The Unmanageable Consumer*, London: Sage Publications.

Gaisford, J.D., Tarvydas, R., Hobbs, J.E. and Kerr, W.A. (2002) Biotechnology piracy: rethinking the international protection of intellectual property. *Canadian Journal of Agricultural Economics*, **50**, 15–34.

Ghaffar, A. and Robinson, G.M. (1997) Restoring the agricultural landscape: the impact of government policies in East Lothian, Scotland. *Geoforum*, **28**, 205–217.

Ghijsen, H. (1998) Plant variety protection in a developing and demanding world. *Biotechnology and Development Monitor*, **36**, 2–5.

Gommes, R., du Guerny, J., Nachtergaele, F. and Brinkman, R. (1997) *Potential Impacts of Sea-Level Rise on Populations and Agriculture*, Rome: FAO.

Goodman, D. and Watts, M.J. (eds) (1997) *Globalising Food: Agrarian Questions and Global Restructuring*, London and New York: Routledge.

Gornitz, V. (1995) A comparison of differences between recent and late Holocene sea-level trends from eastern North America and other selected regions. *Journal of Coastal Research, Special Issue*, **17**, 287–297.

Gornitz, V. and Lebedeff, S. (1987) Global sea level changes during the past century, in D. Nummedal, O.H. Pilkey and J.D. Howard (eds), *Sea Level Fluctuation and Coastal Evolution*, SEPM Special Publication, **41**, 3–16.

Gornitz, V., Lebedeff, S. and Hansen, J. (1982) Global sea level trend in the past century. *Science*, **215**, 1611–1615.

Goudie, A.S. (1993) Land transformation, in R.J. Johnston (ed.), *The Challenge for Geography*, Oxford: Blackwell, pp. 117–137.

Gough, C. and Shackley, S. (2001) The respectable politics of climate change: the epistemic communities and NGOs. *International Affairs*, **77**, 329–345.

Gould, W. (2000) Remote sensins of vegetation, plant species richness, and regional biodiversity hotspots. *Ecological Applications*, **10**, 1861–1870.

Graham, E. and Boyle, P. (2002) Population crisis: from the local to the global, in R.J. Johnson, P.J. Taylor and M.J. Watts (eds), *Geographies of Global Change*, 2nd edn, Oxford: Blackwell, pp. 198–215.

Grainger, A. (1990) *The Threatening Desert*, London: Earthscan Publications Ltd.

Greene, C.R. (2000) US organic farming emerges in the 1990s: adoption of certified systems. *Economic Research Service, Agriculture Information Bulletin, USDA*, no. **770**.

Greenpeace (1997) *Greenfreeze: A Global Revolution in Domestic Refrigeration. A Greenpeace Solution*. A Position Paper prepared for the 9th Meeting of Parties to the Montreal Protocol. London: Greenpeace.

Greenpeace (1998) *The Turning of the Spar*, London: Greenpeace.

Gregory, R.D. and Baillie, S.R. (1998) Large-scale habitat use of some declining British birds. *Journal of Applied Ecology*, **35**, 785–799.

Gregory, R.D., Noble, D.H., Campbell, L.C. and Gibbons, D.W. (1999) *The State of the Nation's Birds*, Sandy, Bedfordshire: RSPB.

Grigg, D. (1980) *Population Growth and Agrarian Change: An Historical Perspective*, Cambridge: Cambridge University Press.

Grimble, R. and Laidlaw, M. (2002) Biodiversity and local livelihoods: Rio plus 10. *Natural Resources Perspectives* No. 73, London: ODI.

Grimble, R., Chan, M.K., Aglionby, J. and Quan, J. (1995) *Trees and Trade-offs: A Stakeholder Approach to Natural Resource Management*. Gatekeeper series No. **52**, London: Sustainable Agriculture Programme, International Institute for Environment and Development.

Grubb, M., Vrolijk, C. and Brack, D. (1999) *The Kyoto Protocol: A Guide and Assessment*, London, UK: RIIA, 342pp.

Grübler, A. (1994) Technology, in W.B. Meyer and B.L. Turner II (eds), *Changes in Land Use and Land Cover: A Global Perspective*, Cambridge: Cambridge University Press, pp. 287–328.

Gutenberg, B. (1941) Changes in sea level, post-glacial uplift and mobility of the earth's interior. *Bulletin of the Geological Society of America*, **52**, 721–722.

Hallam, A. (1984) Pre-quaternary sea-level changes. *Annual Review of Earth and Planetary Science*, **12**, 205–243.

Hannah, L., Lohse, D., Hutchinson, C., Carr, J.L. and Lankerani, A. (1994) A preliminary inventory of human disturbance of world ecosystems. *Ambio*, **23**, 246–250.

Hardin, G. (1968) The tragedy of the commons. *Science*, **162**, 1243–1248.

Hardoy, J.E. and Satterthwaite, D. (1990) The future city, in J.E. Hardoy, S. Cairncross and D. Satterthwaite (eds), *The Poor Die Young: Housing and Health in Third World Cities*, London: Earthscan, pp. 228–244.

Hardoy, J., Mitlin, D. and Satterthwaite, D. (2001) *Environmental Problems in an Urbanizing World: Finding Solutions for Cities in Africa, Asia and Latin America*, London: Earthscan.

Harremoës, P., Gee, D., MacGarvin, M., Stirling, A., Keys, J., Wynne, B. and Guesdes Vaz, S. (2002) *The Precautionary Principle in the 20th Century*, London, UK: Earthscan, 268pp.

Harris, F. (1996) *Intensification of Agriculture in Semi-arid Areas: Lessons from the Kano Close-settled Zone, Nigeria*. Gatekeeper Series no. **SA59**, International Institute for Environment and Development.

Harris, F. (2003) Relying on nature: wild foods in Northern Nigeria. *Ambio*, **32**, 24–29.

Harrison, P. (1993) *The Third Revolution. Population, Environment and a Sustainable World*, London: Penguin.

Hart, J.F. and Mayda, C. (1998) The industrialization of livestock production in the United States. *Southeastern Geographer*, **38**, 58–78.

Hearty, P.J., Neumann, A.C. and Kaufman, D.S. (1999) Chevron ridges and runup deposits in the Bahamas from storms late in oxygen-isotope substage 5e. *Quaternary Research*, **50**, 309–322.

Henao, J. and Baanante, C.A. (1999) *Nutrient Depletion in the Agricultural Soils of Africa*, Washington DC: International Food Policy Research Institute.

Hewitt, K. (1997) *Regions of Risk: A Geographical Introduction to Disasters*, Harlow: Longman.

References

Hoffert, M.I. and 17 others (2002) Advanced technology paths to global climate stability: energy for a greenhouse planet. *Science*, **298**, 981–987.

Hogan, W.J. (ed.) (1995) *Energy from Inertial Fusion*, Vienna, Austria: International Atomic Energy Agency, 457pp.

Hogg, D., Barth, J., Favoino, E., Centemero, M., Caimi, V., Devliegher, W. and Antler, S. (2002) *Comparison of Compost Standards Within the EU, North America and Australasia*, Banbury: Waste and Resources Action Programme (WRAP).

Holdgate, M.W. (1979) *A Perspective on Environmental Pollution*, Cambridge: Cambridge University Press.

Holdgate, M.W. (1993) The sustainable use of tropical coastal resources – a key conservation issue. *Ambio*, **22**, 481–482.

Holdgate, M.W. (1994) Ecology, development and global policy. *Journal of Applied Ecology*, **31**, 201–211.

Holliday, F.G.T. (1995) The dumping of radioactive waste in the deep ocean: scientific advice and ideological persuasion, in D.E. Cooper and J.A. Palmer (eds), *The Environment in Question: Ethics and Global Issues*, London: Routledge.

Hooda, P.S., Truesdale, V.W., Edwards, A.C., Withers, P.J.A., Aitken, M.N., Miller, A. and Rendell, A.R. (2001) Manuring and fertilisation effects on phosphorus accumulation in soils and potential environmental implications. *Advances in Environmental Research*, **5**, 13–21.

Horta, K. (2000) Rainforest: biodiversity conservation and the political economy of international financial institutions, in P. Stott and S. Sullivan (eds), *Political Ecology. Science, Myth and Power*, London: Arnold, pp. 179–202.

Houghton, R.A. (1996) Converting terrestrial ecosystems from sources to sinks of carbon. *Ambio*, **25**, 267–272.

Hulme, M. and Barrow, E.M. (1997) Introducing climate change, in M. Hulme and E.M. Barrow (eds), *Climates of the British Isles: Present, Past and Future*, London, UK: Routledge, pp. 1–7.

Hunter, J.R. (2002) *Note on Relative Sea Level Change, Funafuti, Tuvalu*. Unpublished paper, Antarctic Co-operative Research Centre, Hobart, 25pp.

Huq, S., Ali., S.I. and Rahman, A.A. (1995) Sea-level rise and Bangladesh: a preliminary analysis. *Journal of Coastal Research, Special Issue*, **14**, 44–53.

Iaquinta, D. and Dreschler, A. (2001) More than the Spatial Fringe: an application of the periurban typology to planning and management of natural resources. Paper presented to *the Rural-Urban Encounters: Managing the Environment of the Peri-Urban Interface*. Development Planning Unit, University of London, London, 9–10 November.

Ilbery, B.W. (2002) Geographical aspects of the 2001 outbreak of foot-and-mouth disease in the UK. *Geography*, **87**, 142–147.

Intergovernmental Panel on Climate Change (IPCC) (1991) *The Seven Steps to the Vulnerability Assessment of Coastal Areas to Sea-Level Rise – Guidelines for Case Studies*. IPCC Coastal Zone Management Subgroup, 24pp.

Intergovernmental Panel on Climate Change (IPCC) (2000) *Emissions Scenarios. A Special Report of Working Group III of the Intergovernmental Panel on Climate Change*, Cambridge, UK: Cambridge University Press, 599pp.

Intergovernmental Panel on Climate Change (IPCC) (2001a) Climate change 2001: synthesis report, in R.T. Watson and the Core Writing Team (eds), *Contribution of*

Working Groups I, II, and III to the Third Assessment Report of the Intergovernmental Panel on Climate Change, Cambridge, UK: Cambridge University Press, 398pp (http://www.ipcc.ch/).

Intergovernmental Panel on Climate Change (IPCC) (2001b) Climate change 2001: mitigation, in B. Metz, O. Davidson, R. Swart and J. Pan (eds), *Contribution of Working Group III to the Third Assessment Report of the Intergovernmental Panel on Climate Change*, Cambridge, UK: Cambridge University Press, 700pp (http://www.ipcc.ch/).

Intermediate Technology Development Group (ITDG) (2003) (http://www.itdg.org/annual_review_2003)

International Centre for Local Environmental Initiatives (ICLEI) (2002) *Local Governments' Response to Agenda 21: Summary Report of Local Agenda 21 Survey with Regional Focus*, Canada: International Centre for Local Environmental Initiatives.

International Energy Outlook (IEO) (2002) http://www.eia.doe.gov/

International Environmental Technology Centre (IETC) (1996) *International Source Book on Environmentally Sound Technologies for Municipal Solid Waste Management*, Tokyo: UNEP Technical Publications Series.

International Solid Waste Association (ISWA) (2002) *Industry as a Partner for Sustainable Development: Waste Management International*, Copenhagen: Solid Waste Association, UK and United Nations Environment Programme.

International Union for the Conservation of Nature (IUCN)/WWF/INEP (1980) *World Conservation Strategy*. Gland, Switzerland: IUCN.

International Union for the Conservation of Nature (IUCN) (1994) http://www.unepwcmc.org/protected_areas/categories/index.html

Irwin, E.G. and Geoghegan, J. (2001) Theory, data, methods: developing spatially explicit economic models of land use change. *Agriculture, Ecosystems and Environment*, **85**, 7–23.

Jacobi, P., Kjellén, M. and Castro, Y. (1998) *Household Environmental Problems in São Paulo: Perceptions and Solutions from Centre to Periphery*. Urban Environment Series Report, **5**, Stockholm: Stockholm Environment Institute. Available at http://www.sei.se

Jelgersma, S. and Tooley, M.J. (1995) Sea-level changes during the recent geological past, in *Journal of Coastal Research*, special issue No. **17**, Holocene Cycles, Climate, Sea levels and Sedimentation, pp. 123–139.

Jiang, Y., Kirkman, H. and Hua, A. (2001) Megacity development: managing impacts on marine environments. *Ocean and Coastal Management*, **44**, 293–318.

Jodha, N.S. (1986) Common property resources and rural poor in dry regions of India. *Economic and Political Weekly*, 5 July 1986, 1169–1181.

Johannessen, L.M. and Boyer, G. (1999) *Observations of Solid Waste Landfills in Developing Countries: Africa, Asia and Latin America*, Washington: The World Bank.

Johnson, N. and Ditz, D. (1997) Challenges to sustainability in the US forest sector, in R. Dower, D. Ditz, P. Faeth, N. Johnson, K. Kozloff and J.J. MacKenzie (eds), *Frontiers of Sustainability: Environmentally Sound Agriculture, Forestry, Transportation and Power Production*, Washington DC: World Resources Institute, pp. 191–280.

Jones, P.D., Osborn, T.J. and Briffa, K.R. (2001) The evolution of climate over the last millennium. *Science*, **292**, 662–667.

Kaldis, P.E. (2002) The Greek fresh-fruit market in the framework of the Common Agricultural Policy. Unpublished PhD thesis, University of Coventry.

References

Kaluwin, C. (2001) Adaptation policies – addressing climate change impacts in the Pacific region, in B.J. Noye and M. Grzechnik (eds), *Sea-Level Changes and their Effects*, Singapore: World Scientific Publishing, pp. 273–291.

Kandeh, H.B.S. and Richards, P. (1996) People as conservationists: querying neo-Malthusian assumptions about biodiversity in Sierra Leone. *Africa*, **66**, 90–103.

Kennett, J.P., Cannariato, K.G., Hendy, I.L. and Behl, R.J. (2002) *Methane Hydrates in Quaternary Climate Change*, American Geophsyical Union, Special Publication, **54**.

Kjellén, M. and McGranahan, G. (1997) *Urban Water – Towards Health and Sustainability*, Stockholm: Stockholm Environment Institute. Available at *http://www.sei.se*.

Knight, J. and Song, L. (2000) *The Rural-Urban Divide; Economic Disparities and Interactions in China*, Oxford: Oxford University Press.

Knox, J.W. and Weatherfield, E.K. (1999) The application of GIS to irrigation water resource management in England and Wales. *The Geographical Journal*, **165**, 90–98.

Koc, M., MacRae, R., Mougeot, L.J.A. and Walsh, J. (eds) (1999) *Hunger Proof Cities: Sustainable Urban Food Systems*, Ottawa: International Development Research Centre.

Koziell, I. (2001) *Diversity not Adversity. Sustaining Livelihoods with Biodiversity*, London: International Institute for Environment and Development and Department for International Development.

Kreuger, A.O., Schiff, M. and Valdès, A. (eds) (1992) *The Political Economy of Agricultural Pricing Policy*, Baltimore: Johns Hopkins University Press for the World Bank, 4 vols.

Kuznesof, S., Tregear, A. and Moxey, A. (1997) Regional foods: a consumer perspective. *British Food Journal*, **99**, 199–206.

Ladurie, L.R. (1972) *Times of Feast, Times of Famine: A History of Climate Since the Year 1000*, London, UK: Allen & Unwin.

Lamb, H.H. (1977) *Climate Present, Past and Future: Volume 2 Climatic History and the Future*, London, UK: Methuen & Co., 835pp.

Lambin, E.F. (1994) *Modelling Deforestation Processes: A Review*, Luxembourg: European Commission.

Lambin, E.F. (1997) Modelling and monitoring land-cover change processes in tropical regions. *Progress in Physical Geography*, **21**, 375–393.

Lambin, E.F., Turner, B.L., Geist, H.J., Agbola, S.B., Angelsen, A., Bruce, J.W., Coomes, O.T., Dirzo, R., Fischer, G., Folke, C., George, P.S., Homewood, K., Imbernon, J., Leemans, R., Li, X., Moran, E.F., Mortimore, M., Ramakrishnan, P.S., Richards, J.F., Skånes, H., Steffen, W., Stone, G.D., Svedin, U., Veldkamp, T.A., Vogel, C. and Xu, J. (2001) The causes of land-use and land-cover change: moving beyond the myths. *Global Environmental Change*, **11**, 261–269.

Lang, T. (1998) BSE and CJD: recent developments, in S. Ratzan (ed.), *The Next Cow Crisis: Health and the Public Good*, London: UCL Press, pp. 67–85.

Lang, T., Barling, D. and Caracher, M. (2001) Food, social policy and the environment: towards a new model. *Social Policy and Administration*, **35**, 538–558.

Lappé, M. and Bailey, R. (1999) *Against the Grain: the Genetic Transformation of Global Agriculture*, London: Earthscan.

Lario, J., Zazo, C., Dabrio, C.J., Somoza, L., Goy, J.L., Bardají, T. and Siva, P.V. (1995) Record of recent Holocene sediment input on spit bars and deltas of southern Spain. *Journal of Coastal Research, Special Issue*, **17**, 241–246.

Last, F.T. (1987) Introduction, in P.J. Coughtrey, M.H. Martin and M.H. Unsworth (eds), *Pollutant Transport and Fate in Ecosystems*. Special Publication Series of the British Ecological Society Number **6**, Oxford: Blackwell Scientific, pp. 1–4.

Laupepa, P.K. (2002) High tides and low scientific standards. *Pacific Magazine*, April 2002, pp. 6–7.

Law, B.E. and Curtis, J.B. (2002) Introduction to unconventional petroleum systems. *AAPG Bulletin*, **86**, 1851–1853 (and references therein).

Leach, M., Mearns, R. and Scoones, I. (1997) Environmental Entitlements: A Framework for Understanding the Institutional Dynamics of Environmental Change. *IDS Discussion Papers*, No. **359**, Sussex: Institute of Development Studies.

Leatherman, S.P. and Beller-Simms, N. (1997) Sea-level rise and small island states: an overview. *Journal of Coastal Research, Special Issue*, **24**, 1–16.

Leatherman, S.P. and Nicholls, R.J. (1995) Accelerated sea-level rise and developing countries: an overview. *Journal of Coastal Research, Special Issue*, **14**, 1–14.

Lee-Smith, D. and Memon, P.A. (1994) in A.G. Egziabher, D. Lee-Smith, D.G. Maxwell, P.A. Memon, L.J.A. Mougeot and C.J. Sawio (eds), *Cities Feeding People: An Examination of Urban Agriculture in East Africa*, Ottawa: International Development Research Centre, pp. 67–84.

Le Heron, R.B. (1988) Food and fibre production under capitalism: a conceptual agenda. *Progress in Human Geography*, **12**, 409–430.

Lewcock, C. (1995) Farmer use of urban waste in Kano. *Habitat International*, **19**, 225–234.

Lipton, M. (2001) Reviving global poverty reduction: What role for genetially modified plants? *Journal of International Development*, **13**, 823–845.

Lomborg, B. (2001) *The Skeptical Environmentalist*, Cambridge, UK: Cambridge University Press, 515pp.

London Dumping Convention (1972) Convention on the Prevention of Marine Pollution by Dumping of Wastes and Other Matter. http://sedac.ciesin,org/pidb/texts/marine.pollution.dumping.of.wastes.1972.html (accessed 3 February 2003).

Lucas, R.M., Honzak, M., Curran, P.J., Foody, G.M., Milne, R., Brown, T. and Amaral, S. (2000) Mapping the regional extent of tropical forest regeneration stages in the Brazilian Legal Amazon using NOAA AVHRR data. *International Journal of Remote Sensing*, **21**, 2855–2881.

Lutz, W. (2002) Interview for Home Planet. BBC Radio 4, 1 January 2002.

Lynch, K. (1995) Sustainable Urban food supply for Africa. *Sustainable Development*, **3**, 80–88.

Lynch, K. (2002) Urban agriculture, in V. Desai and R. Potter (eds), *Arnold Companion to Development Studies*, Edward Arnold.

Lynch, K. (in press) *Rural–Urban Interaction in the Developing World*, London: Routledge.

Lynch, K., Binns, T. and Olofin, E.A. (2001) Urban agriculture under threat – The land security question in Kano, Nigeria. *Cities*, **18**, 159–171.

MacKellar, L., Lutz, W., McMichael, A.J. and Subrke, A. (1998) Population and climate change, in S. Rayner and E. Malone (eds), *Human Choice and Climate Change vol 1: The Societal Framework*, Washington: Batelle Press, pp. 89–193.

References

MacMillan, G. (1994) *At the End of the Rainbow? Gold, Land and Society in the Brazilian Amazon*, London: Earthscan Publications.

McCarthy, J.F. (2000) The changing regime: forest property and *reformasi* in Indonesia, in M. Doornbos, A. Saith and B. White (eds), *Forests: Nature, People, Power*, Oxford: Blackwell, pp. 89–128.

McGranahan, G. and Satterthwaite, D. (2000) Environmental Health or Ecological sustainability? Reconciling the brown and green agendas in urban development, in C. Pugh (ed.), *Sustainable Cities in Developing Countries*, London: Earthscan, pp. 73–90.

McGranahan, G. and Satterthwaite, D. (2002) The environmental dimensions of sustainable development for cities, in G. Robinson (ed.), *Geography. Sustainable Development Special*, **87**(3), 213–226.

McKenna, M.K.L., le Heron, R.B. and Roche, M.M. (2001) Living local, growing global: renegotiating the export production regime in New Zealand's pipfruit sector. *Geoforum*, **32**, 157–166.

McLanaghan, S.R.B. (2002) *Delivering the Landfill Directive: The role of new & emerging technologies*. Report for the Strategy Unit: 0008/2002. Cumbria: Associates in Industrial Ecology.

McMorrow, J. and Talip, M.A. (2001) Decline of forest area in Sabah, Malaysia: relationship to state policies, land code and land capability. *Global Environmental Change*, **11**, 217–230.

McNeill, J., Alves, D., Arizpe, L., Bykova, O., Galvin, K., Kelmelis, J., Migot-Adholla, S., Morrisette, P., Moss, R., Richards, J., Riebsame, W., Sadowski, F., Sanderson, S., Skole, D., Tarr, J., Williams, M., Yadav, S. and Young, S. (1994) Toward a typology and regionalisation of land-cover and land-use change: report of working group B, in W.B. Meyer and B.L. Turner II (eds), *Changes in Land Use and Land Cover: A Global Perspective*, Cambridge: Cambridge University Press, pp. 55–72.

Maddox, J. (1985) Avoiding recurrent catastrophes. *Nature*, **315**, 453.

Main, H. (1995) The effects of urbanisation in rural environments in Africa, in T. Binns (ed.), *People and Environment in Africa*, Chichester: Wiley, pp. 47–57.

Mainguet, M. and Da Silva, G.G. (1998) Desertification and drylands development: what can be done? *Land Degradation and Development*, **9**, 375–382.

Mandanipour, A. (1998) *Tehran: The Making of a Metropolis*, Chichester: John Wiley & Sons.

Manne, A.S. and Rutherford, T.F. (1994) International trade in oil, gas and carbon emission rights: an intertemporal general equilibrium model. *Energy Journal*, **15**, 57–76.

Mannion, A.M. (2002) *Dynamic World: Land-Cover and Land-Use Change*, London: Arnold.

Marsden, T.K. (1997) Creating space for food: the distinctiveness of recent agrarian development, in D. Goodman and M.J. Watts (eds), *Globalising food: agrarian questions and global restructuring*, London and New York: Routledge, pp. 169–191.

Marsden, T.K. (1999) Rural futures: the consumption countryside and its regulation. *Sociologia Ruralis*, **39**, 501–520.

Martens, P., Kovats, S., Nijhof, S., de Vries, P., Livermore, M., Bradley, D., Cox, J. and McMichael, T. (1999) Climate change and future populations at risk of malaria. *Global Environmental Change*, **9**, S89–S107.

Mason, C.F. (1996) *Biology of Freshwater Pollution*, 3rd edn, Harlow: Longman.

Mather, A. (1993) Introduction, in A. Mather (ed.), *Afforestation, Policies, Planning and Progress*, London: Belhaven Press, pp. 1–12.

Mather, A.S., Needle, C.L. and Fairburn, J. (1998) The human drivers of global land cover change: the case of forests. *Hydrological Processes*, **12**, 1983–1994.

Mather, P.M. (1999) Land cover classification revisited, in P.M. Atkinson and N.J. Tate (eds), *Advances in Remote Sensing and GIS*, Chichester: Wiley, pp. 7–16.

Meadows, D.H., Meadows, D.L., Randers, J. and Behrens, W.W. (1972) *The Limits to Growth*, London: Earth Island Ltd.

Medina, M. (1997) *Informal Recycling and Collection of Solid Wastes in Developing Countries: Issues and Opportunities*. UNU/IAS Working paper no. 24. Tokyo: The United Nations University Institute of Advanced Studies.

Merrit, J.Q. and Jones, P.C. (2000) Science and environmental decision making: the social context, in D. Sumner and M. Huxham (eds), *Science and Environmental Decision Making*, Harlow: Pearson Education Ltd.

Merson, J. (2000) Bioprospecting or bio-piracy: intellectual property rights and biodiversity in a colonial and postcolonial context. *OSIRIS*, **15**, 282–296.

Michaelson, K.L. (1981) *And the Poor Get Children: Radical Perspectives on Population Dynamics*, New York: Monthly Review Press.

Middelton, N. (1999a) *The Global Casino*, 2nd edn, London: Arnold.

Middelton, N. (1999b) All in the genes. *Geographical*, **71**, 51–55.

Miller, G.T. (1998) *Living in the Environment*, 10th edn, Belmont, California: Wadsworth Publishing Co.

Milliman, J., Broadus, J. and Gable, J. (1989) Environmental and economic implications of rising sea level and subsiding deltas: the Nile and Bengal examples. *Ambio*, **18**, 340–345.

Ministry of Agriculture, Fisheries and Food (MAFF) (2000) *England Rural Development Plan 2000–2006*, London: MAFF.

Ministry of Foreign Affairs, Republic of Belarus (1994) *Belarus*, Minsk: Ministry of Foreign Affairs, Republic of Belarus.

Mitchell, B. (2002) South Pacific Sea Level and Climate Monitoring Project. Printout of Powerpoint Presentation given at the South Pacific Forum, Suva, Fiji, 13 March 2002.

Mitchell, J.F.B. and Karoly, D.J. (2001) Detection of climate change and attribution of causes, in J.T. Houghton, Y. Ding, D.J. Griggs, M. Noguer, P.J. van der Linden and D. Xiaosu (eds), *Climate change 2001: the scientific basis*. Working Group I Report to the Intergovernmental Panel on Climate Change, Third Assessment, Cambridge, UK: Cambridge University Press, 944pp.

Moore, N.W. (1967) A synopsis of the pesticide problem, in J.B. Cragg (ed.), *Advances in Ecological Research*, **4**, London: Academic Press, pp. 75–129.

Moran, K., King, S.R. and Carlson, T.J. (2001) Biodiversity prospecting. Lessons and prospects. *Annual Review of Anthropology*, 505–526.

Mörner, N.-A. (1981) Revolution in Cretaceous sea-level analysis. *Geology*, **9**, 344–346.

Morris, C. and Young, C. (1997) Towards environmentally beneficial farming? An evaluation of the Countryside Stewardship scheme. *Geography*, **82**, 305–316.

Mortimore, M. and Adams, W.M. (1999) *Working the Sahel: Environment and Society in Northern Nigeria*, London: Routledge.

References

Mortimore, M. and Tiffen, M. (1995) Population and environment in time perspective: the Machakos Story, in T. Binns (ed.), *People and Environment in Africa*, Chichester: John Wiley, pp. 69–89.

Mortimore, M. and Harris, F.M.A. (in press) Do small farmers' achievements contradict the nutrient depletion scenarios for Africa? *Land Use Policy*.

Moss, B. (1979) Alarm call for the Broads. *Geographical Magazine*, **52**, 47–50.

Moss, B. (1998) *Ecology of Fresh Waters; Man and Medium, Past to Future*, 3rd edn, Oxford: Blackwell Science.

Moss, B., Stansfield, J., Irvine, K., Perrows, M. and Phillips, G. (1996) Progressive restoration of a shallow lake: a 12 year experiment in isolation, sediment removal and biomanipulation. *Journal of Applied Ecology*, **33**, 71–86.

Murdoch, J. and Miele, M. (1999) Back to nature: changing worlds of production in the food sector. *Sociologia Ruralis*, **39**, 465–483.

Murphree, M.W. (1993) *Communities as resource management institutions*. Gatekeeper series No. **36**, Sustainable Agriculture Programme, London: International Institute for Environment and Development.

Mvena, Z.S.K., Lupanga, I.J. and Mlozi, M.R.S. (1991) *Urban Agriculture in Tanzania: A Study of Six Towns*. The Sokoine University of Agriculture and the International Development Research Centre, Morogoro.

Myers, N. (1983) Tropical moist forests: over-exploited and under-utilised? *Forest Ecology and Management*, **6**, 59–79.

Myers, N. (1996) The world's forests: problems and potentials. *Environmental Conservation*, **23**, 156–168.

Myers, N., Mittermeier, R.A., Mittermeier, C.G., da Fonseca, G.A.B. and Kent, J. (2000) Biodiversity hotspots for conservation priorities. *Nature*, **403**, 853–858.

Nakiboglu, S.M. and Lambeck, K. (1991) Secular sea level change, in R. Sabadini, K. Lambeck and E. Boschi (eds), *Isostasy, Sea-Level and Mantle Rheology*, Dordrecht, Kluwer, pp. 237–258.

National Environment Commission, Royal Government of Bhutan (1998) *The Middle Path. National Environment Strategy for Bhutan*. Government of Bhutan.

National Society for Clean Air and Environmental Protection (NSCA) (2002a) *Comparison of Emissions from Waste Management Options*, Brighton: NSCA.

National Society for Clean Air and Environmental Protection (NSCA) (2002b) *Relative Impacts of Transport Emissions in Recycling*, Brighton: NSCA.

Neumann, A.C. and MacIntyre, I. (1985) Reef response to sea-level rise: keep-up, catch-up or give-up. *Proceedings of the 5th International Coral Reef Congress*, **3**, 105–110.

Newby, H. (1991) One world, two cultures: sociology and the environment. *BSA Bulletin Network*, **50**, 1–8.

Nicholls, R.J. and de la Vega-Leinert, A. (2000) Overview of the SURVAS Project, in N. Mimura and H. Yokoki (eds), *Proceedings of APN/SURVAS/LOICZ Joint Conference on Coastal Impacts of Climate Change and Adaptation in the Asia-Pacific region*, Kobe: Asia-Pacific Network for Global Change Research, pp. 3–9.

Nicholls, R.J., Hoozemans, F. and Marchand, M. (1999) Increasing flood risk and wetland losses due to global sea-level rise: regional and global analyses. *Global Environmental Change*, **9**, S69–S88.

North, R. and Gorman, T. (1990) *Chickengate: An Independent Analysis of the Salmonella in Eggs Scare*, London: Health and Welfare Unit, Institute of Economic Affairs.

Noss, R.F. (1999) Assessing and monitoring forest biodiversity: a suggested framework and indicators. *Forest Ecology and Management*, **115**, 135–146.

Nunn, P.D. (1994) *Oceanic Islands*, Oxford: Blackwell.

Nunn, P.D. (1999) *Environmental Change in the Pacific Basin: Chronologies, Causes, Consequences*, London: Wiley.

Nunn, P.D. (2000) Illuminating sea-level fall around AD 1220–1510 (730-440 cal yr BP) in the Pacific Islands: implications for environmental change and cultural transformation. *New Zealand Geographer*, **56**, 46–54.

Oksanen, M. (1997) The moral value of biodiversity. *Ambio*, **26**, 541–545.

O'Neill, B.C. and Oppenheimer, M. (2002) Dangerous climate impacts and the Kyoto Protocol. *Nature*, **296**, 1971–1972.

O'Riordan, T. (1981) *Environmentalism*, 2nd edn, London: Pion.

O'Riordan, T. (2000) Environmental science on the move, in T. O'Riordan (ed.), *Environmental Science for Environmental Management*, Harlow: Prentice-Hall, pp. 1–27.

O'Riordan, T. (ed.) (2001) *Globalism, Localism and Identity: Fresh Perspectives on the Transition to Sustainability*, London, UK: Earthscan.

Pacione, M. (2001a) Land use in cities of the developed world. *Geography*, **86**, 97–119.

Pacione, M. (2001b) The internal structure of cities in the Third World. *Geography*, **86**, 189–209.

Pacione, M. (2001c) The Future of the City – Cities of the Future. *Geography*, **86**(4), 275–286.

Pacione, M. (2001d) *Urban Geography: A Global Perspective*, London: Routledge.

Park, C.C. (1987) *Acid Rain: Rhetoric and Reality*, London and New York: Routledge.

Parry, M. (2001) Climate change: where should our research priorities be? *Global Environmental Change*, **11**, 257–260.

Parry, M.L. (1992) Agriculture as a resource system, in I.R. Bowler (ed.), *The Geography of Agriculture in Developed Market Economies*, Harlow: Longman, pp. 207–238.

Parry, M.L., Arnell, N.W., McMichael, A.J., Nichols, R.J., Martens, P., Kovats, S., Livermore, M., Rosenzweig, C., Iglesias, A. and Fischer, G. (2001) Millions at risk: defining critical climate change threats and targets. *Global Environmental Change*, **11**, 181–183.

Parry, M.L., Rosenweig, C., Iglesias, A., Fischer, G. and Livermore, M.T.J. (1999) Climate change and world food security: a new assessment. *Global Environmental Change*, **9**, S51–S68.

Pearson, P.N. and Palmer, M.R. (2000) Atmospheric carbon dioxide concentrations over the past 60 million years. *Nature*, **406**, 695–699.

Pederson, T. (2000) Climate change fore and aft: where on earth are we going? *IGBP Newsletter*, **44**, 3–4.

Peltier, W.R. (1998) Postglacial variations in the level of the sea: implications for climate dynamics and solid earth geophysics. *Reviews of Geophysics*, **36**, 603–689.

Pepper, D. (1986) *The Roots of Modern Environmentalism*, London: Routledge.

Perez, R.T. (2000) Assessment of vulnerability and adaptation to climate change in the Philippines Coastal Resources Sector, in N. Mimura and H. Yokoki (eds), *Proceedings of APN/SURVAS/LOICZ Joint Conference on Coastal Impacts of Climate Change and*

References

Adaptation in the Asia-Pacific Region, Kobe: Asia-Pacific Network for Global Change Research, pp. 95–102.

Peterson, G. (2000) Political ecology and ecological resilience: An integration of human and ecological dynamics. *Ecological Economics*, **35**, 323–336.

Pfaff, A.S.P., Kerr, S., Flint Hughes, R., Liu, S., Sanchez-Azofeifa, G.A., Schimel, D., Tosi, J. and Watson, V. (2000) The Kyoto protocol and payments for tropical forest: an interdisciplinary method for estimating carbon-offset supply and increasing the feasibility of a carbon market under the CDM. *Ecological Economics*, **35**, 203–221.

Phantumvanit, D. and Liengcharernsit, W. (1989) Coming to terms with Bangkok's environmental problems. *Environment and Urbanisation*, **1**, 40–49.

Phillips, P.S., Pratt, R.M. and Pike, K. (2001) An analysis of UK waste minimisation clubs: key requirements for future costs effective developments. *Waste Management*, **21**, 389–404.

Pierce, J.T. (1990) *The Food Resource*, Harlow: Longman Scientific & Technical.

Pimental, D. (1995) Environmental and economic costs of soil erosion and conservation benefits. *Science Magazine*, **267**, 251–259.

Pimm, S.L., Russell, G.J., Gittleman, J.L. and Brooks, T.M. (1995) The future of biodiversity. *Science*, **269**, 347–350.

Poore, D. (1993) The sustainable management of tropical forest: the issues, in S. Rietbergen (ed.), *The Earthscan Reader in Tropical Forestry*, London: Earthscan Publications, pp. 50–71.

Postel, S. (1999) *Pillar of Sand: Can the Irrigation Miracle Last?* New York: Norton.

Potter, C.S. (1999) Terrestrial bio-mass and the effects of deforestation on the global carbon cycle. *Bio-Science*, **49**, 769–787.

Potter, R. and Lloyd-Evans, S. (1998) *The City in the Developing World*, Harlow: Longman.

Potts, D. (1997) Urban lives: Adopting new strategies and adapting rural links, in C. Rakodi (ed.), *The Urban Challenge in Africa: Growth and Management of its Large Cities*, Tokyo: United Nations University, pp. 447–494.

Power, M. (2001) Alternative geographies of global development and inequality, in P.W. Daniels, M. Bradshaw, D. Shaw and J. Sidaway (eds), *Human Geography: Issues for the 21st Century*, Harlow: Pearson Education, pp. 274–302.

Pretty, J. (1995) *Regenerating Agriculture: Politics and Practice for Sustainability and Self-reliance*, London: Earthscan.

Pretty, J. (2001) *Genetic Modification: Overview of Benefits and Risks*. Centre for Environment and Society, University of Essex.

Pugh, C. (2000) Introduction, in C. Pugh (ed.), *Sustainable Cities in Developing Countries*, London: Earthscan, pp. 1–20.

Pugh, D.T. (1987) *Tides, Surges and Mean Sea-Level*, Chichester: Wiley.

Pugh, J. (2002) Local Agenda 21 and the Third World, in V. Desai and R. Potter (eds), *Arnold Companion to Development Studies*, Edward Arnold, pp. 289–293.

Quinn, T.M., Taylor, F.W. and Crowley, T.J. (1993) A 173 year stable isotope record from a tropical South Pacific coral. *Quaternary Science Reviews*, **12**, 407–418.

Rakodi, C. (1997) Introduction, in C. Rakodi (ed.), *The Urban Challenge in Africa: Growth and Management of its Large Cities*, Tokyo: United Nations University.

Rangeley, W.R. (1987) Irrigation and drainage in the world, in W.R. Jordan (ed.), *Water and Water Policy in World Food Supplies*, College Station, Texas: Texas A & M University Press, pp. 29–36.

Rayner, S. (2000) Prediction and other approaches to climate change policy, in R.A. Pielke, D. Saretwitz and R. Byerly (eds), *Prediction: Science, Decision-making and the Future of Nature*, Washington DC, USA: Island Press, pp. 269–296.

Rayner, S., Bretherton, F., Buol, S., Fosberg, M., Grossman, W., Houghton, R., Lal, R., Lee, J., Lonergan, S., Olson, J., Rockwell, R., Sage, C. and van Imhoff, E. (1994) A wiring diagram for the study of land-use/cover change: report of working group A, in W.B. Meyer and B.L. Turner II (eds), *Changes in Land Use and Land Cover: A Global Perspective*, Cambridge: Cambridge University Press, pp. 13–54.

RECOUP (2000) Affordable plastic bottle recycling? Analysis and review of collection, sorting, reprocessing and end-market issues and economics. A report for the New Life for Plastics Project, UK: Peterborough.

Redclift, M. (1991) The multiple dimensions of sustainable development. *Geography*, **76**(330), 36–42.

Rees, J. (1997) Equity and environmental policy. *Geography*, **76**, 292–303.

Rees, W. (1992) Ecological footprints and carrying capacity: what urban economics leaves out. *Environment & Urbanization*, **4**, 121–130.

Rees, W. (1997) Is 'Sustainable City' an Oxymoron? *Local Environment*, **2**, 303–310.

Renner, M. (2002) Breaking the link between resources and conflict, in L. Starke (ed.), *State of the World 2002: A Worldwatch Institute Report on Progress Towards a Sustainable Society*, London: Earthscan Publications Ltd, pp. 149–173.

Repetto, R. and Austin, D. (1997) *The Costs of Climate Protection: A Guide for the Perplexed*, Washington DC: World Resources Institute.

Rhind, D. and Hudson, R. (1980) *Land Use*, London: Methuen.

Riebsame, W.E., Meyer, W.B. and Turner II, B.L. (1994) Modelling land use and land cover as part of global environmental change. *Climatic Change*, **28**, 45–64.

Rigg, J. (1991) *South East Asia a Region in Transition: A Thematic Human Geography of the ASEAN Region*, London: Unwin Hyman.

Rigg, J. (1998) Rural–urban interactions, agriculture and wealth: a southeast Asian perspective. *Progress in Human Geography*, **22**, 497–522.

Rigg, J. (2001) *More than the Soil: Rural Change in Southeast Asia*, Harlow, Prentice Hall.

Ritzer, G. (1996) *The McDonaldization of Society: an Investigation into the Changing Character of Contemporary Social Life*, California: Pine Forge Press.

Roberts, C.M., McClean, C.J., Veron, J.E.N., Hawkins, J.P., Allen, G., McAllister, D.E., Mittermeier, C.G., Schueler, F.W., Spalding, M., Wells, F., Vynne, C. and Werner, T.B. (2002) Marine biodiversity hotspots and conservation priorities for tropical reefs. *Science*, **295**, 1280–1284.

Robinson, D.A. (1999) Agricultural practice, climate change and the soil erosion hazard in parts of southeast England. *Applied Geography*, **19**, 13–28.

Robinson, D.A. and Boardman, D.A. (1988) Cultivation practice, sowing season and soil erosion on the South Downs, England: a preliminary study. *Journal of Agricultural Science, Cambridge*, **110**, 169–177.

Robinson, G.M. (1991) The environment and agricultural policy in the European Community: land use implications in the United Kingdom. *Land Use Policy*, **8**, 95–107.

References

Robinson, G.M. (1994) *Conflict and Change in the Countryside: Rural Society, Economy and Planning in the Developed World*, Chichester: John Wiley and Sons, revised edition.

Robinson, G.M. (1997) Greening and globalizing: agriculture in the new times, in B.W. Ilbery, Q. Chiotti and T. Rickard (eds), *Agricultural Restructuring and Sustainability: A Geographical Perspective*, Wallingford: CAB International, Wallingford, pp. 41–54.

Robinson, G.M. (2001) Nature, society and sustainability, in I.R. Bowler, C.R. Bryant and C. Cocklin (eds), *The Sustainability of Rural Systems*, Dordrecht: Kluwer Academic Publishers, for the International Geographical Union's Commission on the Sustainability of Rural Systems, pp. 35–58.

Robinson, G.M. (ed.) (2002) *Geography*, Sustainable Development Special, **87**, No. 3.

Robinson, G.M. (2003) *Geographies of Agriculture: Globalisation, Restructuring and Sustainability*, Harlow: Pearson.

Robinson, G.M., Tranter, P. and Loughran, R.J. (2000) *Australia and New Zealand: Economy, Society and Environment*, London: Edward Arnold.

Roomratanapun, W. (2001) Introducing centralised wastewater treatment in Bangkok: a study of factors determining its acceptability. *Habitat International*, **25**, 259–271.

Rosegrant, M.W. and Cai, X. (2002) Global water demand and supply projections. Part 2: results and prospects to 2005. *Water International*, **27**, 170–182.

Ross, H., Poungomlee, A., Punping, S. and Archavanikul, K. (2000) Integrative analysis of city systems: Bangkok 'Man and the Biosphere' programme study. *Environment and Urbanization*, **12**, 151–161.

Rossum, S. and Lavin, S. (2000) Where are the Great Plains? A cartographic analysis. *Professional Geographer*, **52**, 543–552.

Roy, P. and Connell, J. (1989) 'Greenhouse': the impact of sea level rise on low coral islands in the South Pacific. *The Research Institute for Asia and the Pacific, University of Sydney, Occasional Paper*, **6**, 55pp.

Royal Commission on Environmental Pollution (2002) http://www.rcep.org.uk/

Royal Society (2001) *The Role of Land Carbon Sinks in Mitigating Global Climate Change*. Policy document **10/01**, London: The Royal Society.

Royer, J.S. and Rogers, R.T. (eds) (1998) *The Industrialization of Agriculture: Vertical Co-ordination in the US Food System*, Aldershot: Ashgate.

Sage, C. (1994) Population and Income, in W.B. Meyer and B.L. Turner II (eds), *Changes in Land Use and Land Cover: A Global Perspective*, Cambridge: Cambridge University Press, pp. 263–286.

Salway, A.G., Murrells, T.O., Milne, R. and Ellis, S. (2002) *UK Greenhouse Gas Inventory, 1990 to 2000*. Annual Report for submission under the Framework Convention on Climate Change. Abingdon, Oxon: AEA Technology, National Environmental Technology Centre.

Sampson, R.N., Wright, L.L., Winjum, J.K., Kinsman, J.D., Benneman, J., Kursten, E. and Scurlock, J.M.O. (1993) Biomass management and energy. *Water, Air and Soil Pollution*, **70**, 139–159.

Satterthwaite, D. (1997) Sustainable cities or cities that contribute to sustainable development? *Urban Studies*, **34**, 1667–1997.

Satterthwaite, D. (2002) Urbanization and environment in the Third World, in V. Desai and R. Potter (eds), *The Companion to Development Studies*, London: Arnold, pp. 262–267.

Sawio, C.J. (1994) Who are the farmers of Dar es Salaam? in A.G. Egziabher, D. Lee-Smith, D.G. Maxwell, P.A. Memon, L.J.A. Mougeot and C.J. Sawio (eds), *Cities Feeding People: An Examination of Urban Agriculture in East Africa*, Ottawa: International Development Research Centre, pp. 25–46.

Scepan, J. (1999) Thematic validation of high-resolution global land-cover data sets. *Photogrammetric Engineering and Remote Sensing*, **65**, 1051–1060.

Schaffer, G. (1980) Ensuring Man's food supplies by developing new land and preserving cultivated land. *Applied Geography and Development*, **16**, 7–24.

Schlosser, E. (2002) *Fast Food Nation: The Dark Side of the All-American Meal*, New York: Perennial.

Schmidt-Bleek, F. (1994) *Carnoulies Declaration of the Factor Ten Club*, Germany: Wuppertal Institute.

Schneider, S. (1988) Planning for climate change (Testimony to the US Senate Full Committee on Energy and Natural Resources, 11 August 1988). *In Future*, **10**, 7–11.

Schumacher, E.F. (1973) *Small is Beautiful*, London: Abacus.

Science & Technology Policy Research (SPRU) (1999) The politics of GM food: risk, science and public trust. SPRU Special Briefing No. **5**, ESRC/University of Sussex, 22pp.

Scoones, I. and Toulmin, C. (1998) Soil nutrient balances: what use for policy? *Agriculture, Ecosystems and Environment*, **71**, 255–268.

Scoones, I. and Toulmin, C. (1999) *Policies for Soil Fertility Management in Africa*, London: Department for International Development.

Seip, H.M., Aunan, K., Vennemo, H. and Fang, J. (2001) Mitigating GHGs in developing countries. *Science*, **293**, 2391–2392.

Sen, A. (1982) *Poverty and Famines: An Essay on Entitlement and Deprivation*, Oxford: Clarendon Press.

Shiva, V. (1997) Biodiversity totalitarianism – IRPs and seed monopolies. *Economic and Political Weekly*, **32**, 2582–2585.

Shoard, M. (1980) *The Theft of the Countryside*, London: Temple Smith.

Simon, J. (1981) *The Ultimate Resource*, Princeton: Princeton University Press.

Simpson, S. and Robinson, G.M. (2001) Agri-environment policies, post-productivism and sustainability: England's Countryside Stewardship Scheme, in F. Molinero, E. Baraja and M. Alario (eds), *Proceedings of the Second Anglo-Spanish Symposium on Rural Geography, University of Valladolid*, July 2000, Spain: University of Valladolid, Section 3.3, 1–16.

Skole, D.L. (1994) Data on global land-cover change: acquisition, assessment and analysis, in W.B. Meyer and B.L. Turner II (eds), *Changes in Land Use and Land Cover: A Global Perspective*, Cambridge: Cambridge University Press, pp. 437–471.

Slaymaker, O. and Spencer, T. (1998) *Physical Geography and Environmental Change*, Harlow: Longman.

Smit, B. and Pilifosova, O. (2001) Adaptation to climate change in the context of sustainable development and equity, in J. McCarthy, O. Canziani, N.A. Leary, D.J. Dokken and K.S. White (eds), *Climate Change 2001: Impacts, Adaptation and Vulnerability*. Contribution of Working Group II to the Intergovernmental Panel on Climate Change Third Assessment. Cambridge, UK: Cambridge University Press, pp. 877–912.

Smit, J. and Nasr, J. (1992) Urban agriculture for sustainable cities: using wastes and idle land and water bodies as resources. *Environment and Urbanisation*, **4**, 141–152.

References

Smith, G.M. and Fuller, R.M. (2002) Land Cover Map 2000 and meta-data at the land parcel level, in G.M. Foody and P.M. Atkinson (eds), *Uncertainty in Remote Sensing and GIS*, Chichester: Wiley, pp. 143–165.

Smith, J., Mulongoy, K., Persson, R. and Sayer, J. (2000) Harnessing carbon markets for tropical forest conservation: towards a more realistic assessment. *Environmental Conservation*, **27**, 300–311.

Smith, J.T., Comans, R.N.J., Beresford, N.A., Wright, S.M., Howard, B.J. and Camplin, W.C. (2000) Pollution: Chernobyl's legacy in food and water. *Nature*, **405**, 141.

Smith, W.D. (1998) Urban food systems and the poor in developing countries. *Transaction of the Institute of British Geographers*, **23**, 207–220.

Soil Association (2000) *Organic Facts and Figures*, Bristol: Soil Association.

Sprackling, M. (1993) *Heat and Thermodynamics*, UK: McMillan Press Ltd.

Stenseth, N.C., Mysterud, A., Ottersen, G., Hurrell, J.W., Chan, K.-S. and Lima, M. (2002) Ecological effects of climate fluctuations. *Science*, **207**, 1292–1296.

Stéphenne, N. and Lambin, E.F. (2001) A dynamic simulation model of land-use changes on Sudano-sahelian countries of Africa, SALU. *Agriculture, Ecosystems and Environment*, **85**, 145–161.

Stern, P.C., Young, O.R. and Druckman, D. (1992) *Global Environmental Change: Understanding the Human Dimensions*, Washington DC: National Academy Press.

Stoddart, D.R. (1973) Coral reefs of the Indian Ocean, in O.A. Jones and R. Endean (eds), *Biology and Geology of Coral Reefs*, New York: Academic Press, **1**, pp. 51–92.

Stoorvogel, J.J., Smaling, E.M.A. and Janssen, B.H. (1993) Calculation of soil nutrient balances at different scales. I Supra-national scale. *Fertiliser Research*, **35**, 227–235.

Stott, P. and Sullivan, S. (eds) (2000) *Political Ecology. Science, Myth and Power*, London: Arnold.

Strategy Unit (2002) *Waste Not, Want Not: A Strategy For Tackling the Waste Problem in England*, London: Cabinet Office.

Straus, L.G. (1996) The archaeology of the Pleistocene-Holocene transition in southwest Europe, in L.G. Straus, B.V. Eriksen, J.M. Erlandson and D.R. Yesner (eds), *Humans at the End of the Ice Age: The Archaeology of the Pleistocene-Holocene Transition*, New York: Plenum, pp. 83–99.

Streif, H. (1995) Quaternary sea-level changes in the North Sea, an analysis of amplitudes and velocities, in P. Brosche and J. Sundermann (eds), *Earth's Rotation from Eons to Days*, Berlin: Springer-Verlag, pp. 201–214.

Stuart, M.D. and Costa, P.M. (1998) *Climate Change Mitigation by Forestry: A Review of International Initiatives*. Policy that Works for Forests and People Series No. **8**, London: IIED.

Swanson, T. (1999) Why is there a biodiversity convention? The international interest in centralized development planning. *International Affairs*, **75**, 307–311.

Sylva, J.E. (1665) *A Discourse of Forest, Trees and the Propagation of Timber*, in M. Wackernagel and W.E. Rees (eds), *Our Ecological Footprint: Reducing Human Impact on the Earth*. The new catalyst bioregional series, **9**, Philadelphia: New Society, p. 48.

Tatem, A.J., Lewis, H.G., Atkinson, P.M. and Nixon, M.S. (2002) Super-resolution land cover mapping from remotely sensed imagery using a Hopfield neural network, in

G.M. Foody and P.M. Atkinson (eds), *Uncertainty in Remote Sensing and GIS*, Chichester: Wiley, pp. 77–98.

Taylor, R. (1990) in N.P.O. Green, G.W. Stout and D.J. Taylor (eds), *Biological Science*, **1**, 2nd edn, Chapter 12, 384.

Ten Kate, K. and Laird, S.A. (2000) Biodiversity and business: coming to terms with the 'grand bargain'. *International Affairs*, **76**, 241–264.

Tett *et al.* (2002) Journal of Geophysical Research – D,

Thomas, E.P., Seager, J.R., Viljoen, E., Potgieter, F., Rossouw, A., Tokota, B., McGranahan, G. and Kjellén, M. (1999) *Household Environment and Health in Port Elizabeth, South Africa*. Urban Environment Series Report, **6**, Stockholm: Stockholm Environment Institute, in collaboration with the South African Medical Research Council and Sida.

Thornes, J.E. and McGregor, G.R. (2002) Cultural climatology, in S. Trudgill and A. Roy (eds), *Contemporary Meaning in Physical Geography*.

Thrupp, L.A. (2000) Linking agricultural biodiversity and food security: the valuable role of agrobiodiversity for sustainable agriculture. *International Affairs*, **76**, 265–281.

Tickell, C. (1996) Economical with the environment: a question of ethics. *Journal of Applied Ecology*, **33**, 657–661.

Tickell, C. (2001) *Making Sense of Sustainable Development: Prospects for the Johannesburg Summit of September 2002*. CEESR Annual Lecture, Faculty of Science, Kingston University.

Tiffen, M. (1995) Population density, economic growth and societies in transition: Boserüp reconsidered in a Kenyan case-study. *Development and Change*, **26**, 31–65.

Tiffen, M., Mortimore, M. and Gichuki, F. (1994) *More People, Less Erosion. Environmental Recovery in Kenya*, Chichester: J. Wiley.

Tivy, J. (1990) *Agricultural Ecology*, Harlow: Longman.

Tooley, M.J. (1994) Sea-level response to climate, in N. Roberts (ed.) *The Changing Global Environment*, Oxford: Blackwell, pp. 172–189.

Tooley, M.J. and Shennan, I. (eds) (1987) *Sea-Level Changes*, Oxford: Blackwell.

Townshend, J.R.G. (1992) Land cover. *International Journal of Remote Sensing*, **13**, 1319–1328.

Townshend, J., Justice, C., Li, W., Gurney, C. and McManus, J. (1991) Global land cover classification by remote-sensing – present capabilities and future possibilities. *Remote Sensing of Environment*, **35**, 243–255.

Trudgill, S. (2001) *The Terrestrial Biosphere: Environmental Change, Ecosystem Science, Attitudes and Values*, Harlow: Prentice-Hall.

Trupin, A. and Wahr, J. (1990) Spectroscopic analysis of global tide-gauge sea level data. *Geophysical Journal International*, **100**, 441–453.

Tucker, C.J., Townshend, J.R.G. and Goff, T.E. (1985) African land-cover classification using satellite data. *Science*, **227**, 369–375.

Turner, B.L. and Meyer, W.B. (1994) Global land-use and land-cover change: An overview, in W.B. Meyer and B.L. Turner II (eds), *Changes in Land Use and Land Cover: A Global Perspective*, Cambridge: Cambridge University Press, pp. 3–10.

Turner, B.L., Kasperson, R.E., Meyer, W.B., Dow, K.M., Golding, D., Kasperson, J.X., Mitchell, R.C. and Ratick, S.J. (1990) Two types of global environmental change:

References

definitional and spatial-scale issues in their human dimensions. *Global Environmental Change*, **1**, 14–22.

Twidell, J. and Weir, T. (1986) *Renewable Energy Resources*, London, UK: E & FN Spon, 439pp.

UN (1986) *Declaration on the Right to Development*. Geneva: Office of the United Nations High Commissioner for Human Rights, Gland, Switzerland.

United Nations (1995) *An Urbanising World: Global Report on Human Settlements 1996*, Oxford: Oxford University Press.

United Nations Conference on Environment and Development (UNCED) (1992) *Agenda 21*, Washington: United Nations Publications. Also at: http://www.un.org/esa/sustdev/agenda21.htm.

United Nations Convention to Combat Desertification (UNCCD) (1994) http://www.unccd.int/convention/text/convention.php/ (accessed 9 February 2003).

UNEP, IUCN, WWF (1991) *Caring for the Earth. A Strategy for Sustainable Living*. Gland, Switzerland: UNEP, IUCN, WWF.

United Nations Environmental Programme (UNEP), (1999) *Our Planet, Hazardous Waste Development Programme*, www.ourplanet.com/ (accessed 3 February 2003).

United Nations Framework Convention on Climate Change (UNFCCC) (1992a) http://unfccc.int/resource/docs/convkp/conveng.pdf.

United Nations Framework Convention on Climate Change (UNFCCC) (1992b) Koyoto Protocol. http://unfccc.int/resource/docs/convkp/convkp/kpeng.html (accessed 10 February 2003).

United Nations Framework Convention on Climate Change (UNFCCC) (1997) Koyoto Protocol. http://unfccc.int/resource/docs/convkp/kpeng.pdf

United Nations Fund for Population (UNFPA) (1996) *The State of World Population 1996: Changing Places: Population, Debt and the Urban Future*, Washington: United Nations Publications. Also at http://unfpas.org/swp/1996/SWP96MN.htm.

United Nations Water Conference (1977) *Water for Agriculture*, New York: UN.

Urban Agriculture Network (1996) *Urban Agriculture: Food, Jobs and Sustainable Cities*. United Nations Development Programme.

Vajpeyi, D.K. (2001) *Deforestation, Environment, and Sustainable Development: A Comparative Analysis*, Westport, CT: Praeger.

van Beukering, P., Sehker, M., Gerlagh, R. and Kumar, V. (1999) *Analysing Urban Solid Waste in Developing Countries: A Perspective on Bangalore, India*. Working Paper no. 24. London: CREED (The programme of Collaborative Research in Economics of Environment and Development).

van de Plassche, O. (ed.) (1986) *Sea-Level Research: A Manual for the Collection and Evaluation of Data*, Norwich: Geo Books.

van den Bergh, J.C.J.M. and Verbruggen, H. (1999) Spatial sustainability, trade and indicators: an evaluation of the 'ecological footprint'. *Ecological Economics*, **29**, 61–72.

Varley, A. (1994) The exceptional and the everyday: vulnerability analysis in the International Decade for Natural Disaster Reduction, in A. Varley (ed.), *Disasters, Development and Environment*, Chichester: John Wiley, pp. 1–12.

Vaughan, D. (2000) Tourism and biodiversity: a convergence of interests? *International Affairs*, **76**, 283–297.

Vesilind, P.A. and Peirce, J.J. (1983) *Environmental Pollution and Control*, Michigan: Ann Arbor Science, p. 46.

Victor, D.G. (2001) *The Collapse of the Kyoto Protocol*, Princeton and Oxford: Princeton University Press, 178pp.

Vitousek, P.M. (1994) Beyond global warming: ecology and global change. *Ecology*, **75**, 1861–1876.

Vogelmann, J.E., Howard, S.M., Yang, L.M., Larson, C.R., Wylie, B.K. and Van Driel, N. (2001) Completion of the 1990s National Land Cover Data set for the conterminous United States from Landsat Thematic Mapper data and Ancillary data sources. *Photogrammetric Engineering and Remote Sensing*, 650.

von Braun, J., McComb, J., Fred-Mensah, B.K. and Pandya-Lorch, R. (1993) *Urban Food Insecurity and Malnutrition in Developing Countries*, Washington, International Food Policy Research Institute.

von Weizsäcker, E., Lovins, A.B. and Lovins A.H. (1998) *Factor Four Doubling Wealth – Halving Resource Use*, London: Earthscan.

Wackernagel, M. (1998) The ecological footprint of Santiago de Chile. *Local Environment*, **3**, 7–25.

Wackernagel, M. and Rees, W.E. (1996) *Our Ecological Footprint: Reducing Human Impact on the Earth*. The new catalyst bioregional series, **9**, Philadelphia: New Society Publishers.

Wackernagel, M., Onisto, L., Bello, P., Linares, A.C., Falfan, I.S.L., Mendez, J., Guerrero, A.I.S. and Guerrero, M.G.S. (1997) Ecological footprints of nations: how much nature do they use and how much nature do they have? Commissioned by the Earth Council of Costa Rica for the Rio+ 5 Forum. Toronto: International Council for Local Environmental Initiatives.

Wackernagel, M., Schulz, N.B., Deumling, D., Linares, A.C., Jenkins, M., Kapos, V., Monfreda, C., Loh, J., Myers, N., Norgaard, R. and Randers, J. (2002) Tracking the ecological overshoot of the human economy. *Proceedings of the National Academy of Science*, **99**, 9266–9271.

Warrick, R.A. and Oerlemans, J. (1990) Sea level rise, in J.T. Houghton, G.J. Jenkins and J.J. Ephraums (eds), *Climate Change: the IPCC Scientific Assessment*, Cambridge: Cambridge University Press, pp. 257–281.

Waste and Resources Action Programme (WRAP) (2001) *The WRAP Business Plan; Creating Markets for Recycled Resources*, London: Waste and Resources Action Programme.

Wathern, P., Young, S.N., Brown, I.W. and Roberts, D.A. (1986) The EEC Less Favoured Areas Directive: implementation and impact on upland land use in the UK. *Land Use Policy*, **3**, 205–212.

Wathern, P., Young, S.N., Brown, I.W. and Roberts, D.A. (1988) Recent upland use change and agricultural policy in Clwyd, North Wales. *Applied Geography*, **8**, 147–163.

Watkins, K. (1997) Globalization and liberalization: implications for poverty, distribution and inequality. *UNDP Human Development Reports*, no. **32**. New York UNDP.

Watson, R.T. and the Core Writing Team (2001) *Climate Change 2001: Synthesis Report*, Geneva: IPCC.

Wayerhaeuser, G.H., Jr. (1998) The challenge of adaptive forest management: aren't people part of the ecosystem too? *Forest Chronicles*, **74**, 865–870.

References

Whatmore, S.J. (1995) From farming to agribusiness: the global agro-food system, in R.J. Johnston, P.J. Taylor and M. Watts (eds), *Geographies of Global Change: Remapping the World in the Late Twentieth Century*, Oxford: Blackwell, Oxford, pp. 36–49.

Whatmore, S.J. and Thorne, L. (1997) Nourishing networks: alternative geographies of food, in D. Goodman and M.J. Watts (eds), *Globalising Food: Agrarian Questions and Global Restructuring*, London and New York: Routledge, pp. 287–304.

Wilkinson, G.G. (1996) Classification algorithms – where next? in E. Binaghi, P.A. Brivio and A. Rampini (eds), *Soft Computing in Remote Sensing Data Analysis*, Singapore: World Scientific, pp. 93–99.

Wilson, G.A. (2001) From productivism to post-productivism...and back again? Exploring the (un)changed natural and mental landscapes of European agriculture. *Transactions of the Institute of British Geographers*, new series, **26**, 103–120.

Woodcock, C.E. and Gopal, S. (2000) Fuzzy set theory and thematic maps: accuracy assessment and area estimation. *International Journal of Geographical Information Science*, **14**, 153–172.

Woodroffe, C.D., Thom, B.G. and Chappell, J. (1985) Development of widespread mangrove swamps in mid-Holocene times in northern Australia. *Nature*, **317**, 711–713.

Woodward, J., Place, C. and Arbeit, K. (2000) Energy resources and the environment, in W.G. Ernst (ed.), *Earth Systems: Processes and Issues*, Cambridge, UK: Cambridge University Press, pp. 373–401.

Woodwell, G.M. (1995) Biotic feedbacks from the warming of the Earth, in G.M. Woodwell and F.T. Mackenzie (eds), *Biotic Feedbacks in the Global Climate System: Will the Warming Feed the Warming?* New York: Oxford University Press, pp. 3–21.

World Commission on Environment and Development (WCED) (1987) *Our Common Future*, Oxford: Oxford University Press.

World Resources Institute (1996) *World Resources 1996–97: A Guide to the Global Environment. The Urban Environment*, Washington, World Resources Institute. Also at: http://www.wri.org/wr-96–97/96tocfull.htm.

World Summit on Sustainable Development (WSSD) (2002) www.johannesburgsummit.org/html/documents/wehab_papers.html.

World Wide Fund for Nature (WWF) (2002) *Seven Reasons Why WWF Opposes the Spanish National Hydrological Plan, and Suggested Actions and Alternatives*, Godalming: WWF.

Wynne, B. (1991) After Chernobyl: science made too simple? *New Scientist*, **129**, 44–46.

Wyrtki, K. (1990) Sea level rise: the facts and the future. *Pacific Science*, **44**, 1–16.

Yamada, K., Nunn, P.D., Mimura, K., Machida, S. and Yamamoto, M. (1995) Methodology for the assessment of vulnerability of South Pacific Island countries to sea-level rise and climate change. *Journal of Global Environmental Engineering*, **1**, 101–125.

Zerda-Sarmiento, A. and Forero-Pineida, C. (2002) Intellectual property rights over ethnic communities' knowledge. *International Social Science Journal*, **54**, 99–121.

Index

Note: B after page number indicates material in boxes; Figures in *italic* and tables in **bold**; For acronyms and abbreviations consult pages 277–279.

Global Environmental Issues.   Edited by Frances Harris
© 2004 John Wiley & Sons, Ltd   ISBNs: 0-470-84560-0 (HB); 0-470-84561-9 (PB)

Index

Index

Cities *(continued)*
  Developing world
    And the brown agenda 208–209, **209**
    Challenge to maintain food security
      218–219
    Need for a new urban agenda 209–210
    Need for transparent and participatory
      government 226
  Development of institutional and human
    capacities of 225–257
  Ecological footprints 202
  Efficiency of urban environments 226–227
  Environmental management issues 227, 248
  Growing collaboration between 200–201
  Historic, small 196
  Potential environmental advantages or
    opportunities from 203–204
  Rapid growth of 195, 196, *199*, 200, 201, 223
    Energy production 202
    In the poorest countries 204
    Incorporating smaller ones 196
    Mega-cities 208
    Urban growth 204–205
    Urbanisation 205–206
    World's largest 199, **199**
  Successful, basic goals for 226
  Transfer of the environmental burdens 195,
    207
  Urban utopias of the future 223, **224**
  Vulnerability to hazards 215
  Waste disposal 236–237, 250
  See also Sustainable cities/settlements
Civil unrest, food riots 218–219
Clean development mechanism (CDM) 35B,
  156–157
Clean production processes 42
Clean water, access to 210–214
Climate
  And society 22–24
  Future changes in 26–28
  Has never been stable 21
  Observed changes in 25–26
  Our experiments with 28
Climate change 43, 266, 273
  And energy 181–185
  A result of build ups of greenhouse gases
    183
  And sustainable development 41–42
  Contemporary and future, causes, rate and
    significance of 21–22

Direct impacts of 29–33
  Adaptation 31–33
  Human-induced 26–27
  Indirect effects 33–36, 39
  May increase soil erosion in Western Europe
    123
  Past variation have occurred very quickly
    183
  Political debate 14
  Reasons for 26
  Social aspects of impacts 32B
  Typology of people's attitudes to 38, *38*
  See also Global warming; Greenhouse gases
Climate Change Levy, UK 34, 185
Climate change policies
  Integration with development plans 42
  Shaping of 36–38
    Global nature of the challenge 37–38
Climate management, global, future of
    36–41
  An optimised benefit/cost analysis 39–40
  A risk framework 40–41, *40*
  Precautionary principle 39
  Shaping climate policy 36–38
Climate mitigation policies 36, 41–42, 43
Climate thresholds, critical, identification of
    40–41
Climatic variables, changes in 28
Coal 167, 245
Coastal areas
  Adapting to sea-level rise 60, 62B, 275
  And recent sea-level rises 51–53
  Flooding problems for large cities 215
  Pressured environments 101–102
  See also Bangladesh
Coastal protection 60
Coastal retreat 60
Coffee trade 146
Common Agricultural Policy (CAP)
    130–131
  Less Favoured Areas (LFA) scheme 130
  Supported intensive farming 130
  Uplands adversely affected 130
Common Methodology for Assessing
    Vulnerability to Sea-Level Rise 58, 269
Common-property resources 7
Composting 249, 255
Conference on Human Settlements
  HABITAT I (Stockholm 1972) 223
  HABITAT II (Istanbul 1996) 223

310

Index

Index

Index

Index

Index

Index